THIS LAND

A GUIDE TO CENTRAL NATIONAL FORESTS

THIS LAND

A GUIDE TO CENTRAL NATIONAL FORESTS

Robert H. Mohlenbrock

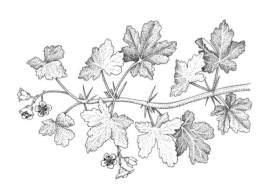

UNIVERSITY OF CALIFORNIA PRESS

Berkeley   Los Angeles   London

This book is dedicated to Mark, Wendy, and Trent, who accompanied their mother and me to the national forests until they went their separate ways.

University of California Press, one of the most distinguished university presses in the United States, enriches lives around the world by advancing scholarship in the humanities, social sciences, and natural sciences. Its activities are supported by the UC Press Foundation and by philanthropic contributions from individuals and institutions. For more information, visit www.ucpress.edu.

University of California Press
Berkeley and Los Angeles, California

University of California Press, Ltd.
London, England

Library of Congress Cataloging-in-Publication Data
Mohlenbrock, Robert H., 1931–
    This land : a guide to central national forests / by Robert H. Mohlenbrock.
        p.   cm.
    Includes bibliographical references and index.
    ISBN 0-520-23982-2 (pbk. : alk. paper)
    1. Forest reserves — Middle West (U.S.) — Guidebooks. 2. Middle West (U.S.) — Guidebooks. I. Title.

    SD428.A2M545 2005
    333.75'0977--dc22

                                                                      2005000741

Manufactured in Canada
10   09   08   07   06   05
10   9   8   7   6   5   4   3   2   1

The paper used in this publication meets the minimum requirements of ANSI/NISO Z39.48–1992 (R 1997) (*Permanence of Paper*). ♾

Cover: Lichen along the Continental Divide, Medicine Bow National Forest, Wyoming. Photograph by Scott T. Smith.

CONTENTS

Plates follow page 176

FOREWORD

As the Civil War came to an end, the United States found itself positioned to become a leader among nations. A country of immigrants with a rich endowment of natural resources, America was already a land of opportunity, but the young nation lacked the cultural marks of achievement that characterized its Old World counterparts. Europe had great temples, cathedrals, and museums filled with artifacts. Asia had great dynasties that embodied its long and glorious past. Though short on history, America did have a powerful national spirit that was expressed especially well through its abundant and bountiful land, much of which was public domain, west of the Mississippi—the great frontier. Historian Frederick Jackson Turner referred to this land as the "greatest free gift ever bestowed on mankind."

Yet around the time of the Civil War, a number of influential Americans were becoming increasingly concerned that some of the country's public lands were being plundered. Although homesteading had served the nation well in bringing territory under effective national control, it had become clear that some of the public domain lands had to be set aside as a legacy for all Americans.

In 1864, Henry David Thoreau called for the establishment of "national preserves" of virgin forests, "not for idle sport or food, but for inspiration and our own true re-creation." That same year, President Abraham Lincoln signed legislation granting Yosemite Valley and the Mariposa Big Tree Grove to the state of California to hold forever "for public use, resort, and recreation." Also in the 1860s, Frederick Edwin Church painted *Twilight in the Wilderness*, which inspired artists to capture on canvas the grandeur of the American landscape.

In 1891, President Benjamin Harrison created the nation's first forest reserve, the 1.2-million-acre Yellowstone Park Timber Land Reservation, just south of Yellowstone National Park. Today this area comprises Shoshone and Teton National Forests. Before his term ended, President Harrison proclaimed another 13 million acres of forest reserves in the West, laying the foundation for a National Forest system. However, the reserves were little more than lines drawn on a map, or "paper parks," without managers, regulations, or budgets.

The National Academy of Sciences established a National Forestry Commission in 1896, which issued a report that became the blueprint for forest policy emphasizing that the federal forest reserves belonged to all Americans and should be managed for them and not for any particular class. It went on to say that "steep-sloped lands should not be cleared, the grazing of sheep should be regulated, miners should not be allowed to burn land over willfully, lands better suited for agriculture or mining should be eliminated from the reserves, mature timber should be cut and sold, and settlers and miners should be allowed to cut only such timber as they need."

Just before the close of the nineteenth century, President Grover Cleveland established another 21 million acres of forest reserves, and in 1897, the Forest Management Act, or Organic Act, was passed, which specified that forest reserves were "to improve and protect the forest, or for the purpose of securing favorable conditions of water flows, and to furnish a continuous supply of timber for the use and necessities of citizens of the U.S."

Although Lincoln was perhaps the first president to see the "people's land" as a legacy to preserve for posterity, President Theodore Roosevelt thrust the nation into its first conservation movement. Roosevelt acted aggressively to expand federal forest reserves and to establish the first national wildlife refuges and national monuments. In 1905, with Gifford Pinchot at the helm, the USDA Forest Service was established. Secretary of Agriculture James Wilson, in 1905, directed Forest Service Chief Pinchot to manage the national forests "for the greatest good for the greatest number for the long run." Thanks to such visionary leaders in the nineteenth and early twentieth centuries, today we have 192 million acres of national forests that are owned by the people and are the birthright of all American citizens.

The pages that follow in this volume by Robert H. Mohlenbrock, a distinguished botanist, natural historian, and conservationist, are an account of the national forests of the West. Mohlenbrock has spent more than 40 years visiting and working in all of the 155 national forests. State by state, Professor Mohlenbrock describes each of the region's national forests in detail, including their size, location, access routes, basic geology, hydrology, and biota, as well as things to see and do.

He describes the trails that take visitors to wilderness areas and features of special interest and concern. He discusses rare and endangered species, notable historical landmarks and events, and sites of scenic beauty. *This Land: A Guide to Central National Forests* contains a wealth of information presented in clear and concise language. It adeptly conveys the sense of awe that characterizes our national forests. In the end, this volume will help us and future generations understand and appreciate the wealth of this land and remind us of the importance of being responsible stewards of the people's land today and for future generations.

Mike Dombeck
Chief Emeritus of the U.S. Forest Service

PREFACE

My family and I began visiting the national forests in 1960, and we have spent all of our vacations and an enormous amount of days in them, eventually visiting each of the 155 national forests. We soon discovered that they contain millions of acres of habitats and scenery that are nearly on a par with those found in national parks and national monuments. Many of these marvelous areas in national forests are little known, and we had to do considerable research to find the most exciting and beautiful areas. I have tried to provide information for these areas in this book.

Because I am a professional botanist, my family and I spent considerable time in areas known as Research Natural Areas, which are part of a national network of ecological areas designated in perpetuity for research and education, to maintain biological diversity on National Forest System lands, or both. Although many of the Research Natural Areas are in remote areas that are not accessible to the ordinary person, others are more accessible. Special permission may be required to visit a few of them. Some of my favorites are included in this book.

When you begin your exploration of a national forest, you will need to obtain an up-to-date forest map showing where the major roads and back roads are located. In most forests I have not listed the forest road and forest highway numbers since these change occasionally and are often vandalized so that visitors to the forest may have difficulty even finding the forest service road markers. The reader is advised to obtain the latest forest service map for the national forest from the district ranger stations or from the forest supervisor's office. While you are at the ranger stations, you can usually pick up several brochures describing trails and other points of interest. Forest fires which may have occurred while this book was in press may have altered the forests to some extent.

In 1984, I received a call from Mr. Alan Ternes, then the editor of *Natural History* magazine, published by the American Museum of Natural History in New York. Mr. Ternes asked if I would be interested in writing a monthly column for the magazine about some areas in the national forests that I particularly liked. An area did not necessarily need to be pretty, but it should have a biological or geological story to tell. My first article appeared in the November 1984 issue of *Natural History* magazine in my This Land column. The articles that I published in *Natural History* that pertained to national forests in the central United States accompany some of the national forest descriptions in this book. I am grateful to *Natural History* magazine and its editor, Ellen Goldensohn, for allowing me to republish these articles in this book.

I am indebted to several forest personnel, some of whom directed me to some little known areas, and others who read drafts of the manuscript of their particular national forest. Any errors which may have crept into the book are strictly my own. I am also grateful to Blake Edgar of the University of California Press who suggested this series of books, and to Mr. Scott Norton who has worked untiringly as my editor.

Whatever your interest in the out-of-doors, you will undoubtedly be able to enjoy our national forests, which are truly a unique American treasure.

INTRODUCTION

During the rapid development of the United States after the American Revolution, and during most of the twentieth century, many forests in the United States were logged, with the logging often followed by devastating fires; ranchers converted the prairies and the plains into vast pastures for livestock; sheep were allowed to venture onto heretofore undisturbed alpine areas; and great amounts of land were turned over in an attempt to find gold, silver, and other minerals.

In 1875, the American Forestry Association was born. This organization was asked by Secretary of the Interior Carl Schurz to try to change the concept that most people had about the wasting of our natural resources. One year later, the Division of Forestry was created within the Department of Agriculture. However, land fraud continued, with homesteaders asked by large lumber companies to buy land and then transfer the title of the land to the companies. In 1891, the American Forestry Association lobbied Congress to pass legislation that would allow forest reserves to be set aside and administered by the Department of the Interior, thus stopping wanton destruction of forest lands. President Benjamin Harrison established forest reserves totalling 13 million acres, the first being the Yellowstone Timberland Reserve, which later became the Shoshone and Teton National Forests.

Gifford Pinchot was the founder of scientific forestry in the United States, and President Theodore Roosevelt named him chief of the Forest Service in 1898 because of his wide-ranging policy on the conservation of natural resources.

Pinchot pursuaded Congress to transfer the Forest Service to the Department of Agriculture, an event that transpired on February 1, 1905. He realized that the forest reserves were areas where timber production would be beneficial to the nation and where clear water, diverse wildlife, and scenic beauty could be maintained.

In 1964, the Wilderness Act was passed by Congress, authorizing the setting aside of vast areas that were still in pristine condition. Although the establishment of wilderness areas has preserved some of our most beautiful areas, it has also made these areas off-limits to anyone who is aged or physically handicapped or who just cannot backpack for miles and miles into an area.

Slowly, as people from all walks of life began to use the national forests for recreational purposes, the U.S. Forest Service adopted the multiple-use concept, where timber production, wildlilfe management, conservation of plants and animals, preservation of clear water, maintenance of historic sites, and recreation could be accommodated in the national forests and enjoyed year round. Although camping, picnicking, and scenic driving are the major recreational activities on national forest land, other activities include boating, swimming, fishing, hunting, whitewater rafting, horseback riding, nature study, photography, wilderness trekking, hang-gliding, rockhounding, and winter sports. Special areas are set aside for off-highway vehicle activity.

Today, we have 155 national forests, although to save administrative costs, some have been combined. The U.S. Forest Service administers other areas as well, such as national grasslands, the Columbia River Gorge National Scenic Area, and the Lake Tahoe Basin Management Area.

NATIONAL FORESTS IN ARKANSAS AND OKLAHOMA

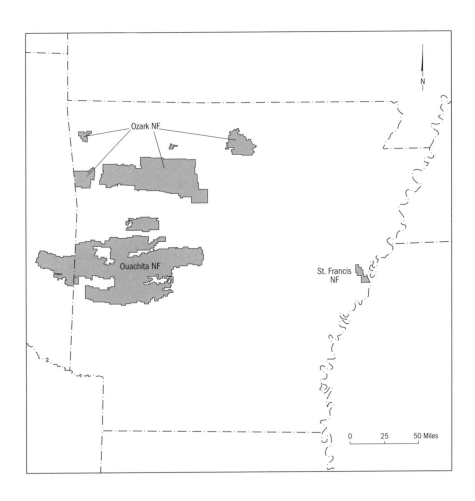

Arkansas has three national forests, and one of them extends into southeast-ern Oklahoma. The Ozark and St. Francis National Forests, both entirely within Arkansas, are administered by one forest supervisor located at 605 W. Main Street, Russellville, Arkansas 72801. The Ouachita National Forest con-tains public lands in both Arkansas and Oklahoma and is administered by another forest supervisor located in Hot Springs, Arkansas. All of these na-tional forests are in the Southern Region of the U.S. Forest Service, Region 8.

Ouachita National Forest

SIZE AND LOCATION: 1,777,478 acres in western Arkansas and southeastern Oklahoma. Major access roads are U.S. Highways 59, 70, 71, 259, 270, and 271, Arkansas State Routes 7, 8, 9, 27, 28, 80, 84, 88, 246, and 250, and Oklahoma State Routes 1, 3, and 63. District Ranger Stations: Booneville, Danville, Glenwood, Jessieville, Mena, Mount Ida, Oden, Perryville, and Waldron in Arkansas, and Hodgen, Idabel, and Talihina in Oklahoma. Forest Supervisor's Office: 100 Reserve, Federal Building, Hot Springs, AR 71901, www.fs.fed.us/oonf/ouachita.

SPECIAL FACILITIES: Boat launch areas; swimming areas; ATV trails; equestrian trails.

SPECIAL ATTRACTIONS: Talimena National Scenic Byway; Arkansas Highway 7 National Scenic Byway; Robert S. Kerr Arboretum and Botanical Area; Indian Nations National Scenic and Wildlife Area; Winding Stair National

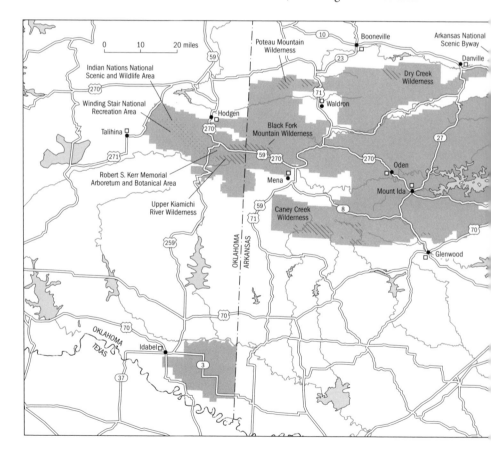

Recreation Area; Beech Creek National Scenic Area; Rustic Road Auto Tour; Crystal Valley Auto Tour; Winona Auto Tour; Shortleaf Pine–Bluestem Grass Ecosystem Management Area.

WILDERNESS AREAS: Poteau Mountain (11,299 acres); Flatside (9,507 acres); Dry Creek (6,310 acres); Caney Creek (14,460 acres); Black Fork Mountain (13,579 acres); Upper Kiamichi River (10,819 acres).

About 500 million years ago, a sea covered the area where the Ouachita Mountains are now located. The sea collected thousands of feet of sediments that became compressed and eventually uplifted to form the Ouachita Mountains. These mountains consist of shale, sandstone, limestone, chert, and novaculite. Lying south of the Arkansas River, the Ouachita Mountains are totally different from the Ozark Mountains that are north of the Arkansas River. The Ouachita Mountain Range extends from near Little Rock, Arkansas, to Atoka, Oklahoma, including many mountains that are unique in that they are oriented in an east-to-west direction. The Ouachitas extend into eastern Oklahoma, but in the southeastern corner of Oklahoma, south of the mountain range, is a very different and smaller unit of the Ouachita National Forest called the Tiak District. This district is in the Gulf Coastal Plain Province, rather than a mountain province.

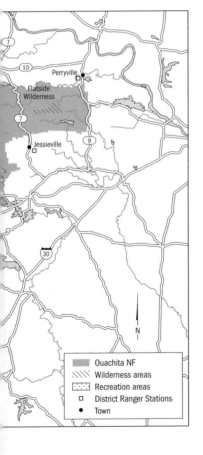

Most of the western third of the Ouachita National Forest that is in the mountain district may be accessed along or just off of the Talimena National Scenic Byway (pl. 1), a 54-mile road that goes through beautiful scenery between Mena, Arkansas, and Talihina, Oklahoma. At either end of the scenic byway the Forest Service has a visitor information station where you may obtain guide books, brochures, maps, pamphlets, and other information. One-and-a-half miles north of the Mena entrance station, the rocks visible to the right are split, forming small cavelike crevices that were used by outlaws for hideouts in the 1920s. These rocks make up Earthquake Ridge, and a seven-mile-long trail winds through the area. Over the next two miles are three worthy overlooks. Blue Haze Vista looks down over the

Ouachita River Valley; Acorn Vista is above the community of Acorn, an early settlement known at one time as Gourd Neck; Round Mountain Vista provides for a broad view over wild, forested areas. The road now stays on the crest of Rich Mountain.

For 13 miles, the area adjacent to the scenic byway to the south has been designated the Middle Rich Mountain Recreation Area, a roadless region up to two miles wide and bisected at the western end by Mill Creek. To the north is the Rich Mountain Botanical Area that stretches for five miles and protects a rich and diverse flora. One unique feature of the flora of Rich Mountain is the elf forest that occurs on exposed south slopes where trees such as blackjack oak and post oak range in height from three to six feet tall. These trees are pruned and shaped by ice storms and the strong south winds, as well as extremes of winter temperatures and poor soil. On the north slopes of Rich Mountain, some of the most rich hardwood forests in Arkansas occur. Here you may find American beech, tulip poplar, Ohio buckeye, basswood, cucumber magnolia, and umbrella magnolia in the canopy. Some of the best natural wildflower and fern gardens may be found on the forest floor. Many picturesque mountain streams are in the area (pl. 2).

Several old homesites from pioneer farms provide nostalgia along the scenic byway. Queen Wilhelmina State Park interrupts the national forest on the scenic byway for two miles. After the road leaves the state park and reenters the national forest, there is a pioneer cemetery. Three-and-a-half miles later, a stone monument indicates the scenic byway has entered Oklahoma and becomes State Route 1. For the next 10 miles, the route is through a narrow corridor between the Upper Kiamichi River Wilderness to the south and the Robert S. Kerr Arboretum and Botanical Area to the north. The wilderness is on the south slope of Rich Mountain but is difficult to traverse because of extremely steep slopes and dense vegetation. The Ouachita National Recreation Trail crosses the wilderness and may be picked up at the state line or at the western end northeast of Big Cedar. The Robert S. Kerr Arboretum and Botanical Area is an outdoor education laboratory, coupled with scientific, cultural, and recreational opportunities, and features several self-guiding interpretive trails.

Just before the scenic byway reaches U.S. Highway 259, Big Cedar Vista provides for a beautiful view through a shortleaf pine forest. U.S. Highway 259 crosses the scenic byway at the western edge of the arboretum. West of the intersection, the scenic byway is in the Winding Stair National Recreation Area. The road descends Spring Mountain and then climbs by way of wide-sweeping curves up Winding Stair. At Emerald Vista is a circular half-mile drive to an overlook over the Poteau River Valley and Cedar Lake. Billy Creek Trail follows a gorgeous mountain stream to the Billy Creek Recreation Area where there is a campground and where fishing in the stream is usually rewarding. The steep Cedar Creek Road branches northward off of the scenic byway and drops rapidly to Cedar Lake where there are two nice campgrounds and access to the Winding Stair Equestrian Trail System.

Both north and south of the Winding Stair National Recreation Area is the Indian Nations National Scenic and Wildlife Area. Although this rugged region consists of a total of 41,409 acres, much of the recreational activities occur in the northern portion where there are several hiking trails, two small fishing lakes, old homesites, and a pioneer cemetery.

At Horsethief Springs, you can see the spring that is enclosed by rock walls that the Civilian Conservation Corps built in the early 1930s. A picnic area and trail are nearby. A side road south eventually winds to the Billy Creek Recreation Area. Within the next five miles along the scenic byway are three fine overlooks.

Just after the scenic byway passes a radio tower, a short circular drive runs to the top of a knob for a 360-degree panoramic view. To the south are Jackfork Mountain and the Potato Hills, the latter unique in that they are the westernmost known occurrence of a mineral known as novaculite. Near the western end of the Talimena National Scenic Byway is the Old Military Road Historic Site and Picnic Area. At this site in 1832, soldiers built a wagon road to connect Fort Smith, Arkansas, with Fort Towson, Oklahoma. A natural spring is nearby. The scenic byway terminates at the West End Visitor Information Center.

Although the Talimena National Scenic Byway provides access to many parts of the area, there are other features both north and south of it that are worthy of exploration. The Beech Creek National Scenic Area lies south of the Upper Kiamichi River Wilderness and is best reached by a side road south off of Arkansas State Route 8 several miles west of Mena. This 7,500-acre area is located between several mountains (Blue Bouncer, Lynn, Cow Creek, Round, and Walnut) in a very scenic setting. Within it are the headwaters of Beech Creek and the entire Beech Creek Botanical Area. Two trailheads at the north side of the scenic area provide access to this beautiful spot. The botanical area contains a rich assemblage of forest trees, with a bountiful understory. At least nine kinds of oaks, three kinds of hickories, and three kinds of elms are present, along with American beech, cucumber magnolia, umbrella magnolia, Ohio buckeye, sugar maple, red maple, black walnut (fig. 1), and more than 40 other species of trees. More than 60 kinds of shrubs and woody vines have been recorded from here.

North of the Talimena National Scenic Byway are the Shut-In Mountain Turkey Hunting Area; Lake Hinkle; Blue Moon Wildlife Demonstration Area; Shortleaf Pine–Bluestem Grass Ecosystem Management Area; and two

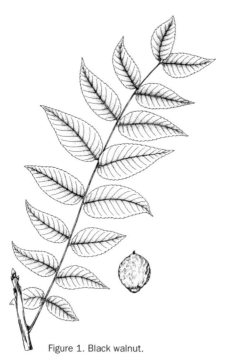

Figure 1. Black walnut.

wildernesses, Black Fork Mountain and Poteau Mountain.

The Shut-In Mountain Turkey Hunting Area, bordered on its western side by Clear Fork of the Ouachita River, is one of the best places in the country for wild turkeys. Although part of huge Lake Hinkle is on private property, most of the eastern half of the lake is in the national forest, with Little River Campground at its northern tip. The Blue Moon Wildlife Demonstration Area has been designed to show wildlife management techniques and is dotted with several large ponds. If you drive Buffalo Road south of Waldron, you come to the 155,000-acre Shortleaf Pine–Bluestem Grass Ecosystem Management Area, one of the best birding areas in Arkansas. Not only are dozens of common species of birds seen here, but rarer ones such as the endangered red-cockaded woodpecker, Bachman's sparrow, prairie warbler, brown-headed nuthatch, summer tanager, scarlet tanager, northern bobwhite, and yellow-breasted chat may also be observed.

Black Fork Mountain Wilderness stretches from near the west end of Shut-In Mountain in Arkansas into Oklahoma. Comprising 13,579 acres, this wilderness extends for 12.5 miles east to west but only two miles north to south. The entire wilderness is along the crest of Black Fork Mountain with several areas of massive, exposed sandstone. Many steep rock flows, known as rock glaciers, descend from the ridge. The forest is primarily dry, with shortleaf pine and several species of oaks dominant. In extremely dry soil are forests of dwarf blackjack oaks. Various roads completely encircle this wilderness, and the Black Fork Mountain Trail penetrates the eastern side of the wilderness.

North of Arkansas State Route 28 and southeast of the town of Hartford is the Poteau Mountain Wilderness that consists of two small but separate units. Several east-west-trending ridges and clear, rocky streams make up the wilderness. Rustic Road Auto Tour is an exceptionally scenic route that goes along the southern border of both units of the wilderness. Although there are no designated hiking trails, old logging roads make for fairly easy access to the wilderness.

The central part of the Ouachita National Forest lies between U.S. Highway 71 and Arkansas State Route 7. This region consists of a series of more or less parallel east-to-west-running mountains. From north to south in the national forest, the major mountains in the Ouachita Range are Bee, Dutch Creek, Mill Creek, Fourche, Missouri, Caddo, and Cossatot. Northeast of Bee Mountain is the small Dry Creek Wilderness with impressive sandstone bluffs rising above Dry Creek. A prominent rock column known as Chimney Rock has broken away from a nearby sheer sandstone cliff. A trail alongside Dry Creek crosses the wilderness. Forest personnel report a dense black bear population in the wilderness.

Arkansas State Route 80 separates Bee Mountain from Dutch Creek Mountain. The particularly beautiful Dutch Creek Scenic Area is about five miles south of the Dry Creek Wilderness. A dirt road runs along the ridge of Dutch Creek Mountain and also follows the southern border of the scenic area.

Several roads zigzag through the Mill Creek Recreation Area. At the southeastern corner of the mountain is the small but extremely picturesque Blowout Mountain Scenic Area. U.S. Highway 270 passes through the scenic area just east of Big Brushy Campground. The Ouachita National Recreation Trail also crosses the scenic area. About 15 miles northeast of Blowout Mountain, on the east side of State Route 27, is Irons Fork Scenic Area, where the mountain stream is lined by outstanding sandstone cliffs.

U.S. 270 and State Route 88 on either side of Fourche Mountain bring you to massive Lake Ouachita where the Forest Service, the State of Arkansas, and the Army Corps of Engineers have camping, boating, and picnicking facilities at several points around the lake.

From the Womble Ranger Station southeast of Mount Ida, you may drive the 21-mile-loop Crystal Vista Auto Tour. Along the way is the 700-acre Ouachita Seed Orchard where trees are grown for reforestation projects in the national forest. The auto route comes to Lil Blue Phantom Mine where quartz was mined from 1985 to 1990. You may still pick up pieces of quartz in the area. The road then stays alongside Twin Creek to the Crystal Mountain Scenic Area that features an old-growth forest of pine and hardwood species. A side road south from the scenic area goes to Collier Spring shelter. Two miles father west on the Crystal Vista Auto Tour is Crystal Vista. It has a marvelous view of Crystal Mountain to the south. A trail leads you to a reclaimed quartz mine where rockhounding is permitted. The auto tour then circles back to the starting point at the Womble Ranger Station.

Between State Routes 370 and 246 is a pretty section of the Ouachita National Forest that includes the Missouri Mountains, the Caddo Mountains, and the Cossatot Mountains. The Little Missouri National Wild and Scenic River tumbles through this region south of the Missouri Mountains. Also in the area is the Athens–Big Fork Hiking Trail that is one of the most rigorous in the national forest as it crosses eight mountains along its 10-mile course from north to south. The northern starting point for this trail is along Big Fork, about 3.5 miles east of Wolf Pen Gap. In a mile, the trail crosses the Little Missouri River and then winds, ascends, and descends its way south to its southern terminus a few miles east of the Shady Lake Recreation Area. Several excellent vistas occur along the way, particularly the Eagle Rock Vista and the overlooks on Brush Heap Mountain and Hurricane Knob. The Athens–Big Fork Trail follows the route of an old postal trail that was established more than 100 years ago.

Near the south end of the Athens–Big Fork Trail is the intersection with the Viles Branch Trail. If you take the Viles Branch Trail to the west, you come to the Shady Lake Recreation Area in about four miles. Here you can camp, fish, and swim, or simply relax in the pleasant surroundings. Shady Lake lies just outside the southeastern corner of Caney Creek Wilderness that is rugged with narrow ridges and steep slopes. Caney Creek and Short Creek flow through the wilderness. Along both creeks are fine stands of bottomland hardwood trees and, in slightly drier areas, giant American beech trees. Sandstone outcroppings and occasional granite rocks provide habitat for even drier-loving trees such as blackjack oak, shortleaf pine, and red cedar. Several trails take you through the wilderness, including the Tall Peak Trail that originates at the Shady Lake Recreation Area.

In addition to the usual forest types found in the Ouachita National Forest, some specialized habitats exist, particularly in the Mount Ida area. Groundwater seeps along Meyers Creek keep boggy areas wet, and rocky openings on south-facing slopes that are referred to as glades have an interesting assemblage of plants.

The easternmost section of the Ouachita National Forest is north of Hot Springs and between State Routes 7 and 9. State Route 7 is a scenic byway, the northern portion of which is in the Ozark National Forest. In the Ouachita National Forest, the scenic byway runs from Fourche Junction at the east end of Nimrod Lake south to the Iron Springs Campground. It is an extremely curvy road through forests of shortleaf pine. The lower part of the scenic byway follows sparkling Trace Creek. South Fourche Campground is midway along the route.

The best way to explore this part of the national forest is to drive the 27-mile gravel Winona Auto Tour that runs from State Route 7 to State Route 9.

Beginning on State Route 7 about two miles north of Iron Springs Campground, the auto tour circles past the Alum Creek Experimental Forest and climbs to Oak Mountain Vista after about seven miles. The 4,659 acres of the experimental forest were set up to monitor precipitation, air temperature, barometric pressure, and other elements and their effects on shortleaf pine and hardwood forests. Looking southward from Oak Mountain, you see various rock formations of sandstone, limestone, novaculite, shale, chert, and white quartz. The Ouachitas are one of the best places in the world to see and find high quality white and clear quartz crystals. Here the scenic byway follows the southern boundary of Flatside Wilderness, named for a 1,550-foot rock tower called the Flatside Pinnacle.

If you stay on the auto tour, you come to Crystal Mountain where some of the best veins of white quartz occur. The crystals vary from clear to milky white. Surface collecting of the crystals is permitted, but if you wish to dig for them, you need a permit from the ranger station in Jessieville. Just east of Crystal Mountain is an overlook with a view of Lake Winona in the distance. Near the west end of Lake Winona is a stand of extremely tall trees preserved in the Lake Winona Research Natural Area.

In about three more miles, the Crystal Vista Auto Tour is intersected by a forest road to the north. Take this to the Flatside Pinnacle and hike the two-tenths-mile trail to the top where there is an incomparable view of the countryside. The auto route then comes to North Fork Pinnacle. This rock formation has the foundation of an old fire tower at its summit. In another two miles, a short side road leads to Lake Sylvia and a campground. State Route 9 is just beyond.

A large, isolated part of the Ouachita National Forest is north of Broken Arrow, Oklahoma, and entirely in that state. Although large Broken Bow Lake is within this portion of the Ouachita National Forest, none of the recreation facilities around the lake is managed by the Forest Service. The Ouachita National Forest maintains a campground and boat launch area along Mountain Fork a few miles south of the lake.

A completely different and isolated part of the Ouachita National Forest, known as the Tiak District, is entirely within the state of Oklahoma east and southeast of Idabel. This small area has elements of the Gulf Coastal Plain and features a couple of unique botanical areas. Kull Campground is situated at a small lake in the western part of the Tiak, and the Bokhoma Campground is nestled in a pleasant wooded area on the east side. Much of the area west of Ward Lake and along Push Creek is swampy.

Ouachita Mountains

Largely contained in Ouachita National Forest, the Ouachita Mountains extend more than 200 miles across western Arkansas and southeastern Oklahoma. To the north, the broad Arkansas Valley separates them from the Ozark Plateau, whereas to the southeast lies the Gulf Coastal Plain. Fifty to 60 miles wide, the range consists mainly of narrow, steep-sloped ridges that run east to west, separated by valleys or basins. Spring-fed streams run parallel to the ridges or occasionally cut across them, forming spectacular ravines with picturesque waterfalls. The highest crest, between 2,850 and 2,900 feet, belongs to Rich Mountain, near the Arkansas-Oklahoma state line.

The Ouachitas are almost entirely composed of sedimentary rock—sandstone, shale, chert, limestone, and conglomerate. Geologists have a field day in the mountains because of all the folding, faulting, overturning, and compression of the strata. Botanists also have a field day with the variety of plant communities and the more than a dozen endemic species, that is, plants found in no other place in the world. The Ouachitas is the largest east-west range of mountains in the United States, and the plant communities on its north-facing slopes are distinct from those that face south. (For most U.S. mountain ranges, in which the ridges run north to south, the contrast is between east- and west-facing habitats.)

Dominating the warmer, south-facing slopes are forests of southern red oak, blackjack oak, shortleaf pine, and mockernut hickory, with an abundance of smaller trees and shrubs, including persimmon, flowering dogwood, redbud, and sassafras. Cooler, north-facing slopes are inhabited by white oak and shagbark hickory, often joined by basswood and cucumber magnolia. Dense, junglelike vegetation grows beneath the trees. On the driest ridgetops, forests of stunted oaks and hickories are usually found, sometimes with mature trees only three or four feet tall.

Viewed from the numerous roads that pass by and through the Ouachita Mountains, the pine–oak forests seem monotonous. But if you hike into the mountains, small habitats with special plant communities and endemic plants become apparent. Many owe their existence to the geological forces that molded the landforms and exposed different types of rock. Rare plants are mostly concentrated where the underlying rock is Arkansas novaculite, an often whitish, fine-grained, gritty material high in silica. Novaculite is brittle, breaking down into large blocks that eventually fragment into smaller particles. The Indians who occupied the Ouachitas used it for making many of their stone implements.

Another rock that is home to rare plants is Mazarn shale, whose black or green layers are sometimes interleaved with layers of sandstone and limestone. Sandstone overhangs, known as rock-shelters, are common in the Ouachitas, and there are several rock-strewn openings, or glades, including novaculite glades and Mazarn-shale glades.

I spent a day a few miles west of Hot Springs examining two of the special plant communities in the Ouachitas. One, a little less than an acre in size, was a soggy, nearly boglike habitat adjacent to Meyers Creek. Groundwater that seeps from the flanks of the surrounding rock strata and drains across the terraces into the creek feeds the habitat. I did not attempt to enter the seep, because it is easy to sink in above the knees. But I observed many of the wetland plants by curling my arm around tree trunks and leaning out as far as I could, or by edging my way out along strategically positioned logs.

A few trees, including sycamore, green ash, and sweet gum, have established a foothold in the seep and provide dense shade for the understory. Shrubs include swamp azalea, brookside alder, and spicebush. Many kinds of wetland grasses and sedges grow in the seep, several of them rare for this part of the country. Wildflowers include grass-leaved lily, yellow bellwort, water horehound, and blue lobelia. Royal fern, lady fern, and sphagnum also abound.

Immediately bordering the seep is a drier woods with umbrella magnolias, witch hazels, and American hollies. Beneath these trees grows a rich assortment of wildflowers and ferns, including crane fly orchid, blue-leaved goldenrod, heart-leaved aster, broad beech fern, and Christmas fern. Partridgeberry, an evergreen with nearly circular leaves that is popular in terraria, creeps over much of the ground.

A few miles to the west, on a wooded, south-facing slope overlooking the Ouachita River, lies Fulton Branch Glade. Surrounded by a dry forest of willow oaks, water oaks, black hickories, and shortleaf pines, the glade has only a sparse tree cover of red cedar, blackjack oak, post oak, winged elm, dwarf hackberry, and one-flowered hawthorn. Most of the trees are eerily draped with long, gray green lichens. Beneath these trees, much of the ground consists of exposed Mazarn shale.

Because the terrain is rocky and subject to direct sunlight, moisture is limited. Among the plants that can survive summers here are prickly pear cactus and succulents such as pink sedum, American agave, and flower-of-an-hour. More surprising, because of their rarity in this part of the Southeast, are wild hyacinth, green milkweed, black quillwort, Barbara's buttons, Engelmann's milk vetch, slender spiderwort, and two endemics—Arkansas cabbage (a large-leaved plant in the mustard family) and Hubricht's bluestar. This arid habitat stands in stark contrast to the wetland community of Meyers Creek Seep.

Tiak

The pointy southeastern corner of Oklahoma, a triangle of land squeezed between Arkansas and Texas, is bounded on the south by the meandering Red River, on the east by a surveyor's straight border, and on the north by the Little River (a tributary of the Red River). This region offers a number of wetlands of a kind you might not expect to encounter here. Some have been incorporated into the relatively new Little River National Wildlife Refuge. Others are found in the Tiak Ranger District, an isolated portion of the Ouachita National Forest, which lies mainly in Arkansas.

I recently toured the area, after first receiving an orientation from ranger Robert Bastarache, of the Tiak Ranger Station, which is housed in the post office in the town of Idabel. Heading southeast from town along State Highway 3, I parked beside the road between the tiny villages of Bokhoma and Tom and made my way along a small tributary of Parker Creek. After walking a short distance through a forest of loblolly pine, oaks, and hickories, I entered rather marshy terrain. False nettle, with its tight spheres of tiny green flowers, began to appear, and soon there were several other marsh species— a beggar's tick with three-lobed leaves; two smartweeds, one with pink flowers and one with white; and mock bishop's-weed, a plant that resembles a delicate Queen Anne's lace. Wetter areas provided excellent habitat for lizard's-tail.

Suddenly the ground was spongy; sphagnum was underfoot. I had come to the edge of McKinney Bog, a peat bog in southeastern Oklahoma—unexpected here because these habitats are rare this far south. Because peat bogs are not good places for hiking (and hikers are not good for the fragile plant community), I poked around the edge of this one. I was rewarded by seeing yellow fringed orchid, small green wood orchid, and grass-leaved ladies' tresses (this is also an orchid, with small white flowers in a spiral spike).

Returning to my van, I headed northward, at first retracing part of my route along Highway 3. I noticed several scissor-tailed flycatchers—an Oklahoma trademark—perched on overhead wires. I then turned off the highway and followed smaller roads to a dead end at the edge of a forest. From there I hiked a short distance through an upland woods to Goodwater Glade, the largest pristine limestone glade in Oklahoma. Glades are rocky, open areas within otherwise forested terrain. In this case, the rocky ground and the dry, northwest- and west-facing exposures discourage the growth of woody plants. The glade lies about 75 feet above Little River to the north and Goodwater Creek to the west. A mesic, or moist, forest grows down along the creek, whereas the upland woods adjacent to the glade are dominated by white oak, chinquapin oak, and mockernut hickory.

Although lacking trees, Goodwater Glade hosts a diversity of wildflowers, including a tiny, endangered one called golden glade cress. This is the only place in the world where this little member of the mustard family lives. (For a long time, a very similar plant found near the east Texas community of San Augustine was thought to be the same species, but botanists now regard them as distinct.) Golden glade cress is a dwarf annual with all its leaves clustered in a ring, or rosette, at the base of the plant. From the center rise flower stalks up to four inches tall, each bearing several lemon-yellow flowers. The plant blooms in early March. By late April the seed pods have formed their nearly spherical seeds, and by early May the plant withers and is gone. Thus, by the time the heat of summer is drying up the glade, the golden glade cress has

Figure 2. Bald cypress swamp.

completed its life cycle. In fall, some of the seeds germinate into new plants, forming new rosettes that lie dormant during winter months.

I left Goodwater Glade early in the afternoon, heading back toward Idabel and then north on U.S. Highway 70 to the other side of the Little River. This route took me through a section of the Little River National Wildlife Refuge. To explore the refuge more thoroughly, however, I had to double back and reenter it on smaller roads. The refuge does not yet have visitor facilities and formal trails, but 10 miles of unpaved roads wind through several wetland habitats. With care, you can walk off the roads into some of these areas, which are much more typical of wetlands farther south and southeast. The refuge is home to bald cypress swamps (fig. 2) and even alligators, which are found nowhere else in Oklahoma. The alligator snapping turtle and the four-toed salamander also live in the refuge.

Wet sloughs (narrow bodies of open water) often lead into the cypress swamps. One of the wildlife refuge roads crosses a slough, providing an opportunity to view its plant life. Growing in the standing water is southern wild rice (a six-foot-tall grass that towers over clumps of soft rush) and several kinds of sedges. The refuge also contains extensive bottomland forests of loblolly pine and broad-leaved deciduous trees. The flora is diverse, but because these forests are subject to spring floods, they have relatively few spring wildflowers. One of several ponds is easily accessible along U.S. Highway 70. Spatterdock, an aquatic plant related to the water lily, is common; it has two-inch-wide, yellow club-shaped flowers. In summer, the large, cream-colored flowers of water lotus, rare for this part of the country, stand on stout stalks above the water.

Bogs contain numerous rare species. In addition to the yellow fringed orchid, the small green wood orchid, and the grass-leaved ladies' tresses, others that may be found in or near bogs are green adder's mouth orchid, southern twayblade, large whorled pogonia, autumn coralroot, Indian pipe, and American pinesap. All but the last two are orchids.

Glades are rocky, relatively treeless areas. Goodwater Glade has some blue ash, cedar elm, and red cedar. Golden glade cress and two other miniature mustards—thread-leaved bladderpod and a golden mustard plant known as *Selenia*—grow there. Other plants in this glade are often found in prairies in the Midwest. The common grass is little bluestem, whereas among the wildflowers are orange puccoon, purple coneflower, yellow coneflower, purple prairie clover, and two types of bluets.

Mesic woods harbor bitternut hickory, sugarberry, green ash, and the rather uncommon nutmeg hickory and swamp chestnut oak. The canopy is so dense during summer that little sunlight penetrates to the forest floor. As a result, most of the wildflowers in this habitat bloom in spring, before the trees have put out all their leaves. Among them are wild blue phlox, Solomon's seal, the spectacular red-and-yellow Indian pink, and bee balm.

Upland woods are composed principally of white oak, chinquapin oak, mockernut hickory, red mulberry, wild black cherry, and black gum. The woods also have an ample midcanopy layer of small trees such as red buckeye, hop hornbeam, redbud, and rough-leaf dogwood. Spring-flowering wildflowers include hairy phlox and goat's rue, the latter a hairy-leaved member of the pea family with pink and cream flowers. In summer, yellow crownbeard and two species of sunflower bloom on the forest floor.

Cypress swamps, besides hosting bald cypress, are home to swamp red maple, water hickory, and water elm. Buttonbush, Virginia sweet spire, and storax, all with attractive white flowers, fill in the shrub layer. When trees die and topple into the murky water, they often form a suitable place for wildflowers to take root. Species that commonly grow on the half-submerged logs in the Little River National Wildlife Refuge include three different kinds of beggar's-ticks, a pink St. John's wort, false nettle, pinkweed (a type of smart-weed), and a small white aster.

Sloughs have southern wild rice, beaked rush, soft rush, and several kinds of sedges, all of which grow in standing water. The muddy banks support meadow beauty, cowbane, bur reed, monkey flower, and two kinds of hibiscus, or rose mallows. All of these bloom in summer. Fall brings the flowering of Letterman's ironweed and a white aster. The only woody plant common along the sloughs is the shrubby swamp privet.

Bottomland forests are a mixture of loblolly pine and broad-leaved deciduous trees. Common are green ash, cherrybark oak, pecan, sweet gum, and musclewood. Overcup oak is at the edge of its range here. American beauty-berry and deciduous holly are frequently encountered shrubs. Giant cane, a type of bamboo, is plentiful, whereas sensitive fern and netted chain fern are found occasionally. One of the few wildflowers that bloom in spring is jack-in-the-pulpit; summer flowers include white mock bishop's-weed, fringed yellow loosestrife, blue waterleaf, purple false dragonhead, and the fire-red cardinal flower.

Ozark National Forest

SIZE AND LOCATION: Approximately 1,200,000 acres in northwestern Arkansas north of Russellville and Van Buren and one unit south and west of Russellville. Major access routes are Interstate 40, U.S. Highways 71 and 412, and State Routes 5, 7, 16, 21 (Ozark Highlands Scenic Byway), 22, 27, 59, 113, 123, 220, and 309. District Ranger Stations: Clarksville, Hector, Jasper, Mountain View, Ozark, and Paris. Forest Supervisor's Office: 605 W. Main Street, Russellville, AR 72801, www.fs.fed.us/southernregion/oonf/ozark.

SPECIAL FACILITIES: Boat launch areas; swimming beaches; ATV trails.

SPECIAL ATTRACTIONS: Sylamore National Scenic Byway; Arkansas Highway 7 National Scenic Byway; Ozark Highlands National Scenic Byway; Mount Magazine National Scenic Byway; Pig Trail National Scenic Byway (State Route 23); Blanchard Springs Caverns; Big Piney Creek Wild and Scenic River; Buffalo National Wild and Scenic River; Mulberry National Wild and Scenic River; Alum Cove; Mount Magazine.

WILDERNESS AREAS: Hurricane Creek (15,307 acres); East Fork (10,688 acres); Upper Buffalo (12,018 acres); Richland Creek (11,801 acres); Leatherwood (16,838 acres).

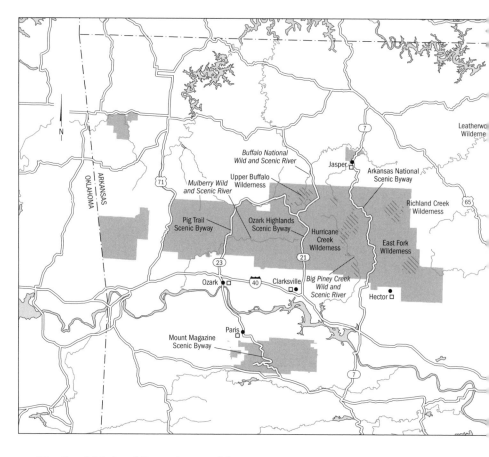

The Ozark National Forest is one of the most scenic regions in the southern United States, and it contains a wide diversity of natural features. Sheer rock cliffs occur along some of the rivers and streams, and rocks that have been eroded into unusual formations are scattered throughout the forest. You can see waterfalls, natural springs, deep canyons, natural lakes, and the best cavern in any national forest. Three features set the Ozark National Forest apart from any other: the Boston Mountains, Blanchard Springs Caverns, and Mount Magazine. The large central part of the Ozark National Forest includes the Boston Mountains, whereas five isolated units, including Blanchard Springs Caverns and Mount Magazine, occur around the periphery of the forest.

The Boston Mountains range across western Arkansas between Harrison and Yellville to the north and Clarksville to the south. Although many roads provide easy access to the area, there are still wild areas where roads either do not exist or have been closed because of the five wilderness areas in the forest. Three national scenic byways pass through the Boston Mountains from

MISSOURI
ARKANSAS

Sylamore National
Scenic Byway

Blanchard Springs
Caverns

Mountain View

0   10   20 miles

Ozark NF
Wilderness areas
District Ranger Stations
Towns

the north to the south. The Arkansas Highway 7 National Scenic Byway provides access to the eastern part of the mountains; the Ozark Highlands National Scenic Byway travels through the central section; and the Pig Trail National Scenic Byway penetrates the western side.

Arkansas Highway 7 National Scenic Byway is actually in two parts, the northern route is in the Ozark National Forest, and the southern route is in the Ouachita National Forest, the two segments separated by the 40-mile-wide Arkansas River valley. The northern entrance of the scenic byway into the Ozark National Forest is 10 miles south of Jasper along the eastern side of Henderson Mountain. In fewer than five miles, there is a forest road that leads back northwest to the Alum Cove Natural Bridge Recreation Area, with one of the most significant natural features in the forest. A natural bridge, carved out of sandstone through millions of years of erosion, is the centerpiece for this highly scenic area. The impressive natural bridge is 130 feet long and about 20 feet wide, with beautiful American beech, tulip poplar and huge-leaved umbrella magnolia trees in the surrounding forest. A 1.1-mile-long interpretive trail points out many of the features in the area. Interesting rock formations and large, cavelike rooms weathered out of the rocky cliffs dot the trail. American Indians used these overhangs for shelters, and the rare French's shooting star occurs in the deep shade under them.

The scenic byway gradually climbs through dense forests to the Fairview Campground, surrounded by the oldest pine plantation in the Ozark National Forest. The Ozark Highlands National Recreation Trail crosses the byway and may be joined here. Hiking this trail westward for five miles brings you to the Hurricane Creek Wilderness through which wild Hurricane Creek flows. The trail follows this creek through the wilderness, at one time going beneath the magnificent Hurricane Creek natural arch, a bridge 60 feet long and 10 feet wide. Hurricane Creek is full of boulders, and hiking the trail is sometimes treacherous with slippery, wet rocks. Hundred-foot-tall cliffs of sandstone and limestone periodically line the creek. The wilderness teems with wildlife, including black bears. The trail passes near two old cemeteries and several abandoned homesites from scores of years ago.

The Ozark Highlands National Recreation Trail east from Fairview Campground curls below the southern end of the Richland Creek Wilderness in about 12 miles and then skirts the eastern edge of the wilderness to Richland Creek Campground. This densely forested wilderness is penetrated by several picturesque streams, some of which are bordered by high, rocky cliffs. A mile-long palisade of cliffs stands above Richland Creek near its junction with Long Devil's Fork Creek. The forests in the wilderness are dominated primarily by several species of oaks and hickories, with flowering dogwood and redbud (fig. 3) in the colorful midcanopy. North outside of the wilderness on different roads are two rock formations worth going out of your way to see. Sam's Throne is along State Route 123 south of the community of Mount Judea; Stack Rock is on a county road south of Eula.

One mile south of the Fairview Campground at Sand Gap is an intersection with State Route 123. To explore the Ozark National Forest thoroughly, driving both eastward and westward from the scenic byway is necessary. State Route 123 to the west follows the southern boundary of Hurricane Creek Wilderness to Haw Creek Falls where Haw Creek drops picturesquely over a rocky bluff. A handicap-accessible trail is available from the campground to the falls. The Ozark Highlands National Recreation Trail also can be accessed from the campground. Big Piney National Wild and Scenic River is nearby, and historic Fort Douglas is just north of the falls. State Route 123 then turns south and exits the national forest in about 12 miles before reaching the small community of Hagarville.

State Route 16 to the east from Sand Gap provides an easy drive across to Richland Creek Wilderness, passing Pedestal Rocks Scenic Area where the cliffs have been eroded into an amazing group of pedestal-shaped rocks. A short hiking trail forms a maze around some of them. About three miles farther east, the road comes to a T. State Route 16 goes south, but after one mile, a side road to the east goes to Falling Water Falls, another sparkling cascade, particularly after a rain. The road to the north from the T follows the wilderness boundary of the Richland Creek Wilderness.

From Sand Gap, the scenic byway continues southward past Sollys Knob and climbs to Rotary Ann observation area where there is a good overview of the surrounding forested terrain. A picnic facility is located here. The road winds back and forth for several miles to Winter Hollow Overlook where there are more panoramic vistas. Just beyond is Moccasin Gap Campground. Before the scenic byway leaves the Ozark National Forest, County Road 21 to the west has two more forest options. Where the byway forks to the right, it climbs Newton Hill to Piney Overlook where you look westward over the vast Big Piney River watershed. The left fork goes to Long Pool Recreation Area situated along Big Piney Creek. Here you may camp, picnic, swim,

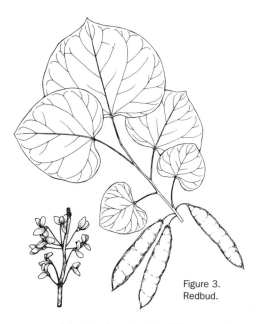

Figure 3.
Redbud.

canoe, kayak, or fish for several species of bass and sunfish.

A wild part of the southeastern section of the Ozark National Forest lies to the east of State Route 7 and is best reached by State Route 27 from the Bayou Ranger Station in Hector. This section includes Bayou Bluff, Buckeye Hollow, Brock Creek Lake, and East Fork Wilderness. Bayou Bluff refers to the scenic bluffs along the Illinois Bayou. You may fish or wade in the water of the bayou at the small campground here. Brock Creek Lake, a 35-acre lake, is a favorite for fishermen. Between Bayou Bluff and Brock Creek Lake is wonderful Buffalo Hollow where there are seven waterfalls.

State Route 27 follows the western edge of East Fork Wilderness, named for the East Fork of Illinois Bayou that flows through it. Several picturesque creeks in the wilderness flow into Illinois Bayou, and there are magnificent sheer sandstone cliffs along some of them. Serviceberry hangs from the top of the cliffs, producing its inch-wide white flowers in March before the leaves expand. White oak, red oak, and shagbark hickory are common in the upland woods, whereas the unusual overcup oak borders two ponds in the wilderness. Although there are several marked trails in the wilderness, the best scenery can be seen by hiking into some of the side canyons.

State Route 21 is the middle route through the Boston Mountains and is known as the Ozark Highlands National Scenic Byway. The 35-mile drive stays mostly on the upper slopes of the Boston Mountains, sometimes reaching some of the highest elevations in the region. The scenic byway begins south of the National Park Service's Buffalo National River Recreation Area outside the forest boundaries. Immediately upon entering the Ozark National Forest, the road is adjacent to Upper Buffalo Wilderness. The headwaters of the Buffalo National Wild and Scenic River (pl. 3) are in the wilderness. The river has carved deep ravines through the entire length of the wilderness, with dense forests on the steep slopes and ridgetops. Because this area was sparsely populated until 1948, any trail you take probably passes abandoned homes, stone fences, rusty farm machinery, and an old cemetery or two. State Route 215 branches off the scenic byway to the west, following the Mulberry National Wild and Scenic River. At Wolf Pen and Highbank, there are boat ramps where you can put your canoe or kayak in the water. The Mulberry River is rated class III white water.

The Ozone Campground on the scenic byway is on the site of an old Civilian Conservation Corps camp and is under a nice pine grove. The Ozark Highlands National Recreation Trail passes through the area.

The third designated scenic byway is called Pig Trail (State Route 23), which has so many twists for its 19-mile length that it resembles the curly tail of a pig. At the north end of this scenic byway is the White River where canoeing and kayaking are popular. From the White River, the scenic byway follows the West Flemming Creek to Cherry Bend where the Ozark Highlands National Recreation Trail crosses. Cherry Bend has interesting scenery. If you hike the trail eastward, make your way through Hanging Rock Hollow and then swing southward to the Mulberry National Wild and Scenic River. If you hike the trail a half-mile to the west, you come to old rock house ruins. The scenic byway then passes between Hornbeck Hollow and Wellcave Hollow. Near the Cass Job Corps Center, County Road 83 to the east comes to the Redding Campground in about three miles. This is a pleasant site along the Mulberry River, and there is an 8.5-mile loop trail to Spy Rock and back through a lush forest.

A road west from the small settlement of Turner's Bend goes to an interesting geological formation known as Bee Rock and then to Gray's Spring Picnic Area. The scenic byway climbs over Bend Hill and exits the national forest eight miles north of Ozark.

Horsehead Lake is at the southern end of the Boston Mountains section of the national forest, five miles north of Interstate 40 between Clarksville and Ozark. This is a serene 98-acre lake with campsites, picnicking facilities, boat ramps, and a swimming beach.

The Ozark National Forest west of the Pig Trail National Scenic Byway is centered around spectacular White Rock Mountain. The mountain rises high above the vast valleys that surround it. When you climb to the summit, there are large, flat, sandstone boulders you may sit on and ponder the wild forested terrain that surrounds you in every direction. Forest Service cabins are available for rent at White Rock Mountain. By driving three miles south of White Rock Mountain you come to attractive Shores Lake where you may camp, picnic, fish, swim, or boat. The Ozark Highlands National Hiking Trail also crosses the White Rock Mountain area. The Shores Lake Spur hiking trail also connects White Rock Mountain with Shores Lake.

Four miles west, off of State Route 215, is an area you do not want to miss. Devil's Canyon has some of the most spectacular rock cliffs in the Ozark National Forest, with high rock bluffs rising above Mill Creek. The creek is a clear stream fed by natural springs. It occasionally disappears into the ground, reappearing a short distance downstream. During summer, the rocky bottom of the creek has intermittent pools of water. Several side canyons jut off of Mill Creek, each very scenic, but the most spectacular one is Devil's Canyon. The cliffs are a montage of red, yellow, and brown that changes as the sun moves across the sky. Some of them rise 100 feet straight up, whereas others have treacherous talus slopes. The surrounding forest consists mostly of white oak, red oak, black oak, mockernut hickory, shagbark hickory, and bitternut hickory, with black walnut, American elm, wild black cherry, basswood, and black gum present. Sycamore and sweet gum are common along the creek.

An isolated part of the Ozark National Forest lies west of U.S. Highway 71 and extends to the Oklahoma state line. The main feature of this section is Natural Dam where a rock barrier across Mountain Fork Creek has impounded a large, sparkling pool of water. Pine Mountain and its lookout tower is a short distance southeast of Natural Dam. Devil's Den State Park is at the northeast corner of this section of the national forest, and the scenic Butterfly Hiking Trail makes a long loop into the national forest from the park.

Blanchard Springs Caverns (pl. 4) is the largest cave system in any national forest and contains some of the finest rooms and formations found anywhere. The Dripstone Trail is a guided and lighted trail through the Cathedral Room and Coral Room where you can see draperies, flowstones, columns, a coral pond, hundreds of stalagmites and stalactites, and other formations with fanciful names. The longer Discovery Trail permits you to see the natural entrance to the caverns as well as a giant flowstone, the Ghost Room, and Salamander Pool. The caverns can be reached by taking State Route 14 from Mountain View to Allison and then continuing on State Route 14 west to the Blanchard Springs Caverns turnoff. This stretch of State Route 14 is the western segment of the Sylamore National Scenic Byway. A short distance west of the turnoff is another side road to Gunner Pool Campground, situated next to a small lake. High bluffs surround the area, and north Sylamore Creek flows here. Across the creek is another fantastic pedestal-like rock called Standstone. Continuing west on State Route 14 in five miles is a side road to Barkshed Campground, and in two more miles a side road climbs over Push Mountain where there is a lookout tower. North of this is the wild Leatherwood Wilderness.

Leatherwood Wilderness, the largest in Arkansas, contains the crystal-clear Leatherwood Creek that has probably the thickest stand of a shrub called leatherwood in the country. Although the leaves of this species are similar to those of many other shrubs, the wood and twigs are so tough that they are difficult to cut. The shrubs bear rather obscure pale yellow flowers in early spring. The wilderness has many V-shaped ravines with exceptionally steep sides. The tops of the ridges are so dry that dense forests of tall trees do not grow there. Instead, the ridgetops are almost gladelike in appearance, with short, gnarly trees widely spaced. Common plants in this scrub woods are red cedar, post oak, blackjack oak, and winged elm, with farkleberry the leading shrub. Wildflowers include prickly pear cactus, American agave, goat's rue, and hairy wild petunia.

The northern part of the Sylamore National Scenic Byway is State Route 5. Near West Livingston Creek is Partee Springs, a crystal-clear spring that emits thousands of gallons of water each day. As the scenic byway proceeds northward and crosses Sugarloaf Creek, it climbs to City Rock Bluff, a limestone cliff, where there is a fantastic view down the White River. Two miles north, the scenic byway leaves the national forest at the small community of Calico Rock. You may hike much of this area by taking the 15-mile Sylamore Creek Hiking Trail from Allison to Barkshed Campground, passing Blanchard Springs Campground along the way.

One district of the Ozark National Forest lies south of Interstate 40 and a few miles west of Russellville. Although this district encompasses Rich Mountain, Huckleberry Mountain, Chickalah Mountain, and Mount Magazine, it is the last that is unique because of its flora and fauna. The mountain rises 2,400 feet above the Arkansas River, which is 16 miles to the north. Pennsylvania sandstone is the bedrock of Mount Magazine, although there are layers of shale. Plant communities include an upland deciduous forest, an evergreen forest, and a glade-type grassland. The Mount Magazine National Scenic Byway is a 25-mile route between Paris and Havana. This road passes 160-acre Cove Lake and Campground and has a branch road to the summit of Mount Magazine. Cove Lake has a boat ramp and swimming area.

Fifteen miles east of Mount Magazine is Spring Lake Campground, situated along a picturesque 83-acre lake on Chickalah Mountain. The campground has a boat ramp and swimming beach, as well.

A very small unit of the Ozark National Forest is in the Henry Koen Experimental Forest about 15 miles south of Harrison. The 720-acre Experimental Forest on the south bank of the Buffalo National River was set up in 1960 to develop forest management techniques. Today, researchers use it for studies in the Boston Mountains. Signs along a handicap-accessible interpretive trail point out more than 40 species of trees and shrubs. A picnic facility is also available.

Another small portion of the Ozark National Forest is 13 miles west of Fayetteville. Lake Wedington is the major attraction here. Not only may you camp, fish, swim, boat, or hike, you may also rent a cabin. A long hiking trail from Lake Wedington winds northward to Twin Mountain and ends at the Illinois River. The river features class II and III rapids.

The Ozark National Forest, and particularly the Boston Mountains, is known for the presence of two endangered bats—the Indiana bat and the gray bat. Indiana bat colonies hibernate in several caves scattered in the national forest. A large colony of gray bats was discovered in 1970 in the Ozark National Forest, and there are a few sites for this rare species. Several of the bat caves have been gated. Four species of rare cave bats and four species of tree bats also live in the Ozark National Forest.

Alum Cove

My first visit to Alum Cove, a rocky ravine in Arkansas's Ozark National Forest, was prompted in 1977 by a telephone call from Paul Redfern, a botanist from Southwest Missouri State University. Paul told me he had just found French's shooting star, a rare and delicate wildflower, growing beneath a sandstone overhang at Alum Cove. Paul knew I would be interested in his discovery, because until then, this plant had been found only at a few places in southern Illinois. In fact, French's shooting star, originally discovered in 1871, was the only flowering plant that grew on my home turf and nowhere else in the world. Despite the possibility of losing an exclusive claim, I was excited that this little plant might be living in the wild about 250 miles from its nearest colonies in Illinois.

I rushed to Alum Cove, where I verified that the plant Paul Redfern had discovered was indeed French's shooting star. At the same time, I was introduced to a beautiful natural area with significant vegetation communities. The geological setting is that of a rocky basin carved into a massive sandstone escarpment that includes several arches, overhangs, and a natural bridge whose 40-foot span rises more than 30 feet above a usually dry, rocky streambed. John David McFarland III, an Arkansas geologist, describes the Alum Cove rocks as crossbedded quartz sandstones sandwiched between layers of shale. He attributes the formation of the stone bridge and other features to the erosion of the sandstone: groundwater and surface water gradually remove iron oxide, the cementing material that holds the grains of sand together, allowing wind and gravity to do their work.

In the lush ravine bottom, the dominant trees are the American beech, with its broad crown, unlobed leaves, and smooth, light gray bark; and the sweet gum, with its pointed crown, roughened, dark bark, and five-lobed, star-shaped leaves. Umbrella magnolia, with two-foot-long leaves in pendulous clusters, adds an Appalachian element. The trees in the ravine cast dense shade over a sparse shrub layer and a thick carpet of spring wildflowers. Sure to catch the eye is the rattlesnake plantain orchid, not because of its flowers, which are tiny and white, but because of its dark green leaves patterned with bright white veins. Stranger yet are beechdrops, six-inch-tall plants that look more like twigs stuck in the ground because they have no leaves and contain no chlorophyll. They get their entire nutrient supply from the roots of beech trees, to which they attach themselves with long, underground runners. Despite their unusual appearance, these twigs bear tiny purple flowers in September and October and are therefore flowering plants.

Upland from the ravine bottom, the predominant trees are oaks—red, white, black, and scarlet. Joining them is mockernut hickory, whose large nuts resemble (mock) those of the very tasty shagbark hickory but house disappointingly little food material. The trees in this slope community are spaced farther apart than those in the ravine bottom, permitting more sunlight to penetrate and creating an overall drier habitat. Drier and even more open conditions prevail on the ridgetops above Alum Cove. Oaks still dominate, but black gum becomes an important tree species. Here and there, shortleaf pine makes an appearance.

The ravine, slope, and ridgetop communities at Alum Cove are typical of the surrounding Boston Mountains, but the sandstone overhangs provide niches where plants with special ecological requirements may develop. Such microhabitats are the spawning grounds for new species and undoubtedly gave French's shooting star its beginning.

The plant was discovered by George Hazen French, the second biologist ever to join the faculty of Southern Illinois University. On the fresh spring morning of May 6, 1871, French bicycled to a sandstone ravine about 10 miles south of Carbondale, entering a densely shaded forest dominated by sugar maple and white oak. He followed along the base of a 60-foot-high, east-facing sandstone bluff, which was undercut here and there by erosion to form overhangs. Under one of these overhangs, sheltered from the sun's rays, French discovered his shooting star.

French was undoubtedly excited by his find, because he knew that the common shooting star, the only other shooting star in southern Illinois, was a plant that grew out in the open in rather dry woods or even prairies. The plants under the overhang were a little smaller than the common shooting star, and the leaves, instead of tapering gradually to the base, were cut abruptly into a distinct leaf stalk. But French was puzzled because he could find absolutely no difference between the flowers of this plant and the flowers of the shooting stars out in the open.

Botanists who classify plants generally follow an unwritten rule that differences in flower structure may constitute a valid reason for naming a new species but differences in leaf shape generally do not, because leaves tend to be influenced more by local environmental conditions. When French sent some of his specimens to George Vasey, the national botanist in Washington, D.C., Vasey acknowledged that what French had found was different all right, but not different enough to make it a new species. Vasey considered the plant to be a variation of the common shooting star, describing it as *Dodecatheon meadia* var. *frenchii.*

In 1932, however, when botanist Per Axel Rydberg ran across French's specimens of shooting star in the dried plant collection at the New York Botanical Garden, he reopened the controversy. He decided it should be recognized as a species and accordingly changed the Latin name to *Dodecatheon frenchii.*

After John Voigt and I joined the botany faculty at Southern Illinois University in the 1950s, we became interested in French's shooting star and visited the many sites in southern Illinois where the plant had been found since its initial discovery. Our interest was piqued further when a respected botanist in Wisconsin, who had never seen French's shooting star in the wild, suggested that it was neither a good species nor even a good variety, but that its peculiarities could be attributed to its adverse, shaded environment. He proposed that if French's shooting star were to be transplanted to an open habitat, it would eventually assume the characteristic leaves of the common shooting star, and vice versa.

To test this theory, John Voigt devised a transplant experiment, which we carried out. We moved a few French's shooting star plants to the ridge above a sandstone overhang and took some common shooting stars from the ridge and relocated them back under the overhang. Although the experiment was repeated several times, the common shooting star never survived beneath the overhang. The French's shooting star plants on the ridge survive to this day, 50 years later, and still have their telltale stalked leaves.

Meanwhile, Ladislao Olah joined the botany faculty at Southern Illinois University in 1962 as a cell biologist, and he too became interested in the shooting star problem. In examining the chromosomes of the two kinds of plants, Olah discovered that French's shooting star had twice as many chromosomes (44) as the common shooting star. This genetic evidence weighs heavily on the side of those who consider French's shooting star to be a distinct species.

Following the unexpected discovery of French's shooting star at Alum Cove in northwestern Arkansas, this species has been found under a single overhanging bluff in southern Indiana, at one setting in southeastern Missouri, and at a few places in central and eastern Kentucky. Its widespread but sporadic occurrence remains a mystery for biologists to unravel.

Blanchard Springs Caverns

Geological formations of varied shapes and, sometimes, gigantic size make Blanchard Springs Caverns one of the most unusual attractions administered by the national forest system. The caverns are a product of the same forces that molded the topography in Arkansas's Ozark National Forest.

The caverns had their beginning perhaps 70 million years ago after an ancient seabed uplifted. The seabed consisted of limestone, or calcium carbonate, derived mostly from the shells of small marine organisms. Rain fell on the newly exposed limestone, entering small cracks created by frost action and earth movements. Because it incorporated carbon dioxide from the atmosphere, the rainwater was actually a weak solution of carbonic acid, which converted some of the limestone to water-soluble calcium bicarbonate. As a result, the limestone began to dissolve away.

Figure 4.
Blanchard
Springs.

Caverns are usually formed only where dissolved limestone is carried off by rapidly moving water. In effect, nearly every limestone cave is (or was) an underground waterway. Sinkholes (bowl-shaped depressions where water drains into the earth) often develop at the surface in association with caverns. Sometimes the sinkhole opening into a cave is small and becomes clogged so that the sink fills, forming a natural pond. In other cases, the opening is a vertical shaft into the cave through which water enters, eventually joining an underground river that flows between the layers of limestone. This is what happened at Blanchard Springs millions of years ago, with the water surfacing as a gushing spring several miles away (fig. 4).

After thousands of years, during which the rushing water carved underground chambers, the river shifted into a deeper channel. Dripping water entered the drained chambers, beginning a process of deposition that eventually built up stalactites, stalagmites, and other formations known collectively as dripstones, or speleothems. Later the river shifted again, leaving a second, younger set of chambers.

Large speleothems take millions of years to form. As droplets of water laden with dissolved calcium bicarbonate evaporate on the ceiling of a cave, they leave solid nuclei of limestone. Where this process is repeated at a single point, a stalactite begins to grow. At first, water runs through a small opening in the center of the dripstone, depositing calcium carbonate at the tip. Such a hollow stalactite is called a soda straw. Eventually, the soda straw becomes plugged and the water trickles down the outside, creating the stalactite's characteristic icicle shape. Because the water source for a stalactite is concentrated at a point on the cave ceiling, the stalactite is usually circular in cross section. Where a crack in the ceiling results instead in an irregular distribution of water, thin sheets of dripstone, known as draperies, develop.

Water dripping off the end of a stalactite and striking the cave floor also has a chance to evaporate, creating a stalagmite. Stalagmites never have a central opening, have rounded tips, and have large, irregularly shaped bases caused by the splash. When a stalactite and a stalagmite grow to meet each other, a column forms. The largest such column in Blanchard Springs Caverns is 65 feet tall.

Sometimes dripstone builds up in sheets that cover walls and older formations. This feature is called flowstone. Along underground pools and streams, intricate coral-like ridges, called rimstone terraces, build up. All of these formations are plentiful along Blanchard Springs Caverns's two underground trails, which were opened to the public in the 1970s. Some are found in impressive hollows, such as the Cathedral Room, a cave 1,150 feet long and 180 feet wide.

Far less apparent than the geological formations is the life in the caverns.

Green plants—mosses and algae—live only near the cave entrance or near the lamps that have been placed along the two trails. Fungi, which need organic matter to grow on, are rare in the caverns.

The animals that make use of Blanchard Springs Caverns range from those that live primarily outside but enter the cave from time to time to those that must spend their entire lives in the cave. Mole crickets are conspicuous among the creatures that freely enter the cave from outside. In contrast, the cave salamander, an orange amphibian with black speckles, rarely ventures outdoors. And far away from the entrance to Blanchard Springs Caverns, where temperatures are uniform and blackness reigns, dwell an assortment of animals so adapted to their secluded habitat that they can never leave it. These include a white cricket, several crustaceans, and the Ozark blind salamander. Because they live in total darkness, these organisms need neither sight nor pigment; in addition to being blind, all are white or transparent.

One animal whose life cycle is heavily dependent on caves such as Blanchard Springs Caverns is the gray bat, an endangered species that lives only in a few southern states. The creature is about five inches long, has a wingspan of up to 13.5 inches, and generally weighs less than half an ounce. Because of disturbance of their habitat, including vandalism in some of their caves, the gray bat population was drastically reduced during the 1980s.

Gray bats hibernate in mixed-sex colonies, choosing their winter caves by early September. After the adults mate, the females enter hibernation immediately. The males join them after several more weeks of feeding to replenish their depleted fat supplies. In the past, as many as 5,000 of these bats hibernated at Blanchard Springs Caverns, but Michael Harvey, a noted authority on Arkansas bats, counted only 150 there in February 1979 and only 33 in December 1985. In 2002, however, the population had exploded to 150,000.

The females emerge from hibernation in late March and become pregnant from the sperm stored in their bodies. Males and juveniles emerge by mid-May. After giving birth to single young in early June, the females select a warm cave for summer, where they congregate into a maternity colony. Males and juvenile females form bachelor colonies in nearby caves. During the migration to summer caves, two to 20 miles away from their winter quarters, the bats' food reserves are at their lowest, and this is when many die.

Blanchard Springs Caverns houses a summer bachelor colony of bats that forage up to 20 miles away each night for insects. Michael Harvey points out that these bats consume about 130 pounds of insects each night, or more than 13 tons during a single summer. This is one of the ways in which the underground caverns make their presence felt in the forest land above.

Mount Magazine

At 2,823 feet, Mount Magazine is Arkansas's highest mountain, towering above the broad Arkansas Valley. Its size, isolation, and outline, which early French hunters thought resembled a *magasin* (barn, or storehouse), have attracted naturalists to this day. The plateaulike summit, located in Ozark National Forest in the southern Ozarks, is about seven miles long, east to west, and less than one mile across. Lining the north side of the mountain are sandstone cliffs that overlook densely wooded ravines more than 1,000 feet below. E. J. Palmer, a botanist from Harvard's Arnold Arboretum, commented in 1927, "Nowhere else in this unusually picturesque part of America have I seen anything that approaches [Mount Magazine] for variety, beauty, and fine views."

The north rim is dense with shagbark hickory, red oak, chestnut oak, basswood, and sugar maple. A trail from Cameron Bluff, the most prominent cliff, leads over the edge of the mountain to a broad ledge, or bench, where some of the largest specimens of black walnut, white oak, red oak, black gum, and white ash are found in the United States. What lured me to the mountain for a visit in late May, however, were a number of rare plants that live on these north-facing bluffs. I needed to take only a few steps toward the rim from the mountaintop parking area at Brown Springs to notice a plant with purple flowers, the Ozark spiderwort, one of four spiderwort species that have evolved in isolation in the ancient Ozark Mountains.

As I proceeded down toward the first bench, I passed two kinds of trees that the U.S. Fish and Wildlife Service is considering listing as endangered species. One was the Ozark chinquapin, a close relative of the majestic American chestnut, which apparently is affected by the same chestnut blight that has devastated most American chestnuts in the country. The other tree, whose foot-long clusters of flowers hung gracefully from slender branchlets, was the smooth-trunked yellowwood. The future of this member of the pea family is in jeopardy because of its susceptibility to wood-rotting fungi.

Two ferns growing along the shaded north face of the bluffs also attracted my attention. Hay-scented fern, with four-foot fronds, looks like a tropical species but is truly a plant of temperate climates. In Arkansas it is found only on Mount Magazine and one other cliff. A more unexpected find is the much smaller Rocky Mountain woodsia fern. Mount Magazine is the only place it grows between the Appalachian Mountains to the east and the Rockies to the west. The population in west-central Arkansas may be left over from ages ago when this delicate-looking fern had a more continuous range.

But the most enigmatic plant to me was the maple-leaved oak, a small, shrublike tree that I had wanted to see ever since I had learned of its existence a few years earlier. Palmer, who discovered this oak along the north rim of Mount Magazine in 1925, had never before encountered it, and it still has not been found anywhere else. Because some of the leaves of this oak reminded Palmer of the leaves of the sugar maple, he coined the common name maple-leaved oak. The choice of a scientific name was not so easy, however. Was the plant an oak species that had evolved only on Mount Magazine, a hybrid offspring of two other oak species that lived in the area, or a form of another species that developed odd-looking leaves because of the local habitat?

Palmer wanted to describe the maple-leaved oak as a new species, and had he found it in similar habitats on other Arkansas or Missouri mountains, he probably would have done so in an instant. But he hesitated to do so because the tree grew only on one mountain, and it was an oak. Botanists name new species of hawthorn, blackberry, or milk vetch at the drop of a hat. But oaks have the reputation of being highly variable and even "promiscuous," when two different species cross to form a hybrid. A new oak is rarely named unless there is nearly irrefutable evidence of major physical distinctions, such as unique acorns. (The acorns of the maple-leaved oak, incidentally, are viable, unlike the seeds of most plant hybrids. The young plants they produce have leaves that resemble those of the adult plants.)

Palmer ruled out the possibility that the maple-leaved oak was a genetic freak because dozens of the plants lived on the rim of the mountain. He also found it difficult to imagine that any two of the other kinds of oaks then living on or near the mountain could have hybrid offspring with leaves like those of the maple-leaved oak. Palmer eventually decided, without experimental evidence, that he was dealing with an upland variant of the Shumard oak, a large tree found in the Arkansas Valley. He called it *Quercus shumardii* var. *acerifolia,* or the Shumard oak with maplelike leaves.

I do most of my botanical studies in southern Illinois, where the Shumard oak is a common forest tree. My first impression of the maple-leaved oak was that it had little in common with the Shumard. The habitat was wrong, the growth form was stunted, and the leaves, acorns, and buds were not really similar. Based on my own observations, I felt the maple-leaved oak was a distinct species.

One bit of research can be done to try to unravel this oak's identity. Nearly three decades ago, James Hardin and Paul Thomson, working independently at North Carolina State University and Southern Illinois University, respectively, discovered that each species of oak in eastern North America has distinctive hairs on its leaves. As seen through an electron microscope, some hairs are one celled, others multicellular; some are branched, others unbranched; some are perched on a pedestal-like stalk, others not. Interestingly, a hybrid between two different oaks has hairs from each of the parents. When the hairs on the leaves of the maple-leaved oak are investigated with an electron microscope, we should learn whether they match those of the Shumard or of any oaks in the area that might be the parents of a hybrid, or whether they are distinct from all other oaks.

The vegetation and scenery of the north face of Mount Magazine would be enough to please most naturalists, but the mountain offers more. Less than a mile away, across the flat summit, the south rim of the mountain presents an entirely different habitat. The arid conditions on the southern face result in gnarled and stunted trees. The post oaks that grow in the thin, parched, sandy soil are between 150 and 200 years old but stand less than 50 feet tall, or two-thirds the normal height. Where moisture is inadequate for trees, prairie grasses and wildflowers take over. The bluff's of the south rim are irregular in outline and often form series of stair-step ledges, where thick-stemmed prickly pear cacti are able to survive droughty conditions by storing up large quantities of water.

Near the western end of the flat-topped summit, a scattering of wet pockets have been filled in by growths of sphagnum. These boggy depressions are thought to be left over from an ancient time when the sea covered the entire area. As the ocean finally receded to its present position, most of the coastal plain vegetation retreated with it. But the small bogs on the summit of Mount Magazine have remained through the millennia, and several of the plants that live there are coastal plain species. The yellow fringed orchid and the brilliant blue-flowered bottle gentian are the most spectacular, but the rarest of all the plants on Mount Magazine is the inconspicuous dwarf pipewort, whose narrow, two-inch-long leaves are barely overtopped by minute spherical gray flower clusters. This species has been found only a half-dozen times, in western Arkansas, Oklahoma, and Texas.

St. Francis National Forest

SIZE AND LOCATION: 20,946 acres in eastern Arkansas, between Marianna and Helena, encompassing a part of Crowley's Ridge. Major access route is State Highway 44. District ranger station: Marianna. Forest Supervisor's Office: 605 W. Main Street, Russellville, AR 72801, www.fs.fed.us/oonf/ozark.

SPECIAL FACILITIES: Boat launch areas; swimming beaches.

SPECIAL ATTRACTIONS: Crowley's Ridge; St. Francis National Scenic Byway.

WILDERNESS AREAS: None.

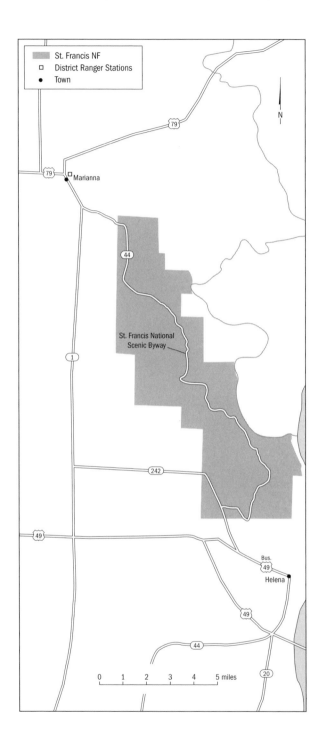

Although the St. Francis National Forest was established in 1960 for areas in Lee and Phillips counties, Arkansas, it is now administered with the Ozark National Forest. Most of the St. Francis National Forest is associated with Crowley's Ridge, a narrow landform of upland ridges that rise more than 150 feet above flat lowlands and floodplains. Crowley's Ridge extends for nearly 200 miles, from southeastern Missouri almost to Helena, Arkansas, on the Mississippi River. In the St. Francis National Forest, the ridge is not more than nine miles wide. The ridge is not rocky but consists of windblown soil, or loess, sand, and gravel that came from earlier channels of the Mississippi River that flowed between the Ozark Mountains and Crowley's Ridge. As the river has changed its channel several times, the erosion of the ridge has created several ravines and exposed sediments of varying ages. On top of the ridge are dry and mesic forests and some open, sandy areas. Because of the relative isolation of Crowley's Ridge from other upland landforms, an interesting flora has developed that is not seen in other parts of Arkansas. In addition, the St. Francis National Forest includes low areas on its eastern side between the ridge and a river system that includes the Mississippi, the St. Francis, and the L'Anguille Rivers.

On top of Crowley's Ridge in the St. Francis National Forest are two large bodies of water, the 625-acre Bear Creek Lake near the north end and the 420-acre Storm Creek Lake near the south end. Bear Creek, Sugar Creek, and Storm Creek are the major streams that drain the ridge. The St. Francis National Scenic Byway traverses the ridge from north to south, a distance of approximately 20 miles; either end of the scenic byway is paved, but the section between the two lakes is a good gravel road. Gravel roads also run along the base of Crowley's Ridge on both the west side and the east side.

The summit of the ridge, which maxes out at an elevation slightly more than 300 feet above sea level, supports a dry upland forest that has been timbered in the past in most places. As one proceeds downslope, particularly on the east side facing the Mississippi River, a more moist, or mesic, forest is present. The mesic forests are reminiscent of those of the Appalachian Mountains and are not found elsewhere in Arkansas. The drier sites support white oak, northern red oak, black oak, shagbark hickory, mockernut hickory, white ash, black gum, and persimmon, whereas the mesic forests have tulip poplar, American beech, sugar maple, and an occasional cucumber magnolia.

One of the unique habitats is open sand where broomsedge grass is often dominant. The unpalatable stems of this grass overwinter in an orange yellow color. Plants that are almost always confined to sandy terrain include sandbur, which is a very prickly grass, narrow rushfoil, which is a type of croton, sand crab grass, dotted wild bergamot, slender knotweed, jointweed, and stiff tick trefoil. Near Horner's Neck Lake at the southeastern corner of the national forest are swamps where bald cypress, tupelo gum, pumpkin ash, water elm, water hickory, and swamp cottonwood occur. Buttonbush is common throughout the swamps.

Two miles southeast of Marianna is the northern tip of the St. Francis National Scenic Byway, which begins as State Route 44. The district ranger station at the beginning of the scenic byway is a good place to obtain maps, brochures, and information. The byway circles around the west side of Bear Creek Lake. At the southern tip of the lake are campgrounds, picnic facilities, boat ramps, and swimming areas. The lake is an excellent fishing spot for catfish, crappie, bass, and bream, and a three-mile-long trail follows the shoreline around the lake. Maple Flats Campground is a very short distance southeast of the lake and is in the woods.

The scenic byway between Bear Creek Lake and Storm Creek Lake stays along the eastern edge of the ridge. Just beyond the Horn Lookout Tower site, a side road drops off the ridge to the tiny community of Phillips Bayou, nestled along the St. Francis River. The scenic byway continues southward along the edge of Crowley's Ridge where there is a nice overlook just before coming to Storm Creek Lake Recreation Area where there is a campground, boat launch area, and swimming area. The road swings around the southeastern side of the lake and then curves to the west as it drops off Crowley's Ridge to West Helena.

From Phillips Bayou, a road below the ridge stays in the national forest. At first, the road is next to the St. Francis River, but after it passes Beaver Pond, it is one mile west of the Mississippi River. This floodplain contains Horner's Neck Lake and its associated swampland. One-and-a-half miles past the lake, the road exits the national forest.

Crowley's Ridge

Extending nearly 200 miles from southeastern Missouri into the St. Francis National Forest in eastern Arkansas, Crowley's Ridge is a conspicuous, one- to 12-mile-wide feature that sometimes rises more than 200 feet above the flat Mississippi River floodplain. Inland seas, glacial melting, and westerly winds all had a role in its creation.

Twice during the past 50 million years, a shallow sea extended from what is now the Gulf of Mexico up along the Mississippi Valley, depositing sediment through wave and tide action. After the shoreline finally retreated gulfward, weak stream patterns developed on both sides of what later became the ridge. These were carved into deep trenches by glacial meltwater released as the last Ice Age drew to a close, between 14,000 and 12,000 years ago. Subsequently, strong winds picked up fine soil particles from the floodplain and redeposited them along the narrow ridge. This windblown material, known as loess, accumulated up to 50 feet thick. In St. Francis National Forest, nearly half the ridge consists of this buff- or ash-colored soil.

Standing alone like an island, Crowley's Ridge has a flora generally unlike that found in the flatter parts of Arkansas and Missouri, which because of poorer drainage are characterized by huge stands of silver maples and cottonwoods. Most of the vegetation that colonized the ridge and adapted to its varied exposures and moisture patterns apparently came from mountainous areas to the east. White oak, red, oak, and black hickory are the most common trees; shortleaf pine abounds on the drier ridgetops; and American beech and tulip poplar dominate the moist, well-drained lower slopes. More than 1,500 different kinds of flowering plants are found along Crowley's Ridge, many of them concentrated on the rich forest floor beneath the beech trees.

Many of the plants that grow on the ridge are on the extreme western edge of their range. Among them are bartonia, a slender member of the gentian family; the nut sedge, with its shiny, white seeds; the attractive white rein orchid; and climbing schisandra, a rare vine belonging to the magnolia family. But perhaps the most peculiar is beechdrops, one of relatively few parasitic flowering plants.

Three basic types of parasitic flowering plants have been identified. Those in the first group, which includes Indian paintbrushes, louseworts, and the bastard toadflax, possess both chlorophyll and roots in the soil but nevertheless attach themselves to chlorophyll-bearing hosts for additional water and nutrients. Others, such as the mistletoe, similarly manufacture some food material by photosynthesis but obtain all their water through attachment to plants. Beechdrops, however, fall into a third group, along with Indian pipes, coralroot orchids, broomrapes, and dodders. These plants lack chlorophyll and must depend entirely on other plants for water and nutrients.

Unlike many species in this last group, beechdrops survives only if it becomes attached to a specific host, the root of a beech tree. Following germination of a beechdrops seed, the white seedling, only one-tenth-inch long, sends forth a mass of curved and twisted roots, called grapplers, which seem to take hold of every object within their reach. If one of these objects happens to be a beech root, the grapplers wrap themselves tenaciously around it. In response to the attack by the grapplers, the beech root produces large amounts of soft tissue around its circumference, from the point of attachment of the grapplers to the end of the root. Despite this defense mechanism, the parasite easily penetrates the root and begins to direct the flow of nutrients produced by the tree into its own system. Eventually the entire current of sap in the beech root enters the parasite, and the beech root that extends beyond the point of attachment dies.

Although beechdrops seeds germinate in mid-June, the mature plants are conspicuous only from late August until the end of the growing season. Their purple brown, twiglike, branched stems, which bear tiny, brown structures known as scale leaves, stand up to one foot tall. They are usually found a few feet from the trunk of a beech tree, although some grow as much as 30 feet away. These plants do not appear to attach to the host until you carefully dig into the humus and fine soil to locate the point of their attachment.

During late fall, beechdrops produces two kinds of flowers. Those high on the plant are very slender, up to one inch long, and white speckled with purple. They are sterile, having lost the ability to form seeds. The lowermost flowers are tiny and inconspicuous, but are fertile. Why this arrangement has evolved is still a botanical mystery.

NATIONAL FORESTS IN COLORADO

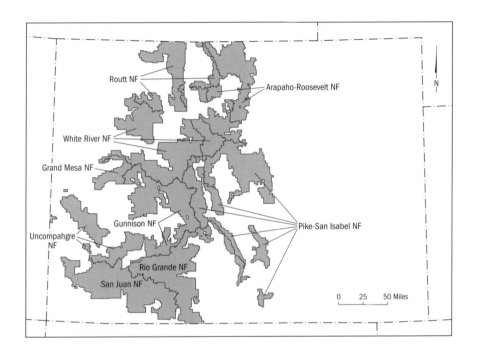

Eleven national forests are scattered throughout the western half of Colorado, all of them in Region 2 of the U.S. Forest Service. To save administrative costs, some of these forests have been combined into three larger administrative units, each comprising about two million acres. But because the forests retain their unique identities, they have been given separate entries in this section.

The Arapaho and Roosevelt National Forests occur in northern and northwestern Colorado and are administered by the forest supervisor at 240 W. Prospect Road, Fort Collins, CO 80526. The Grand Mesa, Gunnison, and Uncompahgre National Forests are in western Colorado; the forest supervisor's headquarters are at 2250 Highway 50, Delta, CO 81416. The Pike and San Isabel National Forests occupy the Front Range of Colorado's Rocky Mountains, extending from Leadville and Colorado Springs south to Trinidad. They are under the administration of the forest supervisor at 1920 Valley Drive, Pueblo, CO 81008.

Arapaho National Forest

SIZE AND LOCATION: Approximately 1,025,000 acres in north-central Colorado west of Denver. Major access routes are Interstate 70, U.S. Highways 6, 34, and 40, State Routes 5, 9, 91, 103, 105, 119 (Peak to Peak National Scenic Byway), and 125. District Rangers Stations: Granby and Idaho Springs. Forest Supervisor's Office: 240 W. Prospect Road, Fort Collins, CO 80526, www.fs.fed.us/r2/arnf.

SPECIAL FACILITIES: Winter sports areas; off-road vehicle areas; boat launch areas.

SPECIAL ATTRACTIONS: Arapaho National Recreation Area; Mount Goliath Natural Area; Guanella Pass National Scenic Byway; Hoosier Ridge; Peak to Peak National Scenic Byway; Elk Park.

WILDERNESS AREAS: Byers Peak (8,913 acres); Eagles Nest (132,906 acres, partly in the White River National Forest); Indian Peaks (73,291 acres, partly in the Roosevelt National Forest); Mount Evans (74,401 acres, partly in the Pike National Forest); Never Summer (20,747 acres, partly in the Routt National Forest); Ptarmigan Peak (12,594 acres); Vasquez Peak (12,956 acres).

Interstate 70 westward out of Denver and Golden is a very scenic highway into the Front Range of the Rocky Mountains. When the highway reaches Idaho Springs, it has come to the eastern edge of the Arapaho National Forest, one of the most beautiful and mountainous of all the national forests. The ranger station at the west end of town is a good place to obtain forest literature and to inquire about forest conditions and activities. From Idaho Springs the highest paved mountain road in the world, the Mount Evans Highway (State Route 105), begins and twists its way to the summit of 14,264-foot Mount Evans. The highway begins alongside Chicago Creek, which is lined during summer by willow shrubs and colorful wildflowers that include cow parsnip, wild geranium, monkshood, larkspur, and many others. At the first major set of switchbacks on the Mount Evans Highway, there is a dirt road that parallels West Chicago Creek to the Hells Hole Trailhead near the West Chicago Creek Campground. The hiking trail follows West Chicago Creek for nearly five miles and into the Mount Evans Wilderness to Hells Hole, a small pond near the base of Gray Wolf Mountain. Staying on the Mount Evans Highway, you curve and climb to the Ponder Point Picnic and Observation Area and then you continue to ascend to Echo Lake and a trailhead. The elevation at Echo Lake is approximately 10,000 feet. The trail from

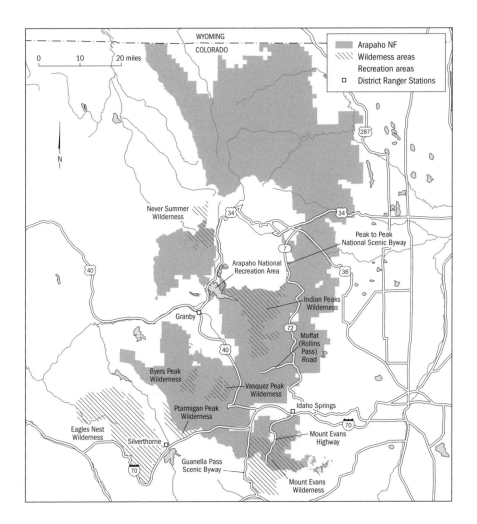

Echo Lake crosses the northeastern corner of Mount Evans Wilderness for about 3.5 miles where it leaves both the wilderness and the national forest.

After negotiating some serious hairpin turns and switchbacks, the Mount Evans Highway reaches 11,000 feet where the Mount Goliath Natural Area is located. The natural area has been preserved since 1950 to protect ancient, windswept bristlecone pines, some of them estimated to be 1,500 years old. Stunted Engelmann spruces occur here and there among the pines. Several species of mountain grasses, such as tufted hair grass, spike trisetum, and sheep fescue, are common, interspersed with occasional plants of yellow-flowered stonecrop, beardtongue, hairy golden aster, and yarrow or milfoil.

One-and-a-half miles beyond the natural area, there is a parking pullout where you may take the half-mile-loop Alpine Garden Trail through the tundralike habitat. The boulders along the way are covered by colorful lichens as well as bright green and nearly black mosses. During July, an array of colorful alpine wildflowers delight the hiker. Most of the wildflower plants are small of stature, and most of them bear tiny flowers. Alpine sandwort is probably the most commonly encountered wildflower along the trail, although golden avens, sky pilot with its blue purple flowers, alpine spring beauty, alp lily, fairy primrose, one-flowered daisy, alpine forget-me-not, and two kinds of dwarf wild clovers are present.

The Mount Evans Highway then continues its climb to the mountain's summit, going along a narrow corridor between two lobes of the wilderness over the top of a glacial cirque. Before starting the last group of switchbacks to the summit, the highway comes next to Summit Lake. The lake is in a beautiful basin surrounded by Mount Warren to the northeast, Mount Spalding to the west, and Mount Evans to the south. Mount Evans's summit is located on the boundary between the Arapaho and Pike National Forests. A short trail to the actual summit leads from the parking lot. An overlook provides unobstructed views in every direction. Abyss Lake lies in a deep ravine between Mount Evans and Mount Bierstadt to the west. Abyss Lake can be reached via a hiking trail from the Burning Bear Campground along the Guanella Pass National Scenic Byway in the Pike National Forest. Bighorn sheep and mountain goats are sometimes seen from the Mount Evans Highway, particularly in the vicinity of Summit Lake, and pikas and golden marmots are regularly seen (fig. 5).

On your return drive back down Mount Evans, you may wish to take the paved State Route 103, which heads east from Echo Lake. This scenic road follows the northern boundary of the wilderness, and there are picnic areas along the way. The highway eventually twists up and across Squaw Pass where it leaves the national forest.

Interstate 70 west from Idaho Springs goes through non–national forest land, but with huge sections of the Arapaho National Forest both to the north and south. At the U.S. Highway 40 exit, you may wish to take this exciting highway as it makes a tortuous climb to Berthoud Pass and then through breathtaking scenery to the towns of Winter Park and Fraser. The highway to Berthoud Pass takes you near several abandoned mines at previously ore-rich veins and the old mining town of Empire, before making a very sharp hairpin turn at Berthoud Falls prior to climbing to the pass. For an exciting drive, take the paved road from the west end of Berthoud Falls and nearby Big Bend Campground. After passing more abandoned mines, follow signs to Jones Pass. This graded dirt road follows along the West Fork of Clear Creek to a point where two tunnels of the Denver Water Board empty into the creek. The road then climbs above timberline to the Continental Divide and Jones Pass and then stays in the alpine tundra for a while before descending through Engelmann spruce, subalpine fir, lodgepole pine, and aspen. The road is closed at the foot of the switchbacks, where you can either retrace your route back to Berthoud Falls or take a long hiking trail along Williams Fork to the Sugarloaf Campground, a distance of a little more than eight miles.

The climb on U.S. Highway 40 to Berthoud Pass (11,316 feet) is very scenic. From the pass there is a 1,500-foot descent to Robbers Roost Campground. After several more miles, just before U.S. Highway 40 leaves the national forest, is the West Portal of the Moffat Tunnel and the historic Moffat, or Rollins Pass, Road that branches off to the east. At one time this road climbed to Rollins Pass on the Continental Divide and then descended to Tolland and eventually Rollinsville to the east. Until a few years ago, it was

Figure 5. Golden marmot.

possible to drive the entire length of this road, but because of dangerous rock slides and the unstable high wooden trestles that spanned deep ravines, the Moffat Road is now closed on the east side of the pass. The United States Army used an existing Indian trail to construct a wagon road over Rollins Pass in 1865. John Rollins improved the road in 1873 and charged a toll to anyone using it. After Berthoud Pass was completed and took most of the traffic across the mountains, David Moffat converted the little-used Rollins Pass Road into a railroad route for the Denver, Northwestern and Pacific Railroad. Because the last few miles on either side of the pass were too dangerous and too expensive to maintain, the railroad was abandoned and the road became an adventure for automobile drivers until it became unsafe for vehicle use. After David Moffat died in 1911, the railroad was bought by

William Freeman who oversaw the construction of the 6.21-mile Moffat Tunnel through the Continental Divide for his railroad, and the tunnel is still in use today. Built from 1923 to 1927, this is the second longest railroad tunnel in the United States, but construction of it cost the lives of 19 workers. Rollins Pass is at the base of the Indian Peaks Wilderness, a wilderness that stretches for 17 miles to the north and contains jagged-peaked mountains, four dozen glacial lakes, and the remnants of several glaciers. A gentle hiking trail from Rollins Pass stays along the Continental Divide in alpine tundra habitat for about three miles to Devil's Thumb Pass and rocky Devil's Thumb Peak. Picturesque Devil's Thumb Lake is at the base of the peak.

The communities of Winter Park and Fraser are good bases for further exploration of the Arapaho National Forest. From Winter Park, the Vasquez Road enters the national forest and follows Vasquez Creek through a narrow notch in the Vasquez Peak Wilderness. Vasquez Peak is on the Continental Divide at the southern border of the wilderness. You may reach it on a trail from the end of Vasquez Road or on the Continental Divide National Recreation Trail either from Berthoud Pass to the east or Jones Pass to the southwest.

Two forest roads run southward from Fraser that dead-end in the Arapaho National Forest. West Creek Road joins a dirt road that follows Fools Creek until it reaches the Vasquez Peak Wilderness. The five-mile length of this road may have more switchbacks on it than any other road of comparable length. St. Louis Creek Road stays near St. Louis Creek almost all the way to St. Louis Lake. From the Byers Creek Campground nestled along St. Louis Creek, there is a dirt road that heads northward and then westward to the small Byers Peak Wilderness. The last part of this road is a series of steep switchbacks best hiked or driven in a four-wheel-drive vehicle to the trailhead. A three-mile hike southward takes you to Byers Peak, or a one-mile hike northward takes you to Bottle Pass at the northern end of the wilderness. Crooked Creek Road, which is also County Road 50, proceeds westward from the north end of Fraser. In about six miles, this road enters the Arapaho National Forest and, after another 13 miles, comes to Horseshoe Campground and the Williams Peak Trailhead at the west side of this part of the national forest. This trail goes southward across private land until the national forest is reentered at the north base of the Williams Fork Mountains.

From the Horseshoe Campground, there is a road that heads south to a fork in the road after about three miles. From either fork, the road is paved. If you take the left fork, this forest road comes to several trailheads and campgrounds. The four-mile trail to Ute Peak (12,298 feet) is the first you come to four miles down the road. In another two miles is the Boardwalk Trail and two campgrounds. Where the forest road ends five miles farther south is the trailhead for the trail along the South Fork of Williams Creek before climbing to Ptarmigan Pass in the Ptarmigan Pass Wilderness. After crossing the pass, you may loop around and climb to the top of 12,458-foot Ptarmigan Peak. If you take the right fork south of the Horseshoe Campground, you ascend immediately to Ute Pass where there are scintillating views. Beyond the pass, the forest road connects with State Route 9.

Fraser is also one of the jumping-off places to explore the northern parts of the Arapaho National Forest. Between Fraser and the small community of Tabernash, the Meadow Creek Road winds around into the southeastern side of the Arapaho National Recreation Area. This region contains Meadow Creek Reservoir, one of five lakes that is encompassed in the Arapaho National Recreation Area. Meadow Creek Reservoir is at an elevation of 10,000 feet and is the most remote lake in the national recreation area. Fishing is good, but only nonmotorized boats may be used.

The other four lakes in the Arapaho National Recreation Area are best reached by taking U.S. Highway 34 from Granby. Lake Granby, the second largest lake in Colorado, has 7,256 surface acres of water and is available for all types of boats, including power ones. Arapaho Bay Road stays near the southern border of Lake Granby to the Arapaho Bay Campground. From here you can camp, see the Roaring Fork Falls, hike the Roaring Fork Trail that goes into a scenic canyon with Marten Peak and Cooper Peak towering above, or access 150-acre Monarch Lake. Monarch Lake was built in 1900 to store logs before they were released by flumes down Arapaho Creek. You can see a steam donkey, which was a steam-operated engine that transported logs from the surrounding forests. The High Lonesome Trail connects Monarch Lake and Meadow Creek Reservoir.

Instead of taking the Arapaho Bay Road, you may stay on U.S. Highway 34, which swerves around the west end of Lake Granby and Shadow Mountain Lake just outside the southern boundary of Rocky Mountain National Park. From the Willow Creek Campground on U.S. Highway 34, there is a dirt road west to an isolated unit of the Arapaho National Recreation Area that contains 756-acre Willow Creek Reservoir and Campground.

Instead of entering Rocky Mountain National Park, you may wish to take the Kawauneeche Road that runs northward within the eastern border of the Arapaho National Forest. On either side of the road are pristine forests in a special management area known as the Bowen Gulch Protection Area. The road finally terminates at the southern end of Never Summer Wilderness where you can take the Bowen Gulch Interpretive Trail. Trails into this wilderness begin from trailheads located within Rocky Mountain National Park. The Bowen Gulch Trail crosses the wilderness to Bowen Pass on the Continental Divide and the border with the Routt National Forest. Bowen

Baker Trail cuts through Bakers Gulch and past Parika Lake to Fairview Mountain.

State Route 125 departs from U.S. Highway 40 three miles northwest of Granby and follows Willow Creek and then Pass Creek to Willow Creek Pass and the Routt National Forest. Three good side roads branch off from State Route 125. Cabin Creek Road begins one mile south of Sawmill Gulch Campground and is a very crooked dirt road that reaches the Continental Divide in about 10 miles. Buffalo Creek Road is a shorter route from Denver Creek Campground to an hourglass-shaped reservoir. A mile-long trail cuts from the west end of the lake to the Continental Divide. The Stillwater Pass Road wanders for many miles across various aspects of the national forest until it intersects the Kawauneeche Road directly west of Shadow Mountain Lake. Numerous jeep trails run off the Stillwater Pass Road.

Georgetown, on Interstate 70, is crowded along Clear Creek and sur-rounded by tall mountains. If you drive up the crooked Guanella Pass Na-tional Scenic Byway behind Georgetown, you enter the Arapaho National Forest at the Clear Creek Campground, which is located at an elevation of 10,000 feet. The road then follows the western edge of the Mount Evans Wilderness, staying close to South Clear Creek and climbing to Guanella Pass. The pass is a great place to walk around in the alpine tundra habitat. During summer months you see myriad alpine wildflowers, most of them tiny. An exception is the remarkable old-man-of-the-mountain, with its three-inch-wide sunflower-like yellow heads. This showy species has a life-style that resembles that of the huge century plants of southwestern deserts in that after taking several years to finally bloom, the entire plant dies. Among the tiny blossoms to be seen at the pass are alpine forget-me-not with five blue petals around a tiny yellow center, alpine avens, whose flower re-sembles a buttercup but whose leaves resemble a fern, American bistort, with spikes of tiny white or sometimes pinkish flowers, alp lily, with six white petals and grasslike leaves, and the very pink alpine primrose. If you are here just as the snow melts, usually by mid-June, you may see the white flower clusters of mountain candytuft, which has leaves covered with a fine powder. Just before the snow begins to fly again in fall over the rocky terrain, the white flowers of arctic gentian open.

When gold was discovered in the Snake River Basin west of the Conti-nental Divide, miners in the Georgetown area needed to get to the deposits of ore. Although only 10 miles from Georgetown, the mines were on the other side of the Continental Divide and the miners had to find a way across the divide. To do this, a road was built across 13,207-foot Argentine Pass. With a four-wheel-drive vehicle and nerves of steel, you can drive to the pass on a road off the Guanella Pass National Scenic Byway, about 2.5 miles before you come to the Clear Creek Campground. Originally a toll road, the Argentine Pass Road was completed in 1871 and is the highest road ever constructed over the Continental Divide. The west side of the pass, in the Pike National Forest, is so rough and steep as it follows a narrow shelf that the Argentine Pass Road was soon abandoned. Today, even the jeep road does not descend the western side, although there is a hiking trail. Just before coming to Argentine Pass, you pass the townsite of Waldorf. At 12,666 feet, the elevation of the post office at Waldorf was the highest in the country around 1870.

If you stay on Interstate 70, heading west out of Georgetown, you climb to the Continental Divide and the engineering marvel of the Eisenhower Tunnel. Before the tunnel was bored through the Continental Divide, travelers had to cross the Continental Divide over Loveland Pass. You may still drive scenic U.S. Highway 6 over Loveland Pass from an exit off Interstate 70 before the Eisenhower Tunnel. As you approach the pass, the road travels through several miles of alpine tundra all around you; it is an unforgettable drive. As you start your descent to the west from the pass, there is the Pass Lake Picnic Area. If you decide to have your picnic here, you may need your sweater or jacket because you are at an elevation of 11,900 feet. On the west side of the pass, the highway eventually circles around the northern shore of Dillon Reservoir before rejoining Interstate 70 at Frisco.

At Frisco, State Route 9 follows the southern end of Dillon Reservoir. A side road goes into the Peninsula Recreation Area that juts into the lake. At the south end of the Blue River Arm of Dillon Reservoir, Swan Mountain Road turns north off of State Route 9 and continues to stay near the reservoir until it joins U.S. Highway 6 just west of Keystone, where there is a developed winter sports area. If you continue south on State Route 9, there is another fine winter sports area at Breckenridge on the eastern side of the Tenmile Range. The Boreas Pass Road begins in Breckenridge and winds its way to the pass and into the Pike National Forest (pl. 5). Before reaching the pass, you come to the historic Bakers Water Tank that has been restored.

State Route 9 follows the course of the Blue River, eventually zigzagging its way to a bleak alpine area at Hoosier Pass. The south side of the pass is in the Pike National Forest.

Just beyond Frisco, Interstate 70 squeezes between Royal Mountain to the east and Wichita Mountain to the west. At the junction of Interstate 70 and State Route 91, the Copper Mountain Ski Area is a very popular winter sports area. State Route 91 southward leaves the Arapaho National Forest after four miles and begins a long but gentle climb to the Climax Molybdenum complex, Fremont Pass, and the Pike National Forest.

At Vail Pass, Interstate 70 leaves the Arapaho and enters the White River

National Forest. A paved bike trail parallels Interstate 70 from a few miles west of Georgetown to Vail Pass and beyond.

State Route 9 northward from Silverthorne to Kremmling follows the Blue River to huge Green Mountain Reservoir. The Forest Service maintains campgrounds and boat launch areas around the lake. Most of the Forest Service land to the west is in the Eagles Nest Wilderness. A paved side road to the east off of State Route 9 climbs over Ute Pass.

A small segment of the Arapaho National Forest sits north of Idaho Springs. The Peak to Peak National Scenic Byway leaves Central City on State Route 119. Although most of the scenic byway is in the Roosevelt National Forest, there are two campgrounds at Pickle Gulch and Cold Springs that are in the Arapaho. Fewer than two miles north of Central City, a well-graded gravel side road follows North Clear Creek past several mines to a small mining settlement at Apex and then to the site of another mining town appropriately named Nugget. Just before the road crosses into the Roosevelt National Forest is a spectacular wetland at Elk Park. Because the underlying rock beneath the wetland is limestone, the soggy wetland is a fen.

One other side trip into the Arapaho National Forest is worth taking. From the first exit off of Interstate 70 west of Idaho Springs, drive to the mining town of Alice. You may hike uphill from town for about one mile to St. Mary's Glacier and the glacial lake below the ice.

Hoosier Ridge

The Continental Divide, the Rocky Mountain watershed that separates west-ward- from eastward-flowing rivers, generally runs from the north to the south. Locally, however, it may run in more of an east-to-west direction. One such place is Hoosier Ridge, in central Colorado, an 11,600- to 13,200-foot-high crest, the slopes of which support tundra vegetation—stunted, cold-adapted species characteristic of high altitudes and high latitudes. Located in Pike and Arapaho National Forests, it is most easily reached by taking Colorado Highway 9 to Hoosier Pass and hiking eastward.

To the west of Hoosier Pass lies the Mosquito Range, consisting of calcareous rock, or limestone. Hoosier Ridge, to the east, is essentially granitic. The most abundant vegetation lies on the moister, north-facing slopes of the ridge, where depressions are snow-covered even in late spring or early summer. Here the dominant plants are tufted hair grass, with its threadlike leaves, and golden avens, a wildflower in the rose family. Other plentiful wildflowers are sky pilot, which is a handsome, blue-flowered member of the phlox family, and the densely tufted, pink-and-white-flowered whiproot clover, a species confined to granitic soil. In small, sheltered areas where the most snow and water accumulate, miniature, stunted trees of subalpine fir and Engelmann spruce grow alongside the equally diminutive gray willow.

South of the Continental Divide, on the warmer but drier south-facing slopes, the same golden avens and a grass known as kobresia prevail at the higher elevations, whereas tufted hair grass grows lower down, punctuated by shrubby, dwarf, gray willows and barren-ground willows. Scattered throughout the plant communities on both sides of Hoosier Ridge are areas of bare rock, bare soil, and permanent snowfields, where vegetation other than mosses and lichens cannot survive.

What makes Hoosier Ridge so significant botanically are some dozen rare alpine species that grow in isolation from their closest relatives. The most celebrated of these is Penland's alpine fen mustard, a three-inch-tall plant discovered on Hoosier Ridge in 1935 by botanist C. William Penland. Since that time, it has been found in a few other areas, all along a 17-mile stretch of the Mosquito Range crest. This rare plant grows in sphagnum-covered fens above 12,500 feet, habitats confined to small, flat ledges kept moist by surrounding, persistent snowfields. Such snowfields exist along the north slopes of this east-to-west portion of the Continental Divide; where the watershed runs from the north to the south, the slopes are more exposed to the drying effects of the prevailing winds.

A tiny plant with minute, white flowers and shiny, heart-shaped leaves borne on slender stalks, Penland's alpine fen mustard is most closely related to Edwards's arctic mustard, found more than 1,000 miles away above the Arctic Circle. Another species in the same genus grows in Asia.

Other rarities include globe gilia, a sweet-smelling member of the phlox family, whose creamy white flowers form in a dense cluster at the top of a six-inch-tall stem. Although this species was first discovered in 1872, it has never been found anywhere in the world except on southern slopes along Hoosier Ridge and in the Mosquito Range.

Sea pink grows on rocky slopes in the Hoosier Ridge tundra at elevations above 12,000 feet. Its spherical clusters of pink flowers rise above a basal tuft of very narrow leaves. The only other places in the world where this species is found are in Canada's Northwest Territories and in Mongolia.

Weber's saussurea grows in the Beartooth Mountains of northeastern Wyoming, in the Belt Mountains of Montana, as well as in the tundra of Hoosier Ridge. It has purple flowers and is protected from the bitter conditions by woolly leaves and bracts. Its closest relative, another kind of saussurea, grows in Saskatchewan.

These and other rare wildflowers can be damaged or destroyed by trampling or other disturbance. Colorado has already acted to designate 925 acres of Hoosier Ridge as a state natural area. Because this zone falls within

National Forest land, however, full government protection will not be assured unless the U.S. Forest Service designates this a Research Natural Area as well.

Elk Park

In the heart of Colorado's Rocky Mountains, west of the nearly deserted mining community of Apex, a rough gravel road climbs steadily through a forest of spruce and pine. After a mile, the road dips into the next valley, and a panorama of the Rockies comes into view, with James Peak rising in the distance. The abandoned Nugget Gold Mine lies on a crest about three-fourths mile to the right, and a small, still-operational gold mine nestles in the hillside below. At the foot of the hill, a beautiful, meadowlike opening known as Elk Park fills the broad valley. Through it runs Elk Creek, which draws elk down from the surrounding mountains to drink and browse on the vegetation. Except for one privately owned corner, Elk Park is managed by the Arapaho National Forest.

Near where the road crosses the creek is a wetland habitat known as a shrub carr, "carr" being a word of Scandinavian origin that means marsh or fen. Shrubs, none more than four feet tall, grow in and along the creek and adjacent rivulets. Flat-leaved willow, short-fruited willow, and dwarf birch are the most common shrubs, and they are often so close together that little else can grow with them. Dwarf birch is unlike most other species of birch, no more than three feet tall with small leaves that have no coarse teeth along the edges. In addition, the bark does not peel off the way it does naturally on other birches. The flat-leaved and short-fruited willows, like most willows, are well adapted to moist habitats, with extensive root systems to provide anchorage in the saturated earth.

Willows grow primarily in temperate and boreal climates; hardly any are tropical. Most have leaves that are considerably longer than they are broad, and all willows form simple flowers without petals. The male flowers each consist of one or two pollen-producing stamens and are crowded into long, drooping structures known as catkins. Each willow has both the male and female variety. Most trees that have flowers without petals, such as oaks, hickories, and birches, depend on the wind as the agent for pollination, but willows are pollinated by insects.

A tuft of soft, silky hairs enables the willow seed to be blown a considerable distance from the parent plant. When the light seeds glide onto suitable, moist soil, they can quickly colonize the area. If a seed happens to fall on water, it may float for a few days before lodging in the soil and germinating.

Several species of dwarf willow grow in the American Arctic and above timberline in the Rocky Mountains. Some are less than a foot tall at maturity, but their woody stems qualify them as shrubs. At the other extreme, some willow trees found along streams at lower latitudes and lower elevations grow more than 60 feet tall. The willows along Elk Creek bristle with branches but seldom exceed four feet.

Although the willows fan out into the carr, here and there layers of soil have built up in small clearings. These have become covered with thick mounds of mosses and harbor a diversity of wildflowers, whose display of blooms peaks in August. The heart-leaved bitter cress, with its pure white flower clusters and deeply cleft green leaves, grows in the wettest areas adjacent to the numerous rivulets. The pink, pendulous flowers of the Rocky Mountain shooting-star contrast with the white blossoms of the marsh marigold. Two succulents, king's crown and rose crown, reach 10 inches above the mosses; their fleshy leaves would seem to be better adapted to desert conditions. Some wildflowers that live among the crowded willows have to grow tall to compete for sunlight. They include blue monkshood, yellow triangle-leaved groundsel, and tall larkspur.

The terrain rises gradually as you follow Elk Creek toward its source, and the soil becomes drier. About 300 yards upstream from the road, the thicket of willows and dwarf birch begins to open up, providing more space for the wildflowers. Soon the willow and birch zone ends, and plants that tolerate drier conditions take over. Milfoil and locoweed abound in the now-pebbly soil, along with scattered, six-foot-tall monument plants. A robust member of the gentian family, the monument plant has inch-wide flowers with four yellow green petals, each marked with small purple spots and bearing two hairy glands that secret nectar to attract insects. This plant is also sometimes known as deer's ears because of the shape of the very fuzzy leaves.

Grand Mesa National Forest

SIZE AND LOCATION: 368,418 acres in western Colorado, its western edge about 15 miles east of Grand Junction. Major access roads are Interstate 70, U.S. Highway 50, and State Routes 65 (Grand Mesa National Scenic Byway) and 330. District Ranger Station: Grand Junction. Forest Supervisor's Office: 2250 Highway 50, Delta, CO 81416, www.fs.fed.us/r2/gmug.

SPECIAL FACILITIES: Winter sports area; motorized trail vehicle areas; boat launches.

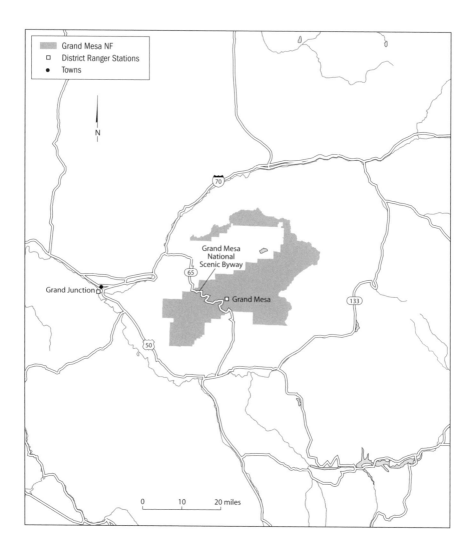

SPECIAL ATTRACTIONS: Grand Mesa National Scenic Byway; Crag Crest National Recreation Trail; Grand Mesa.

WILDERNESS AREAS: None.

One of the largest flat-topped mountains in the world, Grand Mesa's 540 square miles is dotted with 300 lakes and ponds, and the surface of the mesa is not really flat but contains narrow ridges, canyons, and an assortment of pinnacles. The mesa is entirely encompassed in the Grand Mesa National Forest. Although the average elevation of the mesa is 10,500 feet, several areas rise above 11,000 feet, with Leon Peak topping out at 11,327 feet. The national forest lies directly east of Grand Junction, with a small, narrow strip north of State Route 330 and a tiny isolated unit 22 miles south of Grand Junction. The geology of Grand Mesa is complex, with the mesa dating back about 100 million years.

Four vegetation zones exist on the mesa, determined primarily by elevational differences. Above 10,000 feet, the alpine zone is dominated by Engelmann spruce and subalpine fir, although Douglas fir may occur at the lower limits of this zone. This zone receives snow during winter and is subject to frequent rainstorms during summer. Between 8,000 feet and 10,000 feet, the montane zone, aspens are plentiful and are joined by Douglas firs as the dominant trees. The transitional zone occupies the 7,000- to 8,000-foot elevations and is recognized by dense stands of gnarly Gambel's oaks that rarely grow taller than 10 feet. Several shrubby species, such as serviceberry (fig. 6) and mountain mahogany, are common in this zone. Creeks and streams throughout the mesa are usually lined with narrow-leaved cottonwood and chokecherry. The lowest elevation around the periphery of Grand Mesa, usually below 7,000 feet, is more open with trees of piñon pine and juniper.

To get a great orientation of the national forest, drive the Grand Mesa National Scenic Byway. Beginning at the small town of Cedaredge, about nine miles south of the national forest, this highway (which is State Route 65) climbs over a couple of switchbacks before coming to the southern boundary of the national forest. The scenic byway curves past many of the better areas on the Grand Mesa before exiting the national forest near the Powderhorn Ski Area. State Route 65 retains the scenic byway designation all the way to Interstate 70 and then for several miles west on the interstate.

After entering the national forest's southern boundary, the scenic byway follows a wide circular route westward and then northward before coming to Ward Creek Reservoir, where there is a boat launch area and where fishing is reported to be good. From the reservoir the scenic byway climbs about 600 feet to the Grand Mesa Visitor Center. In addition to the usual amenities at the visitor center, there is a half-mile loop interpretive Discovery Trail that is handicap accessible.

If you wish to spend some time exploring the mesa, depart from the scenic byway at the visitor center and take the forest road eastward, quickly following the north shore of Ward Lake, where

Figure 6. Western serviceberry.

there is a fine campground and another boat launch. The forest road then stays along the northern edge of Alexander Lake to a very narrow bit of land separating Alexander Lake to the south and Hotel Twin Lake to the north. At this point, the road forks. You may take the left fork north to Hotel Twin Lake and return to the scenic byway in about two miles, or you may take the right fork for additional points of interest in the national forest. If you choose the latter, the forest road passes Baron Lake and comes to the west end of Eggleston Lake, one of the largest on the mesa. You have another choice at a junction here. South is a very circuitous gravel road that allows you to see and perhaps fish at several lakes until the road comes to the Grand Mesa National Scenic Byway you were on just as you entered the national forest from Cedaredge. Kiser Creek Campground is along the way.

The north route from the junction brings you to the Eggleston Lake boat launch area and, just beyond, the Crag Crest Campground. The campground has the eastern trailhead for the Crag Crest National Recreation Trail, which is a 10-mile loop that allows you to observe several different aspects of the mesa. From the campground the trail climbs sharply to Crag Crest, which is a rocky escarpment with steep drop-offs up to 500 feet on either side of the trail. Caution should be taken, particularly when the trail becomes extremely narrow in places. The trail follows the crest for about six miles before it loops back on a somewhat lower and less spectacular route to the starting point at the Crag Crest Campground. You can also access the Crag Crest National Recreation Trail at its western end from the trailhead just north of Island Lake. If you take the trail during summer, you encounter many wildflowers along the rocky crest, including mountain sorrel, dwarf yellow whitlow grass (a diminutive member of the mustard family), pussy-toes, a yellow-flowered paintbrush, and three species of fleshy-leaved sedums—a rose-pink one known as rose crown, a dark red-to-purple one called king's crown, and a yellow one usually referred to as stonecrop.

The forest road continues beyond the Crag Crest Campground and winds around to four reservoirs, each with a boat launch. After Big Creek Reservoir, the road leaves the national forest after about six miles and eventually comes to Collbran on State Route 330.

From the Grand Mesa Visitor Center, the scenic byway curves around the upper end of Island Lake. At Island Lake is a campground, a boat launch, and nearby, the western trailhead for the Crag Crest National Recreation Trail.

At the Land O' Lakes parking area at the western end of Island Lake is a short interpretive trail. A rough four-wheel-drive road leads southward from the campground and ends at Big Battlement Lake. From here there is a hiking trail to the west. Hiking during summer through a forest dominated by aspens brings you in contact with several attractive wildflowers, including a wild larkspur, blue columbine, meadow rue, waterleaf, heart-leaf arnica, mule's-ears, and two giant herbs, cow parsnip and giant hyssop.

After another three miles north on the Grand Mesa National Scenic Byway is the Lands End Road turnoff, which is a route that eventually drops off the western end of Grand Mesa. After only 1.5 miles, Flowing Park Road leads south from Lands End Road, terminating in about nine miles at the Flowing Park Reservoir. Back on the Lands End Road, the first 10 miles of this road are relatively easy and are very scenic as the road winds its way through forests and meadows. Where Engelmann spruce dominates the forest, there is an assemblage of attractive wildflowers beneath them. Among the showiest are Solomon's seal, twisted stalk, wild candytuft, Jacob's ladder, bluebells, valerian, a purple violet and a yellow violet, and parrot's-head, the latter a type of lousewort whose flowers have a fanciful resemblance to the head of a bird. Shrubs such as snowberry, Oregon grape, and bearberry honeysuckle are usually present beneath the spruces. Kinnikinnick forms a ground cover over much of the forest floor.

As you travel along the Lands End Road, common plants along the roadside include Mariposa lily, nodding wild onion, little sunflower (related to the larger sunflowers but in a different genus), goldeneye, and wild flax. The Lands End Observation Point is at the western edge of Grand Mesa, just before the road descends by many switchbacks that take you on a white-knuckle drive to U.S. Highway 50 in a desertlike habitat at the foot of the mesa.

From a side road off of the Lands Ends Road, outside the national forest, there is a rugged 12-mile hiking trail that reenters the Grand Mesa National Forest and climbs alongside Kannah Creek where it ends at Carson Lake between Lands End Road and Flowing Park Road. It begins in the piñon pine–juniper zone through the scrub oak zone of Gambel's oak and finally to a forest of aspen.

Should you not wish to leave the Grand Mesa in such a wild departure, stay on the Grand Mesa National Scenic Byway. At Skyway Point stop at the breathtaking observation pullout before the byway makes a sharp hairpin turn just before coming to the Mesa Lakes. An interpretive nature trail runs along the shoreline of one of the lakes. The distance between the Land O' Lakes and Mesa Lakes is only four air miles, but the elevation decreases by 1,000 feet. From Mesa Lakes, the scenic byway continues to descend rapidly, losing another 3,100 feet in just more than five air miles. Where the byway leaves the national forest, there is a short side road to the Powderhorn Ski Area.

Numerous openings in the forest on Grand Mesa offer a rich diversity of grasses and wildflowers. These meadows usually contain yellow sneezeweed, golden pea, wild geranium, leafy polemonium, scarlet gilia, mountain parsley phacelia, pink plumes, alumroot, and three kinds of gentians—star, fringed, and small-flowered. Where moisture builds up in shallow depressions in the meadows, monkshood, false hellebore, and snowball saxifrage may occur. In areas that are so wet as to be boglike, marsh marigold, globeflower, mountain death camas, rose paintbrush, and elephant's head occur.

A smaller part of the Grand Mesa National Forest lies north of State Route 330 and the town of Collbran. The western third of this part of the national forest is undeveloped and has no roads; however, a road runs northward from Collbran to Battlement Mesa where you can pick up the Battlement Trail just before the road narrows to a four-wheel-drive road. The Battlement Trail virtually runs the width of the eastern part of this segment of the national forest, passing McCurry Reservoir just south of South Mamm Peak (10,732 feet) and crossing Bald Mountain. The trail then circles north along Middleton Creek and enters the White River National Forest.

Twenty-two miles south of Grand Junction and nine miles south of Glade Park is a 12-square-mile isolated unit of the Grand Mesa National Forest. Known as the Fruita Division, this small forested section has two picnic areas—Fruita and Hay Press—along the forest road that bisects it. Three small Fruita Reservoirs and Black Pine Reservoir are located in this unit. Eighteenmile Lake Road cuts across the extreme southeastern corner of the Fruita Division.

Grand Mesa

The mountain that rises abruptly 6,000 feet above the Colorado and Gunnison River valleys, 25 miles east of Grand Junction, Colorado, looks as if its top has been sheared off by some giant machete. Contained within the Grand Mesa National Forest, this is Grand Mesa, perhaps the largest flat-topped mountain in the country. Not that the 540-square-mile mesa is completely flat; it just lacks any skyward-pointing peaks. Crevices and canyons dot the mesa, particularly around the periphery, and a narrow ridge runs along much of the mesa's length. In addition, more than 300 lakes pock the surface.

The summit of Grand Mesa, which averages 10,500 feet above sea level, supports large, open meadows punctuated by patches of Engelmann spruce. Rocky outcrops harbor fleshy-leaved succulents known as stonecrops and rose crowns. Because the winter snowpack prevails for many months, meadow wildflowers do not begin to bloom until at least late May. Jacob's-ladder, larkspur, and columbine abound. Marsh marigolds grow in continuous colonies in wet seepage areas (pl. 6).

Below the spruce zone, and extending down to about 8,000 feet, aspens grow, often propagated in dense stands by their extensive root systems.

Coralroot orchid and monkshood are two of the showier wildflowers that grow beneath these trees (pl. 7). On the mountain slopes below the aspens is a zone of scrubby vegetation dominated by the low-growing Gambel's oak. More open areas in this zone support the bushy growth of mountain mahogany and sagebrush. Here and there are isolated patches of lupines, beardtongues, and a sunflower-like plant known as mule's-ears (pl. 8).

From a distance Grand Mesa can be seen to consist of a series of nearly horizontal rock layers of various thickness. One hundred thirty million years ago, before these layers were formed, a great sea covered the region. As this sea receded westward, it left behind mud and the shells of countless clams. Eventually these deposits became consolidated into a 5,000-foot layer of gray black shale, which forms the foundation of Grand Mesa.

One hundred million years ago, as mountains began to form to the west of Grand Mesa, streams deposited vast quantities of sand, silt, and clay on the shale. These materials eventually became a 2,000- to 3,000-foot layer of gray brown sandstone, interlaced with coal seams that were formed from ancient decomposed vegetable matter. (Numerous coal mines operate near Grand Mesa today.) Streams poured more sandy sediments into the area, forming the next stratum, more than 2,000 feet thick in some places, of red sandstone and mudstone. Subsequently the land's surface sagged near Grand Mesa. Water filled this huge depression, forming the prehistoric Green River Lake. Sand, silt, and mud combined with the skeletons of fishes, clams, snails, and microorganisms on the lake bottom to form another 2,000-foot-thick layer of gray-and-brown sandstone and gray to brown to black beds of oil shale.

Thirty-five million years ago, with renewed warping of the land surface in the Rocky Mountain region, basins deepened and volcanoes formed. The Grand Mesa and surrounding areas were uplifted a few thousand feet. Then massive amounts of molten rock pushed up as domes into the overlying rocky layers. One of these domes, immediately to the southeast of what is now Grand Mesa, is now the West Elk Mountains. About 10 million years ago, cracks developed near the east end of Grand Mesa. According to geologist Robert G. Young, some lava reached the surface, where it flowed westward down an old stream valley before cooling to form a sheet of basalt. Eventually, as a result of eight or nine different lava flows, the entire surface of Grand Mesa was covered by a cap of lava up to 500 feet thick. Since that time, the sides of the old valley have been entirely removed by erosion, whereas the broad valley bottom, protected by the lava, has become the mountaintop.

But nature was not finished. During the Wisconsin glaciation, which began about 100,000 years ago, a massive ice sheet several hundred feet thick covered most of the top of Grand Mesa. Deposits left behind as this ice melted can be found along the northern and western edges of the mesa. After westerly winds piled up sand and red dust over the top of the mesa, building up a soil layer, a second ice sheet sent tongues of ice down precipitous slopes, adding more deposits. During the past 5,000 years, permanent streams have gradually carved deeply into much of the softer layers, but the protective lava cap has eroded very little, enabling the mountain to retain its tabletop character.

Gunnison National Forest

SIZE AND LOCATION: Approximately 1.7 million acres in western Colorado west of the Continental Divide and surrounding the town of Gunnison. Major access routes are U.S. Highway 50 and State Routes 92, 114, 133, 135, and 149 (Silver Thread National Scenic Byway). District Ranger Stations: Gunnison and Paonia. Forest Supervisor's Office: 2250 Highway 50, Delta, CO 81416, www .fs.fed.us/r2/gmug.

SPECIAL FACILITIES: Winter sports areas; boat launch areas.

SPECIAL ATTRACTIONS: Gothic Research Natural Area; Fossil Ridge Recreation Management Area; Slumgullion Slide.

WILDERNESS AREAS: Fossil Ridge (31,534 acres); Powderhorn (61,510 acres, partly on Bureau of Land Management land); Maroon Bells–Snowmass (181,117 acres, partly in the White River National Forest); Collegiate Peaks (166,938 acres, partly in the White River National Forest); Raggeds (64,992 acres, partly in the White River National Forest); West Elk (176,172 acres); La Garita (128,858 acres, partly in the Rio Grande National Forest).

Gunnison looks like the way a western town is supposed to look, and the landscape surrounding the town epitomizes the west. Much of the land surrounding Gunnison is in the Gunnison National Forest, although two significant areas are administered by the National Park Service—Black Canyon of the Gunnison National Monument and Curecanti National Recreation Area.

The Gunnison National Forest is a region of high mountains, although not as high as those of the San Juan Range. Tranquil lakes, peaceful meadows, and clear mountain streams add to the beauty of the area. The forest still has many wild, roadless areas, but jeep trails that often lead to abandoned

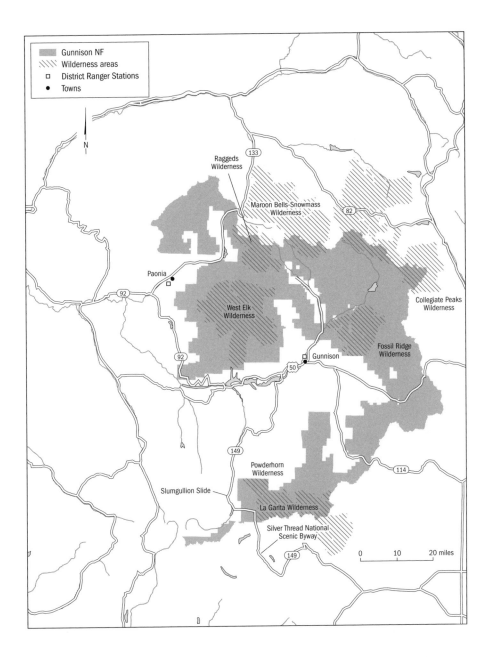

Gunnison NF
Wilderness areas
□ District Ranger Stations
● Towns

N

Raggeds Wilderness

133

Maroon Bells-Snowmass Wilderness

82

Paonia

92

West Elk Wilderness

Collegiate Peaks Wilderness

92

Fossil Ridge Wilderness

Gunnison

50

149

Powderhorn Wilderness

114

Slumgullion Slide

La Garita Wilderness

Silver Thread National Scenic Byway

149

0   10   20 miles

mines crisscross just as much land. The gold strike at Tincup and the discovery of coal at Crested Butte triggered an influx of people seeking a fortune or just seeking work.

The unique feature of the Gunnison National Forest occurs at the southern end of the forest a few miles southeast of Lake City. Soil, rocks, and trees on the flanks of Mesa Seco have been sliding down the side of the mountain for many years. Known as Slumgullion Slide, the movement is due to the spongelike mineral in the soil called montmorillonite. Slumgullion gets its name because, at times, the mineral has a smell that recalls that of a miner's stew, which the miners called slumgullion. The mountainside can be viewed from Windy Point along State Route 149, about eight miles southeast of Lake City. If you have time to linger in Lake City, you may enjoy the ambience of the laid-back town with its Victorian buildings. That part of State Route 149 from Lake City in the Gunnison National Forest to South Fork in the Rio Grande National Forest has been designated the Silver Thread National Scenic Byway. Most of the byway is in the Rio Grande National Forest, with only the 20-mile section between Lake City and Spring Creek Pass in the Gunnison. Beyond Windy Point, the highway climbs over Slumgullion Pass where there is a fine campground. At 11,200 feet, this is one of the highest campgrounds in the country. Eventually, the scenic byway ascends to Spring Creek Pass.

From the Slumgullion Campground there is a great road that follows Mill Creek to the northeast past lovely Deer Lake where there is another campground. The road then continues through a narrow corridor that separates the La Garita Wilderness to the south from the Powderhorn Wilderness to the north. Two miles beyond the Deer Lake Campground, a trail follows the western edge of the Powderhorn Wilderness into Cañon Inferno and then to Devils Lake. This wilderness is said to have the largest alpine mesas in the nation. Three campgrounds are along the corridor between the two wildernesses. The Mineral Creek Trailhead is for a trail that penetrates deep into the La Garita Wilderness between 13,383-foot Baldy Cinco and the towering 14,014-foot San Luis Peak. A branch trail to the east off the Mineral Creek trail takes the hiker over San Luis Pass and into the Rio Grande National Forest a few miles above Creede. The road continues to twist and turn and climb to Los Pinos Pass. Pretty Cebolla Creek Canyon lies to the west of the pass.

The remainder of the Gunnison National Forest lies either north of U.S. Highway 50 on either side of Gunnison or along either side of State Route 114 southeast of Gunnison. State Route 114 branches off of U.S. 50 about eight miles east of Gunnison. After about 22 miles, there is a junction with County Road 20. Regardless of which route you decide to take, you eventually enter the Gunnison National Forest. State Route 114 rises to the Continental Divide, crossing it at North Cochetopa Pass; County Road 20 crosses the divide at Cochetopa Pass. Side roads cross into the national forest from either road.

Monarch Pass on the Continental Divide marks the far eastern edge of the Gunnison National Forest. The pass that U.S. Highway 50 crosses has been in existence since 1956, with an earlier pass, now called Old Monarch Pass, about one mile northwest of the current pass. U.S. Highway 50 on the west side of Monarch Pass descends through the Gunnison National Forest to the small community of Sargents. The highway then leaves the national forest in about three miles at the Tomichi Creek Picnic Area and continues on to Gunnison.

An all-weather road connects Sargents with Marshall Pass, with the usual switchbacks to the top. The descent to the east is in the San Isabel National Forest. Marshall Pass is about eight miles south of Monarch Pass, but is more than 1,000 feet lower.

For an exciting alternate route in the east side of the Gunnison National Forest, take the Old Monarch Pass Road. Although it has a dirt surface, it can be driven in a passenger vehicle with proper caution. This road begins in the San Isabel National Forest, about one mile east of the new pass. Be prepared for many switchbacks as the old highway descends into the Gunnison. The scenery is marvelous. The road reaches Triano Creek at the bottom of the descent where there is a crossroad. The road to the south goes to Sargents in 5.5 miles; the road to the west climbs over Black Sage Pass and then to Waunita Hot Springs at the edge of the national forest. The road north stays near Triano Creek for a short while before starting a gentle climb through beautiful mountain scenery to secluded Snowblind Campground at about 9,800 feet. The road continues to ascend toward the Continental Divide, crossing several picturesque streams. Jeep roads run to either side. As you cross Robbins Creek, you have driven about as far as possible without a four-wheel-drive vehicle. The jeep road climbs over Tomichi Pass and finally to historic Hancock Pass on the Continental Divide. Hancock Pass is more easily reached from the west, however (*see* below).

Parlin is a tiny community on U.S. Highway 50. From Parlin there is a road along Quartz Creek to Pitkin, a town at the confluence of several gulches. Several jeep roads to the north head toward the Fossil Ridge Wilderness. These rough roads are within the Fossil Ridge Recreation Management Area that surrounds the wilderness on three sides. Because the Recreation Management Area is not wilderness, ATVs (All-Terrain-Vehicles) are permitted on its four-wheel-drive roads. The southern part of the area has hiking trails into the wilderness.

The road to Pitkin becomes a dirt road east of town and splits in about one mile. The right fork goes along Middle Quartz Creek to a pleasant campground. The left fork follows the course of North Quartz Creek, eventually climbing over Cumberland Pass. In 1.5 miles, just before the Quartz Campground, another dirt road branches off to the east and climbs to Hancock Pass. The last two miles, which should only be driven in a four-wheel-drive vehicle, are above the timberline. Before reaching Hancock Pass, there are two side roads. One is a jeep road south that quickly climbs over Tomichi Pass and in two miles comes to the townsite of White Pine. From Tomichi Pass there is a superb view into Brittle Silver Basin below. The other road is a passable dirt road north for 2.5 miles to the east portal of the abandoned Alpine Tunnel, at one time the highest railroad tunnel in the country. But when the roof of the tunnel collapsed in 1910, the tunnel was never used again.

Northeast of Gunnison is a huge section of the Gunnison National Forest situated between three wilderness areas. Collegiate Peak Wilderness is at the northeast corner of the national forest, Maroon Bells–Snowmass Wilderness is at the northwest corner, and Fossil Ridge Wilderness is at the southern end of the region.

This large section of the Gunnison National Forest has several roads. State Route 135 north from Gunnison goes to the Almont Campground. From here, the state highway heads northwest to Crested Butte. However, the paved road to the northeast from the campground enters the heart of the national forest to huge Taylor Park Reservoir, passing through scenic Taylor Canyon along the way. Trails at either end of the canyon lead into the Fossil Ridge Wilderness. From the east side of the reservoir are three all-weather roads. The north road gradually climbs along the Taylor River through Taylor Park, staying a short distance south of the Collegiate Peaks Wilderness. From this road is a trail into the wilderness along Red Mountain Creek to Lake Pass. At the Dorchester Campground, the forest road gradually worsens into a jeep road that ascends to Taylor Pass (11,928 feet) and into the White River National Forest. Taylor Lake is a beautiful high mountain lake just below the pass on the south side and just below timberline. At one time Taylor Pass was on a major route to Aspen.

The middle road east from Taylor Park Reservoir cuts across Taylor Park and eventually switchbacks its way to Cottonwood Pass, from which you can get marvelous views in all directions, particularly if you hike a few feet up to the summit from the pass. The drive to Cottonwood Pass is not a difficult one. East of the pass, the road descends into the San Isabel National Forest and to the town of Buena Vista.

The southern road from the reservoir is perhaps the most scenic and rewarding. It curves south along Willow Creek to the mining town of Tincup, passing several abandoned mines along the way. Tincup was a gold town, with perhaps 6,000 residents in 1882. After exploring around the old townsite, follow the old route over Tincup Pass to St. Elmo east of the Continental Divide. The dirt road east from Tincup to the Mirror Lake Campground is passable with a passenger vehicle, but from the campground to the pass, a four-wheel-drive vehicle is needed, unless you choose to hike the two miles above timberline to the pass at 12,154 feet. Another good trip south of Tincup climbs over Cumberland Pass, eventually following Quartz Creek to the town of Pitkin. The pass has less vegetation than other passes in the area and is at an astonishing 12,020 feet, the second highest pass in Colorado that a passenger car can negotiate.

State Route 135 between Gunnison and Crested Butte has two interesting side roads into the Gunnison National Forest. About 17 miles north of Gunnison is the road to Cement Creek Campground, with a natural hot spring nearby. The road continues to follow Cement Creek past the campground for about five miles before it deteriorates into a jeep trail. The jeep trail heads toward Taylor Pass but peters out near an old abandoned mine on the flank of Mount Tilton before reaching the pass.

Those with a four-wheel-drive vehicle, mountain driving experience, and nerves of steel may wish to tackle the Pearl Pass Road, which branches off of State Route 135 two miles before Crested Butte. This road follows Middle Brush Creek for most of the way, but the final ascent to Pearl Pass ranks second only to the Black Bear Road near Telluride as the most difficult of roads. The scenery is breathtaking, with the colorful Elk Range looming before you. Because of loose rocks and the always possible snow slides that cross the road, driving this trail is dangerous.

Deposits of coal primarily brought miners to Crested Butte, now the town caters to tourists and those interested in winter sports. Crested Butte is nestled in a beautiful setting between three wilderness areas that have high mountain peaks. The Maroon Bells–Snowmass Wilderness is north of town, the Raggeds Wilderness northwest, and the huge West Elk Wilderness is to the southwest. One road north of town runs between the Maroon Bells–Snowmass Wilderness to the east and the Raggeds Wilderness to the west. This road is situated east of Gothic Mountains where the Gothic Research Natural Area is located. This high elevation area is on land that ranges from 10,000 feet at the East River to 12,809 feet at the summit of Mount Baldy. Although the 1,490-acre area contains glacial cirques, talus slopes, and high mountain lakes, its main features are the virgin stands of Engelmann spruce and subalpine fir. Beneath the trees is a shrub layer of myrtle-leaved blueberry, bush honeysuckle, wild currant, and red elderberry. Among the colorful wildflowers are heartleaf arnica, curled lousewort, twisted stalk, and avalanche lily. A mile-long hiking trail from the old town of Gothic goes to Judd Falls. North of Gothic are the Avery Peak and Gothic campgrounds. Beyond the campgrounds, a rough but usually passable road climbs to Schofield Pass at the edge of the Maroon Bells–Snowmass Wilderness. The north side of the pass in the White River National Forest descends into Crystal Canyon via a jeep road only and eventually comes to the community of Marble.

About three miles west of Crested Butte and a short distance north of County Road 12 is a small natural gem in the national forest. Mount Emmons Iron Bog is about a three-acre bog at the foot of a steep slope of Mount Emmons. The bog is fed by several highly acidic natural springs that lie above it. The surface of the bog is covered with limonite, an iron oxide that has been deposited by the springs. The bog is the only location for the carnivorous sundew plant in Colorado as well as for two rare species of dragonflies. Mount Emmons Iron Bog is a registered Colorado Natural Area. A few miles west of the bog on County Road 12 is the Lake Irwin Campground. From the campground are several trails that meander past abandoned mines and into the Raggeds Wilderness. This wilderness includes the western end of the beautiful Elk Range (pl. 9). The forested areas are dominated by Douglas fir, ponderosa pine, Engelmann spruce, and subalpine fir, with stands of aspens here and there. Above 11,500 feet, the habitat is alpine tundra with small flowers and rocks everywhere. A popular but difficult trail from Erickson Springs goes down the incredible switchbacks past the Devils Staircase to Horse Ranch Park. A branch trail then climbs over Oh-be-joyful Pass.

A fine day trip can be enjoyed by driving good graded roads over Kebler Pass and Ohio Pass. Between the Lake Irwin Campground and Horse Ranch Park, County Road 12 climbs to Kebler Pass. Near the pass is another road south that comes to Ohio Pass in just two miles. These are two of the prettiest passes in the Gunnison National Forest, with many kinds of colorful wildflowers abundant during summer. These are great places to see blue columbine, Colorado's state flower (pl. 10). The road south of Ohio Pass goes to Gunnison. A 10.5-mile hiking trail into the West Elk Wilderness from Ohio Pass crosses Swampy Pass and Beckwith Pass before curving northward to Horse Ranch Park.

From Horse Ranch Park to Erickson Springs Campground, County Road 12 is one of the prettiest around, passing between the Raggeds Wilderness to the north and the West Elk Wilderness to the south. The road passes Lost Lake Campground, which is situated near a gorgeous wildflower meadow. From the campground there is a loop trail that passes Lost Lake and Dollar Lake and goes to Lost Lake Reservoir. Marcelline Mountain looms above the north side of County Road 12. The Dark Canyon Trail starts at the Erickson Springs Campground and crosses the Raggeds Wilderness.

County Road 12 joins State Route 133 at the south end of Paonia State Recreation Area. Coal Creek Road southbound climbs steadily through a narrow cut in the West Elk Wilderness. Several hiking trails into the West Elk Wilderness are off this road. Because elevations in the wilderness range from 7,000 feet to 13,035 feet on West Elk Peak, several vegetation zones are present, from piñon pine–juniper–Gambel's oak woods at the lower elevations to Engelmann spruce and subalpine fir at timberline. Alpine tundra vegetation occurs where there are no boulder fields.

From the Paonia Reservoir, State Route 133 northward or westward provides access to a large segment of nonwilderness land in the northwest corner of the Gunnison National Forest. If you stay on State Route 133 to the north, you soon climb over McClure Pass and into the White River National Forest and the city of Glenwood Springs. About seven miles north of the north end of Paonia Reservoir is a good graded road westward along East Muddy Creek and then the North Fork of Little Henderson Creek to the West Muddy Guard Station. From here you may either drive south to Pilot Knob, northwest along Gold Creek and into the Grand Mesa National Forest, or circle south to Paonia. Before reaching Paonia, there is a side road to the large Overland Reservoir.

U.S. Highway 50 west of Gunnison goes along the southern boundary of the Gunnison National Forest, passing along the north side of Blue Mesa Reservoir in the Curecanti National Recreation Area administered by the National Park Service. Side roads north of U.S. Highway 50 take you to the edge of the West Elk Wilderness, and then hiking trails disappear into the wilderness. After State Route 92 splits off of U.S. Highway 50 at the west end of Blue Mesa Reservoir, the state highway crosses north to Crawford. From State Route 92 are several roads onto Black Mesa in the Gunnison National Forest.

Slumgullion Slide

South from Gunnison, Colorado, whose wide main street typifies western American towns, State Highway 149 meanders toward the San Juan Mountains, part of the Continental Divide. The road parallels the Gunnison River for a stretch, cuts across dry, rolling hills peppered with sagebrush and snakeweed, and then begins to follow the Lake Fork of the Gunnison River upstream. The entrance to mountain country is signaled when the highway and river pass through the Gate, a narrow gap between two high cliffs. Eventually, the road enters the one-time gold- and silver-mining town of Lake City (altitude 8,800 feet), where a number of Victorian houses give the few streets an old-fashioned look.

Four miles beyond Lake City is Lake San Cristobal, for which the Lake Fork is named. The clear blue waters stretch nearly two miles across and reach a depth of 90 feet. The lake was created about 700 years ago, when a gigantic flow of earth moved down a nearby mountain and dammed the Lake Fork. Heading upland away from the river, State Highway 149 crosses this earthflow, known as Slumgullion Slide. After a great switchback, you can stop at Windy Point, in Gunnison National Forest, to see the slide up close.

The origin of Slumgullion Slide is the summit of Mesa Seco, nearly four-and-a-half miles from the lake and, at 11,400 feet, 2,600 feet higher. Below the summit is a semicircular basin, nearly 4,500 feet in diameter, contained on one side by 400- to 600-foot-tall cliffs of volcanic rock. Probably as a result of the action of gases and superheated water of magmatic origin, much of the rock has been altered to a clay mineral called montmorillonite. This mineral can absorb water and swell to several times its original volume. Swelling of montmorillonite, combined with freezing and thawing, causes rocks and clay within the cliffs to fall and slide into the basin.

At least twice in the past, rock debris and clay built up in the basin to such an extent that the material began to slide down the mountain. Based on radiocarbon dating of buried wood, geologists set the time of the slide that created Lake San Cristobal at about A.D. 1270. This earthflow was initially swift but gradually slowed until it reached the Lake Fork. Readily seen from the air or in aerial photographs, the flow is nearly four-and-a-half miles long and 500 to 2,000 feet wide. It slopes an average of 600 feet per mile.

About 350 years ago, a second flow began from the source basin, taking the same route down the side of the mountain. This flow has proceeded much more slowly, pushing and moving vegetation with it. So far it has moved two-and-a-half miles in a path that ranges in width from 500 to 1,000 feet.

Forty-five years ago, Dwight R. Crandell and David J. Varnes, of the U.S. Geological Survey, examined the new flow and conducted some experiments on it in order to determine its rate of movement and other characteristics. They studied aerial photographs taken from 1939 to 1952 and followed the progress of some of the identifiable trees in the photos. They found that the trees moved downslope at varying speeds. Some, about midway down the new flow, had moved 194 feet in 13 years, an average of 15 feet per year. Nearer the bottom, or toe, of the flow, things slowed down, with some trees having moved only five or six feet during the 13-year period. Similar results were obtained by identifying trees in a 1951 photo and then finding them "on the ground" in 1959.

For further confirmation, in 1958 Crandell and Varnes placed a series of control stakes along and across the earthflow. After two years, those stakes at midflow had moved about 20 feet per year, whereas those at the toe had moved only about 2.5 feet. Time-lapse motion picture photography, used during summer 1960, showed that the toe moved about half-an-inch each day. Crandell and Varnes concluded that even though the flow rate varies over the length of the slide, the mass in each section moves at a constant rate, apparently little affected by seasonal fluctuations of temperature and precipitation. They attribute the variation in flow rate to waves that move downhill through the slide.

The old flow has been inactive so long that the trees established on it—predominantly Englemann spruces and quaking aspens—grow straight. On the new flow, however, movement has disturbed the roots of the trees, and so most of them lean in various directions. Except for the trees, few other plants grow on the new flow. The soil is sticky, spongy, and yellowish. Some say it has a foul odor during the hottest summer days, calling to mind the thick, yellowish, smelly slumgullion stew that local miners used to eat.

Taylor River

In 1853, Capt. John Gunnison led a party of topographical engineers through western Colorado, across the main range of the Rocky Mountains, seeking a middle route for a railroad from the Mississippi River to the Pacific Ocean. The mountains proved too high for a feasible railroad route, but the discovery of gold in 1878 opened up the territory. Although Gunnison was killed by Indians, his name lives on in various landmarks, including Colorado's Gunnison National Forest.

Visitors to the area today can follow an all-season forest road to Taylor Park Reservoir, a huge and popular recreation lake. The road goes past the winter range of a herd of Rocky Mountain bighorn sheep and near their summer range on three mountain peaks. Although the sheep are elusive, they can sometimes be seen grazing during winter months or during their spring and fall migrations.

Rocky Mountain bighorn sheep are sturdy-bodied mammals that graze in mostly open areas of high mountain regions. Adult males, whose creamy white rump patches contrast with their generally gray-and-brown coats, weigh as much as 330 pounds. Their large brown horns curl upward and backward, then down and forward. The horns are not shed but grow throughout the sheep's life, continuing to curl and forming annual ridges. The approximate age of a bighorn sheep can be estimated by counting the ridges. Females are much smaller, lack the contrasting hair color, and have smaller, scarcely curved horns.

In winter, a herd of almost 100 animals congregates in a steep, rocky canyon near the junction of the East and Taylor Rivers, where the elevation ranges between 8,000 and 9,000 feet. Breeding occurs here in November and December. During winter months, the sheep graze primarily on a variety of grasses, although they occasionally nibble on sagebrush, rose bushes, serviceberries, aspens, and willows. The sheep are agile climbers and can move over the rocky terrain with amazing ease. They usually feed during the day, occasionally taking a noontime nap. At night, they seek protection among the rocks.

In late May, when snow begins to melt on the high mountains northeast of their wintering grounds, the bighorn sheep set out on a 14-mile trek along Taylor River to a few mountain peaks west of Taylor Park Reservoir. Serving as the lookout, the dominant ewe cautiously leads the way. She alerts the rest of the herd to any trouble and quickly leads them to safety.

The lambs are born on the way to the summer grounds on the top of Matchless Mountain, the 12,383-foot summit of which is well above timberline. Here the ewes and newborn lambs spend summer, foraging on the vegetation until the first heavy snows blanket the mountaintop. The rams, meanwhile, go to two slightly lower mountain peaks nearby—Rocky Point and Park Cone. Early snows on the mountains may force the sheep to start down to their wintering site earlier than usual, but winter migration normally begins in October.

This seasonal migration took place routinely for years, but early in the 1980s, wildlife biologists observed that the sheep were spending more time than usual on the wintering grounds and were using different migration routes to the summer range. In fact, some did not go to the summer range at all. Concerned that the herd might not maintain itself, range consultant Jim Berry, assisted by zoologist Tom Hobbs, of the University of Colorado, and wildlife manager Tom Henry, studied the Taylor River herd in 1982.

They discovered that lungworm was becoming prevalent among the animals because they were remaining at lower altitudes, where, with the coming of warmer weather, they were exposed to the snail that transmits the parasite. The sheep inadvertently eat the snails that contain the lungworm larvae, and the larvae develop into adult worms in the sheeps' lungs, causing a pneumonia-like disease. In addition, the grasses on which the herd depended were diminishing because of the greater grazing pressure. To create more grassland in the winter range, the Forest Service burned acres of sagebrush in 1982, 1983, and 1984, encouraging the regrowth of forage grasses such as brome and fescue. They also tried salting the summer range area in an effort to entice the animals to migrate to the mountain summits—but without success.

Finally, the study team concluded that the reason the sheep were delaying, altering, or skipping their normal migration to the summer range was because they had lost eye contact with landmarks along their migration route. Through the years, intervening forests of lodgepole pine had grown up, in effect preventing the sheep from seeing where they were going. Instead, they would often wander along Taylor River, where nine campgrounds and the associated human activity often discouraged their progress. Only those sheep that left their wintering ground earlier in spring than usual, while snow still covered the ground, could avoid this human contact.

Jim Paxon, who was the Taylor River district ranger of the Gunnison National Forest, came to the conclusion that if the Taylor River herd was to be saved, the sheep would have to take priority over timber. The decision was made to remove the interfering trees that had grown up in the steep-sided ravines of the winter range. Most of the lodgepole pines were not desirable timber in any case, because they were heavily infested with western mistletoe. The mistletoe had weakened the wood fibers of the pines and distorted their limbs and branches.

The doomed trees, however, were growing on such steep slopes that normal timber-cutting methods could not be used. Early in 1985, the Forest Service made the startling announcement that 1,000 acres of forest would be burned. In May and June, foresters carrying drip torches burned 600 acres of lodgepole pine. The remaining 400 acres, consisting of nearly inaccessible terrain, were burned by helitorch: a 55-gallon drum of gasoline jelly, suspended 15 feet below a helicopter. Golf-ball-sized globs of the jelly, ignited by a spark plug as they flow out of the drum, splash into flames when they hit the ground.

In evaluating the 1985 helitorching, Jim Paxon estimated there was a 70 percent kill of the standing timber, opening up the area and returning it to grassland. Most of the sheep immediately began to use their usual summer migration route. Lodgepole pines, however, have cones that open and shed their seeds only after fire. Since the 1985 burn, seedlings have sprung up everywhere. Paxon says that the Forest Service will reburn every seven years in order to keep the lodgepole pine under control.

Pike National Forest

SIZE AND LOCATION: 1,105,700 acres in the Front Range of the Rocky Mountains, immediately west of Colorado Springs. Major access routes are Interstate 25, U.S. Highways 24 and 285, and State Routes 9, 67, and 126. District Ranger Stations: Colorado Springs, Fairplay, and Morrison. Forest Supervisor's Office: 1920 Valley Drive, Pueblo, CO 81008, www.fs.fed.us/r2/psicc.

SPECIAL FACILITIES: Boat launch areas; motorized off-road-vehicle areas; mountain bike areas; winter sports areas.

SPECIAL ATTRACTIONS: Pikes Peak; Bristlecone Pine Scenic Area; Hurricane Canyon Research Natural Area; Saddle Mountain Research Natural Area; Guanella Pass National Scenic Byway.

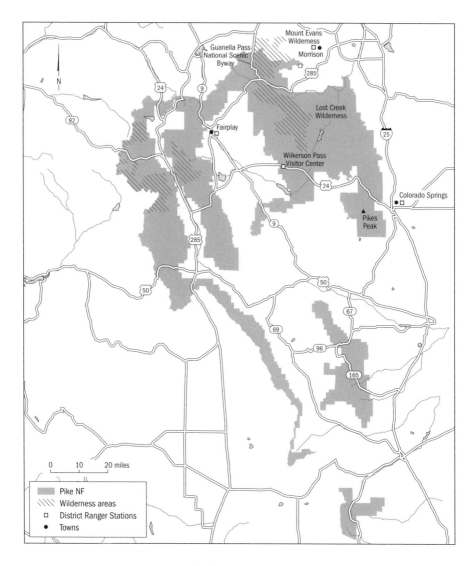

WILDERNESS AREAS: Lost Creek (119,700 acres); Mount Evans (74,401 acres, partly in the Arapaho-Roosevelt National Forests).

When most people think of Colorado, Pikes Peak comes to mind. Located in the Pike National Forest, Pikes Peak is perhaps the most famous mountain in the United States. To travelers heading west, Pikes Peak is the first mountain to be seen soon after passing Limon, Colorado, in the plains. Although Capt. Zebulon Pike was one of the first to record an observation of the peak in 1806, it was ascended by Dr. Edwin James (a botanist) and two colleagues on July 14, 1820. They were part of an expedition under the direction of Maj. Stephen Long.

The 14,110-foot summit of Pikes Peak rises more than 7,000 feet above Colorado Springs. The granite mountain has been carved by glaciers over millions of years. You can reach the summit three ways: hike the Barr Trail, drive the Pikes Peak Highway, or take the cog railroad. The hiking trail begins in Manitou Springs at the Ruxton Trailhead at an elevation of 6,720 feet. From the trailhead it is a 13-mile hike to the summit, gaining an elevation of 7,258 feet. If you are in good physical condition, it is possible to climb to the Summit House and back in one long day. Plan for the temperature at the top to be no warmer than in the 30s.

The Pikes Peak Highway starts at Cascade on U.S. Highway 24 and is open from May to September, from 7 A.M. to 7 P.M. After paying your toll, it is 19 miles to the Summit House. Be sure you and your vehicle are in good condition; otherwise, you may climb the highway via a tour bus from Colorado Springs. The first Pikes Peak road to the summit opened in 1889 as a carriage road, but when the cog railroad was completed in 1891, the carriage road received little use because of the competition. But from 1913 to 1915, Spencer Penrose built the current Pikes Peak Highway. Today there is a Pikes Peak Hill Climb auto race held every July.

Only one mile from the toll booth, after passing through forests of ponderosa pine, Douglas fir, and Engelmann spruce, a pullout at Camera Point allows you to look down Ute Pass to Colorado Springs and the vast plains beyond. With granite boulders along the right-of-way, the highway comes to the Crow Gulch Picnic Area. The highway then climbs and circles the north end of Crystal Creek Reservoir, where fishing for trout is usually good. The paved road soon becomes gravel. Views of the north face of Pikes Peak reveal gullies of snow and ice. After 10 miles on the toll road, the Halfway Picnic Area is at about 10,000 feet. A series of switchbacks, some narrow, bring you to Glen Cove in about three more miles. The cove, in a spruce and fir forest, is a glacial cirque with huge boulders above it. Just beyond, bristlecone pine, limber pine, subalpine fir, and Engelmann spruce are short and gnarly as the highway approaches alpine tundra habitat above the timberline. In a mile is Elk Park where you may rest for a while if you need to. The next two miles is a series of very narrow switchbacks that elevate you another 1,000 feet. Sixteen-mile Turnout is at the top of the switchbacks. If you look to the northwest, you see the Tarryall Range. South of the turnout is a hiking trail through the alpine tundra, with several observation points along the way. After the highway crosses the north ridge of Pikes Peak, there is an overlook into Bottomless Pit. The final approach to the summit is across a bleak, rocky terrain before the flat, 60-acre top is reached. Summit House offers refreshments and a gift shop. The views from the summit are as inspirational today as they were to Katherine Lee Bates in 1893 when she wrote the words to

"America the Beautiful" on her visit during that summer. Looking to the south you see the Spanish Peaks and other mountains in the Sangre de Cristo Range. West are the high mountain peaks along the Continental Divide. To the north, you can see Longs Peak in Rocky Mountain National Park on a clear day, 100 miles away. To the east, Colorado Springs lies at the base, with the Great Plains stretching far into the distance.

For a different aspect of Pikes Peak, take the cog railroad, the world's highest, from Ruxton Avenue in Manitou Springs. The railroad climbs 8.9 miles to the summit, offering different vistas as it ascends. The cog railroad has been in existence since 1891 and there is a charge for the three-hour and 15-minute round trip. Weather permitting, the cog runs from late April to October. Reservations are required.

The Rampart Range Road is a good way to explore the eastern side of the Pike National Forest. Beginning in Manitou Springs, this gravel road reaches the national forest in two miles and immediately climbs by a number of switchbacks into the Rampart Range. The first two side roads terminate at reservoirs outside the forest boundary, but the third side road goes to the Rampart Reservoir Recreation Area. Picnic areas and campgrounds are available around the large reservoir. From the end of the road to Rampart Reservoir is a nice trail that follows West Monument Creek east 3.5 miles to the U.S. Air Force Academy.

Continuing northward on the Rampart Range Road, you come to an extremely crooked dirt side road that eventually switchbacks around the southern end of Mount Herman to Monument Rock. The tower-shaped rock is composed of extremely crumbly sandstone. From Monument Rock the road is paved for three miles to the town of Monument on Interstate 25. An interesting hiking trail runs from the Mount Herman Trailhead into Limbaugh Canyon.

Three miles north of the Mount Herman turnoff is a jeep road east that curves around to an ice cave. Three-and-a-half miles farther along the Rampart Range Road, you see Chimney Rock less than two miles to the east. Several jeep roads branch off to the east before the Rampart Range Road comes to Jackson Creek, and a side road follows the creek all the way out of the national forest to the community of Shamballah on County Road 38. A campground sits 1.5 miles up the Jackson Creek Road. Across the Rampart Range Road from Jackson Creek, there is a hiking trail southward that passes the east side of Turtle Mountain as you follow first Eagle Creek and later Trout Creek before the trail comes out on State Route 67 about 18 miles north of Woodland Park. Beyond the Jackson Creek Road turnoff, the Rampart Range Road comes to a sharp switchback at Topaz Point where there is a magnificent view from the picnic area. A dirt side road goes to Devil's Head Campground from which you may hike 1.5 miles to the historic Devil's Head Lookout Tower on the Devil's Head National Recreation Trail.

The Rampart Range Road continues for nearly four miles to a fine overlook and campground at Flat Rock. The Rampart Range Motorized Vehicle Area is on both sides of the main road and contains 120 miles of off-road-vehicle trails. Beyond Flat Rock, the Rampart Range Road proceeds for another six miles to its northern terminus on State Route 67.

State Route 67 north out of Woodland Hills parallels the Rampart Range Road two to six miles farther to the west. This highway comes within sight of various formations known as Red Rocks, Painted Rock, and Bell Rock. To the east of Painted Rock is the Manitou Experimental Forest. About 22 miles north of Woodland Park, State Route 67 comes to the Platte River at the community of Decker. You may stay on State Route 67 and follow the Platte River north past Skull Rock, which will be on your left. Beyond Skull Rock are several campgrounds before eventually the highway leaves the Pike National Forest at its northeastern corner. Or, from Decker, you may take the paved road westward off of State Route 67. Two campgrounds are available in the first four miles of this road. At the second campground, called Wigwam, State Route 165 branches off northward and comes to the Buffalo Creek Mountain Bike Area near the northern edge of the Pike. The bike area offers 40 miles of mountain bike trails. You may choose to take the gravel road west from Wigwam Campground. This is the Matukat Road, which winds for many miles and goes for a while along the southeastern border of the Lost Creek Wilderness before ending on County Road 77. From the Goose Creek Campground along the Matukat Road are two trails that enter the wilderness. Goose Creek Trail follows Goose Creek north for several miles before leaving the wilderness and coming to the Wellington Lake Road. Goose Creek wanders in and out of huge boulder slides nine different times. Hankins Trail cuts directly across the wilderness from the east to the west, climbing up and over Hankins Pass and following the north slope of South Tarryall Peak before coming to the Tarryall Road. Lost Creek Wilderness is fascinating with its granite spires, variously shaped pinnacles, rocky slopes, deep canyons, and wildflower meadows. The wilderness derives its name from the several times the Lost Creek disappears beneath boulders, only to reappear again. A trail from just north of the Flying G. Ranch Girl Scout Camp off the Matukat Road follows Lost Creek completely across the wilderness, terminating at the Lost Park Campground and Trailhead on the west side of the wilderness. If you stay on the dirt road toward the Lost Creek Trailhead but continue north, the road comes to a nice meadow known as Webster Park, and then the road climbs alongside Cabin Creek to cross Stoney Pass before descending to Wellington Lake. Stoney Pass, not to be confused with Stony Pass in the Rio Grande and San Juan National Forests, is a low pass (8,560 feet) between the Platte River and its North Fork. When you first glimpse Wellington Lake from the pass, you see The Castle rock formation rising high above the lake. The Castle is on the northeastern border of Lost Creek Wilderness.

The Hurricane Canyon Research Natural Area near Manitou Springs preserves a prime example of the original vegetation in the area dominated by Douglas fir, ponderosa pine, Engelmann spruce, and Colorado blue spruce. Gambel's oaks are in the drier areas. The steep-sided Hurricane Canyon has been carved by French Creek. Some elk use the area as a wintering ground. The area is extremely rugged and difficult to maneuver, but the quality of the area may be worth the effort to get there. Two miles up the Barr Trail from the trailhead in Manitou Springs, the natural area lies to the north of the trail.

U.S. Highway 24 west of Lake George is a fine route from which to explore the southern part of the Pike National Forest. West of Lake George, the highway climbs to Wilkerson Pass (9,502 feet), where there is a well-staffed visitor center. Also from Lake George, County Road 96 is a scenic gravel road that stays along the South Platte River all the way to the east end of huge Elevenmile Canyon Reservoir (not in the national forest). Seven Forest Service campgrounds are available along the way.

When you leave Lake George on County Road 96, if you take County Road 61 that branches south off of County Road 96 about one mile south of Lake George, the road circles around Blue Mountain where it becomes County Road 98. This road heads west to its junction with County Road 59. Turn left here and go for 3.5 miles. You see Saddle Mountain to the east. A half-mile hike cross-country brings you to the Saddle Mountain Research Natural Area, a pristine montane grassland with a scattering of Engelmann spruce and quaking aspen. The rocky south slope of Saddle Mountain is dominated by a grassland community of Parry oatgrass and mountain muhly. Needleleaf sedge is a codominant in the meadows on the eastern slopes. Where grazing has had an impact, Parry oatgrass has decreased and been replaced by American vetch, rose pussy-toes, Arizona fescue, mutton grass, and squirrel-tail grass. The area is also a dedicated Colorado Natural Area.

North of Lake George is the gravel Tarryall Road off of U.S. Highway 24. This is a most interesting and scenic route along Tarryall Creek, first going up Tappan Gulch and past Tappan Mountain to the east. After continuing past the south end of Matukat Road, the Tarryall Road skirts the south end of Lost Creek Wilderness.

Tarryall Road continues past the remnants of the old Tarryall community, passes Spruce Grove Campground and the Twin Eagles Trailhead into the

Lost Creek Wilderness, and comes to the Ute Creek Trail that enters the southwest corner of the wilderness. Just before Sugarloaf Mountain to the north, the road becomes paved as far as the Rock Creek crossing. Within a mile, the Tarryall Road leaves the Pike National Forest and winds its way west to U.S. Highway 285 near Como.

The western and northern parts of the Pike National Forest can be explored off of U.S. Highway 285. The historic town of Fairplay is your jumping-off place for these parts of the national forest. Four-and-a-half miles south of Fairplay, County Road 5 branches off west of U.S. Highway 285. In two miles there is a scenic jeep road westward that immediately climbs to Breakneck Pass. This pass road is a one-lane track that goes between Sheep Ridge and Round Hill. Miners used it while exploring for gold and silver in the area during the 1860s. The abundance of aspens in the area makes it extremely attractive during fall. The jeep road continues beyond the pass to the intersection with Browns Pass Road. The main road to Browns Pass is County Road 20 that branches off of U.S. Highway 285 three-and-a-half miles south of Fairplay. If you are not driving a four-wheel-drive vehicle, stay on County Road 5 past the Breakneck Pass Road and swing southward until you reach the South Platte River. At this point, County Road 5 becomes County Road 22 and curves to the north, soon becoming dirt as it climbs to Weston Pass. Where the road curves to the north, Rich Creek joins the South Platte River. A nice hiking trail makes a complete circle, first staying alongside Rich Creek and then following Rough and Tumble Creek back to the trailhead. The complete round trip is about 13 miles.

For many years, Weston Pass was the main route across the mountains from Leadville to South Park and Fairplay. But when Mosquito Pass was completed several miles to the north, Weston Pass became little used. The road to Weston Pass from the eastern side is suitable for passenger cars. The west side, in the San Isabel National Forest, is better traversed by four-wheel-drive vehicles.

From Fairplay, State Route 9 goes into the northwest corner of the Pike National Forest all the way to Hoosier Pass, after which it enters the Arapaho National Forest. Just before reaching Alma Junction, the massive mountain that appears directly in front of you is Mount Bross, and in the distance to the right of Mount Bross is Quandary Peak. At Alma Junction, County Road 12 branches off of State Route 9 and passes Park City with its historic cemetery and several old mine sites. The road splits, with one fork going south of London Mountain and the other fork north of the mountain and up to Mosquito Pass. At 13,188 feet, Mosquito Pass is the highest pass road in the country that is still open, but it is suitable only for four-wheel-drive vehicles. Because it is so high, Mosquito Pass was not used for very long for mining operations. Today, the World Championship Mosquito Pass Burro Race from Leadville to Fairplay is held annually on this road.

Back on State Route 9, Alma is about one mile north of Alma Junction, and County Road 8 takes off from here. Do not miss this. In 1.5 miles you come to Buckskin Cemetery to your right, near which was also a stockade built for protection from Indians. Beyond the cemetery about one mile you may note at the edge of a stream a circular depression, known as an arrastra, that is about six feet across. Ore was brought down the rocky hill to your right and placed in the depression. A large rock chained to a pole was dragged around and around the arrastra by waterpower, pulverizing the rock and freeing the gold particles, which sank to the bottom of the depression and were then recovered. About three miles west of Alma, a side road climbs up to Windy Ridge and the incredible Bristlecone Pine Scenic Area. These bristlecone pines live at an elevation of 12,000 feet, and they must withstand the high winds that give Windy Ridge its name. These trees, many of which resemble living driftwood, are estimated to be about 1,000 years old. Their windswept shape is truly a sight to behold. County Road 8 continues past the Windy Ridge turnoff to Kite Lake and a secluded campground. The lake is located in a glacial cirque. About one mile before reaching Kite Lake, if you look to Mount Bloss, you see a natural red amphitheater that has been carved by nature. A five-mile drive on State Route 9 takes you from Alma to Hoosier Pass and the Arapaho National Forest.

By staying on U.S. Highway 285 north out of Fairplay, you come to the small community of Como. From Como, a dirt road follows the old railroad grade as it climbs to 11,482-foot Boreas Pass on the Continental Divide. When the railroad was built over the pass in 1882, Boreas Pass was the highest railroad pass in the country. The railroad grade was converted into an automobile road in 1937. Boreas was a small settlement at the pass, but little remains of it today. Across the pass, the road is in the Arapaho National Forest as it heads toward Breckenridge.

From the community of Jefferson, about 17 miles northeast of Fairplay on U.S. Highway 285, County Road 35 heads northeast. Take the dirt County Road 54 that branches off of County Road 35 and follow this very twisting and curving road that ascends to Georgia Pass. The pass is relatively flat and open, and there is a spectacular view of Mount Guyot directly west. Do not try to descend the west side of the pass into the Arapaho National Forest unless you have a four-wheel-drive vehicle.

U.S. Highway 285 crosses Kenosha Pass about four miles north of Jefferson where it reenters the Pike National Forest. The Colorado National Recreation Trail crosses the pass. Hiking the trail westward winds your way down

Figure 7. Ptarmigan.

and up several times before arriving at Georgia Pass. The trail to the east eventually cuts across the Lost Creek Wilderness.

At the town site of Webster some 4.5 miles north of Kenosha Pass, County Road 60 is a dirt road that proceeds up Hall Valley to the Hand Cart and Hall Valley campgrounds. This valley is historic because of the number of mines and mining settlements that used to be here. If you take the left fork at the Hall Valley Campground and continue up Hall Valley, you pass the south side of Handcart Peak. A trail continues on to the abandoned Whale Mine. Should you take the right fork at the campground, you proceed up a very rough and narrow road into Handcart Gulch and pass between Handcart Peak on your left and Red Cone Peak on your right, both at elevations exceeding 12,500 feet. After a half-mile of switchbacks, you come to Webster Pass. This approach to Webster Pass from the east can be made only in a four-wheel-drive vehicle. The western descent from the pass in the Arapaho National Forest to the town of Montezuma is passable in a passenger car.

Three miles beyond Webster, the Guanella Pass National Scenic Byway climbs north to Guanella Pass, following the western side of the Mount Evans Wilderness. This is perhaps the most scenic of the accessible mountain pass roads in the Pike. The road stays near Duck Creek the entire way to the pass. Watch for bighorn sheep and the interesting ptarmigan, which has white feathers during winter and brown during summer (fig. 7). Because the east side of the Guanella Pass National Scenic Byway is at the edge of the wilderness, there are no driving routes in that direction, but there are a couple of side trips you could take to the west. One of these leaves the byway at Geneva Park Campground and winds its way along Geneva Creek to Geneva Lake at the foot of Santa Fe Peak. The campground at Geneva Park is in a lodgepole pine forest near the huge Geneva Park mountain meadow. As it climbs toward the pass from Geneva Park, the byway becomes extremely narrow, with steep drop-offs on either side.

Mount Evans Wilderness is centered around the 14,264-foot mountain that is located on the border between the Pike and Arapaho National Forests. The paved road to the summit is in the Arapaho, except for the final switchbacks. Three trails enter the wilderness from Guanella Pass National Scenic Byway. The Rosalie Trail bisects the wilderness diagonally from Guanella Pass to a point just west of the community of Highland Park. The Abyss Trail originates at the Burning Bear Campground and follows Scott Gomer Creek. It ends at Abyss Lake just west of the Mount Evans summit. At 12,500 feet, Abyss Lake lies in a rock-rimmed gorge and has a good population of trout. The third trail leaves from the Whiteside Campground and stays alongside Threemile Creek, passing between Arrowhead Mountain on the left and Spearhead Mountain on the right before intersecting with the Rosalie Trail.

Rio Grande National Forest

SIZE AND LOCATION: Approximately 1,851,000 acres in south-central Colorado. Major access routes are U.S. Highways 160 and 285 and State Routes 15, 17, 114, and 149 (Silver Thread National Scenic Byway). District Ranger Stations: Del Norte, La Jara, and Saguache. Forest Supervisor's Office: 1803 W. Highway 160, Monte Vista, CO 81144, www.fs.ed.us/r2/riogrande.

SPECIAL FACILITIES: Winter sports area; ATV areas; boat launch areas.

SPECIAL ATTRACTIONS: Wheeler Geologic Area; Silver Thread National Scenic Byway; Bachelor Loop Interpretive Historic Tour; Cumbres and Toltec Scenic Railway.

WILDERNESS AREAS: La Garita (128,858 acres, partly in the Gunnison National Forest); Sangre de Cristo (226,420 acres, partly in the San Isabel National Forest); South San Juan (158,790 acres, partly in the San Juan National Forest); Weminuche (488,200 acres, partly in the San Juan National Forest).

The headwaters of the Rio Grande, the nation's third longest river, are at the far western edge of the Rio Grande National Forest. The forest is flanked on the east by the Sangre de Cristo Range and on the west by the San Juans, with the broad San Juan River valley in between. U.S. Highway 285 and State

Route 17 in the heart of the valley separate the two mountain ranges. Throughout the national forest are high peaks, glacial cirques, tarns, reservoirs, clear trout-filled streams, and rugged mountain passes.

Most of the Sangre de Cristo Mountain Range is in the Sangre de Cristo Wilderness. This long, narrow range of high mountains, some of them towering 7,000 feet above the valley below, is approximately 70 miles long and no more than 10 miles wide at its widest point. Lying on the west side of the wilderness, the Rio Grande National Forest part may be reached by a few roads off of U.S. Highway 285 and State Route 17. It takes a four-wheel-drive vehicle to climb over Hayden Pass from Villa Grove, but the views above timberline are super. Hayden Pass is the most northern pass over the Sangre de Cristos and was the major route over the mountains to the mining town of Bonanza. An alpine tundra habitat flourishes at the 10,709-foot pass. A road to the east where U.S. 285 and State Route 17 split goes to the old resort spa of Valley View Hot Springs. Separate trails to Cottonwood Peak, Thirsty Peak, and between Lakes and Electric Peaks follow the crest of the mountains. All of these peaks are above 13,000 feet. Farther south, Hermit Pass and Venable Pass can only be reached by steep trails that originate outside the wilderness. Crestone Peak and Crestone Needle are the most prominent peaks on the crest but are best reached from the San Isabel National Forest.

The eastern edge of the Great Sand Dunes National Monument abuts the Sangre de Cristo Wilderness. From the monument is a jeep trail through a break in the wilderness to Medano Pass and a 2.5-mile hiking trail from the monument's visitor center over Mosca Pass. Just outside the southern boundary of the wilderness is the four-wheel-drive road through Chokeberry Canyon that ends just before reaching several picturesque lakes. In front of you to the east, at the southeast corner of the national forest, is the highest peak in the Sangre de Cristo Range, Blanca Peak, at 14,345 feet.

At the western edge of the San Juan Mountains is the once prosperous silver-mining town of Bonanza. Between 1880 and 1888, this boomtown was a bustling center of activity. A few people reside there today. Four miles southeast and reached only by trail is Hayden Peak, named for the leader of the survey team that used the peak as a triangulation station during 1873 and 1874.

One of the great ways to see the San Juan Mountains is to drive the Silver Thread National Scenic Byway that connects Lake City to the north in the Gunnison National Forest with South Fork in the Rio Grande National Forest. The boundary between the two forests is just south of Spring Creek Pass where the scenic byway, actually State Route 149, crosses the Continental Divide. The byway is named for the silver-mining district through which it passes. The Continental Divide National Recreation Trail crosses Spring

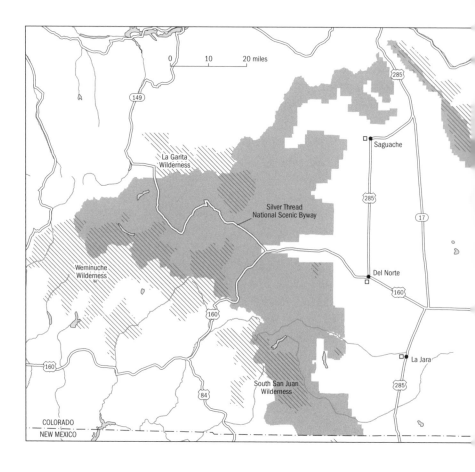

Creek Pass. Although the pass is nearly 11,000 feet above sea level, most of the byway is at the 9,000-foot elevation. From the pass the scenic byway descends along Spring Creek, which is lined by cottonwood and several species of willows. A short side road to the east comes to an overlook for a breathtaking view of North Clear Creek Falls. The highway becomes very curvy as it continues to descend to two campgrounds. Silver Thread Campground has a half-mile loop to a shimmering falls, whereas Bristol Head Campground has an easy 1.5-mile trail along Clear Creek to another falls.

A paved side road to the west comes to the Bristol View Guard Station and then passes through a narrow canyon to Road Canyon Reservoir No. 1. This road then parallels the Rio Grande along the northern boundary of the Weminuche Wilderness. Thirtymile Campground is at the eastern end of the Rio Grande Reservoir where there is the trailhead for a trail along Square Creek into the wilderness. A dirt road stays along the northern edge of the reservoir as far as Last Trail Campground. From here you can hike southward along Ute Creek into the wilderness, hike northward along North Clear

N

Rio Grande NF
Wilderness areas
□ District Ranger Stations
• Towns

Creek to the heart-shaped Heart Lake and to a massive landslide of 1991 along the north side of West Lost Trail Creek, or hike along a closed jeep road for several miles to the headwaters of the Rio Grande. In the vicinity is the townsite of the defunct mining community of Beartown.

After the Bristol View turnoff, the Silver Thread National Scenic Byway enters a broad valley along the Rio Grande, with side roads and then trails to Ruby Lakes at the end of the Fern Creek Trail and to Seepage Lake. The byway follows the Rio Grande as it makes a wide bend around Bristol Head Mountain into the town of Creede. Once a thriving mining community, Creede now is an interesting town catering to tourists. Pick up a booklet at the Creede Ranger Station for the Bachelor Loop Interpretive Historic Tour. This tour loops northward along West Willow Creek past many mines and historic sites and is a good introduction to the gold, silver, lead, and zinc mining days of the past. At the northern tip of the loop road, another road continues for a while along West Willow Creek to another old mine. From here is a hiking trail that climbs to the Continental Divide. San Luis Pass is 2.5 miles south of here. The pass is on the Rio Grande–Gunnison National Forests boundary and the southern edge of the La Garita Wilderness. Trails off the Bachelor Loop Road enter the western side of the wilderness to Halfmoon Pass, Wason Park, and a longer route goes to the Wheeler Geological Area. The La Garita Wilderness contains high mountains that include San Luis and Stewart Peaks, both 14,000-footers. Because of steep talus slopes, much of the area is difficult for the average hiker to get around.

After experiencing Creede, you may continue on the Silver Thread National Scenic Byway as it follows the Rio Grande on a southeasterly course. Eight miles from Creede, and just before coming to Wagon Wheel Gap, there is a paved side road through Spring Gulch to the site of the old Hanson's Mill. From this site is a trail through the La Garita Wilderness or another trail and rough jeep road into a narrow corridor of the wilderness to the incomparable Wheeler Geologic Area. with more unusual pink and white rock formations than you can imagine. Spires and pinnacles are everywhere. Geologists describe the rock as coarse volcanic tuff, and because the rock particles are not firmly compacted, wind and rain have carved these rocks into fantastic shapes. Visitors to the area have given many of these mysterious creations fanciful names such as Cyclops, the Temple, Dante's Lost Souls, Beehives, Pipe Organ, Milk Bottles, and Finger Rocks. The 640-acre area is named for George Wheeler, one of the leaders of an 1874 surveying party in the area. President Theodore Roosevelt recognized the value of the area in 1908 when he named it a national monument, but because of its small size and isolation from other areas under the jurisdiction of the National Park Service, it was given to the Forest Service in 1950.

The Silver Thread National Scenic Byway descends through the narrow canyon at Wagon Wheel Gap where dense stands of Douglas fir, ponderosa pine, and Engelmann spruce dominate the forest and soon comes to the small community of South Fork. The stone structure at the summit of Sentinel Rock west of town is apparently from an old observation lookout point. Although the Weminuche Wilderness is shared with the San Juan National Forest, the northern part is in the Rio Grande National Forest, extending 42 miles from the east to the west and 15 miles from the north to the south.

From South Fork, U.S. Highway 160 traverses southwest, climbing in about 18 miles to Wolf Creek Pass. The pass was not completed for vehicular traffic until 1916. This is one of the most beautiful passes on the Continental Divide where a major highway crosses. At 10,857 feet, the pass is alpine tundra with colorful wildflowers blooming in a very short summer growing season. The popular Wolf Creek Pass Ski Area is near the summit. Between South Fork and Wolf Creek Pass are three interesting side roads off of U.S. Highway 160. Allow a full day for each of these adventures. About one mile south of South Fork is a good gravel road. After passing the South Fork Guard Station, some summer homes, and two campgrounds along Beaver Creek, the road comes to the Beaver Creek Reservoir where there is a boat ramp. Stay on the main gravel road as it continues along Beaver Creek before curving left over switchbacks to secluded Poage Lake, which is great for fishing. Prior to the switchbacks, a dirt road branches eastward and then, after less than one mile, it branches southward. A five-mile stretch of very crooked road leads to the sparkling Crystal Lakes. Beyond the lakes, the road follows a very serpentine route. Within the next seven miles, there are junctions with two gravel roads. No matter which way you choose, you encounter curve after curve and switchback after switchback. At the first junction, the road twists its way to Graybark Mountain. Although there is a radio tower on the summit, the view from here is rewarding. At the second junction, a longer road circles Fuch Reservoir and then winds through beautiful mountain scenery. Where the gravel road becomes a four-wheel-drive road that climbs over Blowout Pass, take a mile-long hike north to the Fitton Guard Station, a tiny log cabin dating back to 1906. From here you can see Poison Mountain to the northeast and Bennett Peak to the southeast.

A second great trip off of U.S. Highway 160 leaves from the Pearl Creek Campground on a dirt road and tries to follow every turn that the adjacent Pearl Creek makes. Many four-wheel-drive roads are off the Pearl Creek Road, but if you stay on the main road, you come to the very tranquil and beautiful Fivemile Park. Beyond the park, the road forks. Take the left fork for three miles to the abandoned buildings of Summitville. This became a boomtown when gold was discovered here in 1870, the first such strike in the San Juan Mountains. Summitville enjoyed a brief revival in 1935 when some of the mines reopened. The road from Summitville continues on to Grayback Mountain. If you return from Summitville back to the fork, take the south road that makes a white-knuckle climb to Elwood Pass on the Continental Divide. The pass was constructed in 1870 to provide a route from Fort Garland on the east and across the Continental Divide to Fort Lewis in Pagosa Springs. This was the major route over the Continental Divide in this area until Wolf Creek Pass was completed. Most passenger vehicles can make it nearly to Elwood Pass from Summitville, but descending down the west side into the San Juan National Forest requires a four-wheel-drive vehicle.

The third side road off of U.S. Highway 160 leaves the highway just south of the Columbine Picnic Area and turns right, first coming to Big Meadows Reservoir and Campground and then sharply heading northward around narrow Shaw Lake before coming to the border of the Weminuche Wilderness. The road then makes another sharp right and follows the edge of the wilderness.

A vast amount of the Rio Grande National Forest is east of the Weminuche Wilderness and south of Elwood Pass and Summitville, extending all the way to the New Mexico state line. To get another aspect of the national forest, approach the national forest from the east from the end of State Route 15 or from the south off of State Route 17. Thirteen miles south of Alamosa and just before LaJara, State Route 15 leaves U.S. Highway 285 and heads due west. At Centro, where State Route 15 makes a sharp right and follows Gunbarrel Road, stay on the westward-heading dirt road, which soon comes to the Alamosa River and the entrance to Rio Grande National Forest. This road comes to the townsite of Jasper, a mining town in its heyday in 1874. The road continues to follow the Alamosa River to the Stunner townsite, another town that grew up overnight when gold was discovered above the Alamosa River around 1880. At Stunner, take the switchback road that climbs to Stunner Pass southward. At the small Mix Lake, one of the highest elevation campgrounds is located at 10,000 feet. The road splits at the campground, the west fork staying close to the edge of large Platoro Reservoir, the east fork dropping into the Conejos River valley and another ghost town known as Platoro. South of Platoro, the road follows the eastern edge of the South San Juan Wilderness to State Route 17, with several trailheads and campgrounds along the way. At the road junction, State Route 17 to the west switchbacks to a fine overlook before following North Fork Creek. The highway then climbs over Cumbres Pass, which is now a grassy meadow following a devastating fire in 1879. A paved road northward from the pass ends at Trujillo Meadow Reservoir and another campground at 10,000 feet. After descending the western side of Cumbres Pass, State Route 17 soon leaves the Rio Grande National Forest and crosses into New Mexico.

State Route 17 to the east continues to parallel the Conejos River, eventually coming to the Mogobe Campground at the edge of the national forest. A few miles beyond is Antonito. In Antonito, you can have another exciting and unforgettable adventure through the southern end of the Rio Grande National Forest by taking the Cumbres and Toltec Scenic Railroad, which makes its way through wild scenery from Antonito, Colorado, to Chama, New Mexico.

East of the La Garita Wilderness and north of U.S. Highway 160 between South Fork and Del Norte, the Rio Grande National Forest stretches northward in nonwilderness land for nearly 50 miles to the Gunnison National Forest. The area has innumerable mountain streams, pointed peaks, narrow gulches, open meadows, and some alpine tundra on the upper peaks. Aspen and ponderosa pine dominate the lower forested areas and Engelmann spruce and subalpine fir are in the higher elevations. It features sites of special interest, as well. A few miles north of South Fork is the old Alder Ranger Station, built in 1910 and still in use by the Forest Service. Eleven miles north of the Alder Ranger Station is the Fremont Campsite in the La Garita Mountains. When the John Fremont expedition was in the La Garita Mountains in 1848, they had to spend Christmas Day at this site following a huge snowstorm. Several members of the expedition lost their lives. The site can be reached by a three-mile trail from the end of a jeep road that follows Embargo Creek northward from U.S. Highway 160 midway between South Fork and Del Norte.

One of the most delightful and scenic but isolated parts of the Rio Grande National Forest is the 15,000-acre Saguache Park at the northwestern corner of the national forest. This is said to be the largest park in Colorado entirely within a national forest. You may best reach it from the north via a forest road off the Cochetopa Pass Road in the Gunnison National Forest. The park is just across the Continental Divide from Salt House Pass. A few miles south of Saguache Park is Stone Cellar Campground and nearby Guard Station. The road actually continues past the campground to a trailhead at the northeastern corner of LaGarita Wilderness. The trail winds past Twin Peaks to Machin Lake at the western edge of the wilderness.

State Route 114 from Saguache is on private land, but with national forest land both to the north and south. About 15 miles west of Saguache, a dirt road southward enters the Rio Grande National Forest and soon comes to a huge natural spring where water continuously emerges from behind rocks. Big Springs Picnic Area is nearby. Where State Route 114 heads north to ascend over North Cochetopa Pass (or North Pass on the forest map), a county road branches due west through Rabbit Canyon to Luders Creek Campground neatly situated along the creek. The road then makes a gentle climb to Cochetopa Pass where a bronze marker indicates an elevation of 10,820 feet. This was another toll road operated by Otto Mears 125 years ago.

North of Del Norte and reached by a series of dirt roads is an opening called the Natural Arch near the top of a mountain. The trip takes a bit of time, but the arch is a unique site and well worth the effort.

Roosevelt National Forest

SIZE AND LOCATION: Nearly 800,000 acres in the northern part of the Front Range west of Boulder and Fort Collins and extending to the Wyoming border. Major access routes are Interstate 70, U.S. Highways 34, 36, and 285, and State Routes 7, 14, 72, and 119. District Ranger Stations: Boulder and Fort Collins. Forest Supervisor's Office: 240 W. Prospect Road, Fort Collins, CO 80526, www.fs.fed.us/r2/arnf.

SPECIAL FACILITIES: Winter sports areas; boat launch areas.

SPECIAL ATTRACTIONS: Peak to Peak National Scenic Byway; Cache la Poudre Canyon; Boston Peak Research Natural Area; Long Lake Fens.

WILDERNESS AREAS: Cache la Poudre (9,238 acres); Comanche Peak (66,791 acres); Indian Peaks (73,291 acres, partly in the Arapaho National Forest); Neota (9,924 acres); Rawah (73,068 acres); James Peak (14,000 acres).

The Peak to Peak National Scenic Byway connects Central City and Estes Park. Except for the southernmost six miles, which are in the Arapaho National Forest, the scenic byway winds and climbs through the eastern side of the Roosevelt National Forest. It is a great highway from which to explore this part of the national forest as well as Rocky Mountain National Park, which is located just west of Estes Park. A few miles north of Central City, the Peak to Peak National Scenic Byway comes to Rollinsville and the eastern entrance to the Moffat, or Rollins Pass, Road. Following the railroad grade over Rollins Pass, a road was maintained for passenger vehicles after the railroad abandoned the pass crossing when the Moffat Tunnel was completed in 1927. Rock slides and the unstable condition of high wooden trestles does not permit one to drive the entire route, however, because the Moffat Road is closed just beyond the Forest Lakes Trailhead below Rollins Pass. Nonetheless, the Moffat Road that is open is very scenic as it follows South Boulder Creek and climbs through mountainous terrain and past several abandoned mines. In order to gain elevation for the final climb over the Continental Divide, the railroad made its way out of the valley by means of giant switchbacks up the side of the mountain. From the Jumbo Picnic Area west of Rollinsville, you can see the rungs of the switchbacks, known as the Giant's Ladder, carved into the mountain in front of you. Less than one mile west of Jumbo Picnic Area is the town of Tolland that was the first stop on the Moffat Railroad west of Denver. Two-and-a-half miles west of Tolland, the Moffat Road makes a sharp hairpin turn. From the hairpin, you can drive the mile-long gravel road westward to get a close look at the East Portal of the Moffat Tunnel. Huge fans are used to force smoke and gas westward out of the tunnel. The Moffat Road then begins its climb toward the Continental Divide, passing the ghost towns of Ladora and Antelope. Ten miles from the East Portal hairpin turn, the road takes a hemispherical route around Yankee Doodle Lake. From the lake you can see James Peak to the south. This is a very scenic spot in the heart of the James Peak Wilderness. The Forest Lakes Trail provides access to the Arapaho Lakes and Crater Lake before reaching the Continental Divide and Rogers Pass. The west side of Rogers Pass is in the Arapaho National Forest. Just beyond, the road passes to the left of Jenny Lake. The lake provided water for the steam locomotives' final four-mile climb to the Continental Divide 800 feet above. The Needle's Eye Tunnel can be seen above Jenny Lake, but a rock slide between the lake and the tunnel forced the closing of the road, although hiking is still an option. Between the Needle's Eye Tunnel and Rollins Pass are the twin trestles that stand a breathtaking 1,000 feet above Middle Boulder Creek. This deep ravine spanned by the trestles is called the Devil's Slide. Constructed in 1904, the trestles cling to the mountainside. Rollins Pass (officially known as Corona Pass) lies a little more than one mile from the twin trestles. At one time the community of Corona on the pass had a restaurant. The foundation of the restaurant may still be seen today.

After Rollinsville on the Peak to Peak National Scenic Byway, the highway comes to Nederland. County Road 130 westward is a good road that paral-

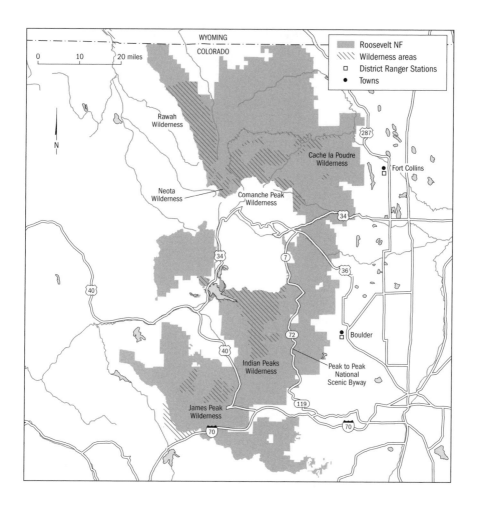

lels Middle Boulder Creek before reaching the remains of Eldora and Hessie. A ski area is near Eldora and a trailhead with two trails is at Hessie. The southernmost trail goes over Woodland Mountain to Skyscraper Reservoir and passes to the west of Chittenden Mountain before coming to a fork. The left fork proceeds due west to Jasper Reservoir and then up across Devil's Thumb Pass on the Continental Divide and into the Arapaho National Forest. The right fork takes a circuitous route around Diamond Lake and joins the Arapaho Pass Trail at the Fourth of July Trailhead at the edge of Indian Peaks Wilderness. You may also drive to this trailhead by staying on the graded road past Hessie. Arapaho Pass is about three miles from the Fourth of July Trailhead. Scenic Lake Dorothy is right on the Continental Divide. After crossing Arapaho Pass, the hiking trail takes a northern course through the Arapaho National Forest to Monarch Lake.

From Nederland, State Route 72 becomes the Peak to Peak National

Scenic Byway because State Route 119 takes a curvy route down into Boulder Canyon and then into the city of Boulder. The scenic byway climbs to Ward where there are paved side roads both to the west toward the Continental Divide and to the east. The road westward is very scenic, first coming to Red Rock Lake and then Brainerd Lake where there is a turnaround and parking area. Red Rock Lake is surrounded by spruces and firs and is nestled below the Indian Peaks of Pawnee, Shoshoni, Navajo, and Arikaree. The lake is at an elevation of 10,345 feet.

A short hike from the parking area at Brainerd Lake brings you to Long Lake, one of the prettiest high mountain lakes in the country. The meadow and forested areas around the lake are without equal. The trail completely encircles Long Lake, and at the western end of the lake another trail begins a climb to lovely Lake Isabelle. All along the way you can see Isabelle Glacier on the Continental Divide. A Forest Service trail branches off so that you may come closer to the glacier, or you may stay on the main trail and climb over Pawnee Pass and into the Arapaho National Forest. The mountain peaks, glacial lakes, glaciers, wildflower meadows, and tundra are as fine as any in the United States. A four-mile trail from Brainerd Lake into the wilderness ascends Mount Audubon.

As you look eastward from Ward, you can see the vast Great Plains far into the distance. The paved road eastward from Ward follows a ridge above Left Hand Creek. Just before reaching Gold Hill, a wetland community is visible in the deep ravine to the south. A scramble down into the ravine finds that the wetland is a floating mat bog, that is, a very watery habitat that bounces up and down as you walk onto it. Several sedges are in the floating mat, as well as three-leaved bogbean, mare's-tail, yellow bladderwort, dwarf bur reed, water smartweed, northern bedstraw, marsh bluegrass, and alpine reed grass. It is a great place to see a high-elevation floating bog, but be careful not to become submerged.

A few miles north of Ward, another side road to the west off of the Peak to Peak National Scenic Byway goes to another interesting wetland known as Beaver Bog. The scenic byway then drops into Peaceful Valley and a campground, which is at a sharp hairpin turn. Take the mile-long dirt road that leads from the hairpin to Camp Dick Campground. From this campground is a four-wheel-drive road and hiking trail along the south side of Middle St. Vrain Creek. At the end of the road, there are two trails into the Indian Creek Wilderness. The Buchanan Pass Trail climbs to Buchanan Pass, the St. Vrain Glacier Trail to the impressive St. Vrain Glacier just below the southern boundary of Rocky Mountain National Park. Should you decide to continue past the hairpin turn, the Peak to Peak National Scenic Byway continues to drop and then climb again to a junction with State Route 7. State Route 7 to the east winds through part of the Roosevelt National Forest, eventually leaving the forest and ending at U.S. Highway 36 at Lyons. State Highway 7 westward is now the Peak to Peak National Scenic Byway. As you travel from Ferncliff to Allenspark to Estes Park on the byway, Rocky Mountain National Park is to the west and the Roosevelt National Forest is to the east. A visitor center is located just across from Lily Lake prior to coming to Estes Park.

From Estes Park, U.S. Highway 36 turns west and is one of the main routes through Rocky Mountain National Park. U.S. Highway 36 to the east drops into Muggins Gulch and out of the national forest in about 20 miles. In addition, U.S. Highway 34 also cuts across Estes Park. To the west it enters Rocky Mountain National Park; to the east it enters the Roosevelt National Forest and magnificent Big Thompson Canyon. The drive through the canyon to the town of Loveland is one of the most scenic and breathtaking canyon drives in the country. Although still beautiful, the canyon never recovered after the devastating flash flood on July 31, 1976, which wreaked havoc and took the lives of at least 139 residents who lived in the canyon and tourists who were fishing in the river. Eventually the canyon highway leaves the national forest at The Narrows just four miles west of Loveland. Prior to reaching The Narrows is the Viestenz Smith Picnic Area from which the Round Mountain National Recreation Trail curves around to the 8,450-foot mountain.

Although Fort Collins lies several miles to the east of the Roosevelt National Forest, it is the base for exploring the northern extremities of the national forest. A few miles north of Fort Collins, State Route 14 branches westward from U.S. Highway 287. At the entrance to the national forest is the Poudre Canyon Information Center where you can obtain brochures, maps, and find out about the recreational activities available along the Cache la Poudre River. From the information center, State Route 14 is a fantastic scenic highway that crosses the entire upper part of the national forest, climbing to Cameron Pass at the western end of the forest. The entire route follows the Cache la Poudre River for 50 miles.

Just west of the information center is the three-mile-long Greyrock National Recreation Trail to Greyrock Mountain. State Route 14 then snakes its way from the Standing Prairie Campground to Kelly Flats Campground, following the northern edge of the Cache la Poudre Wilderness. During this stretch, the highway passes south of Red Mountain and through the well-named Big Narrows to Dutch George Flats where there is a campground. A short distance west is the four-mile-loop Mount McConnel National Recreation Trail into the wilderness that climbs to 8,010-foot Mount McConnel. About half the length of this trail is the self-guided Kreutzer Nature Trail through a forest of ponderosa and lodgepole pines. Black bears, mule deer, and elk are in the wilderness, along with bighorn sheep on the higher peaks.

From Kelly Flats there is a good side road to the south that forks after about five miles near the Bennett Creek Picnic Area. The right fork is the Crown Point Road that dead-ends at the Comanche Peak Wilderness. The Zimmerman Trail follows just inside the wilderness's eastern boundary all the way back to State Route 14. The left fork that branches southward winds its way for several miles to Pingree Park Campus outside the national forest. Along the way are Camman and Badsprings natural springs, a couple of Forest Service picnic areas, and several trailheads into the Comanche Peak Wilderness. One of these trails follows Fall Creek to picturesque Emmaline Lake at the foot of 12,258-foot Fall Mountain. From the Emmaline Lake Trailhead, there is a dirt road to Comanche Reservoir outside the wilderness. A hiking trail from the west end of the reservoir follows Beaver Creek into the wilderness.

By staying on State Route 14 past the Crown Point Road turnoff, you come to the settlement of Rustic where County Road 69 provides one of the best routes to reach the northern part of the Roosevelt National Forest. The road twists and turns, crosses several creeks, and passes small mountains and the site of Manhattan with its old cemetery, before coming to Bellaire Lake where there is a campground at an elevation of 8,650 feet. From the campground it is only three miles on the county road to the community of Red Feather Lakes where there is a major intersection. By continuing north, the road now becomes County Road 179, the Prairie Divide Road, which ultimately crosses the divide at 7,977 feet and continues on until it reaches County Road 80C. Known as the Cherokee Park Road, this road goes through a patchwork of national forest and private land, with one branch reaching the Wyoming state line.

From Red Feather Lakes, the paved road to the east provides access to Dowdy Lake, Wet Lake, Parvine Lake, and Lady Moon Lake, some with campgrounds and trailheads. The side road from Red Feather Lakes to the west is the gravel Deadman Road. This is a more interesting route as it climbs to the 10,269-foot Deadman Hill that rises above the North Fork of Cache la Poudre River. A three-mile-long gravel road to the north dead-ends at the Deadman Lookout. Beyond Deadman Hill, the road joins the Boulder Ridge Road at Four Corners.

Returning to State Route 14 and continuing along the Cache la Poudre River, the Arrowhead Visitor Center is just across the highway from Profile Rock. After the visitor center, the road passes the northern terminus of the Zimmerman Trail, passes a state fish hatchery, and comes to the Sheep Viewing Station and Big Bend Campground. Bighorn sheep are sometimes seen from here. State Route 14 makes a sweeping curve and begins a southwesterly coarse. Just to the west of the Bliss State Wildlife Area is Boston Peak where there is a marvelous wetland on the upper slopes. The wetland is one of the better preserved fens in Colorado and has been designated a Research Natural Area. Sphagnum covers much of the wetland, giving the fen a spongy surface. Although there are scattered trees of Engelmann spruce, subalpine fir, limber pine, and lodgepole pine associated with the fen, it is the sedge and wildflower community together with short willow shrubs that are outstanding. The willows include mountain willow, alpine fen willow, and purple-twig willow, often growing in thick stands. Other shrubby species in the fen are yellow cinquefoil, bog birch, and a species of wild gooseberry. Colorful flowering plants in the fen include four kinds of gentians, yellow Indian paintbrush, elephant's head, small-flowered grass-of-Parnassus, alpine meadowrue, hemlock parsley, bistort, northern bedstraw, and a bog orchid. Although not showy, the dominant vegetation cover consists of several species of sedges.

South of Boston Peak is Sleeping Elephant Campground, which is situated across State Route 14 from Sleeping Elephant Mountain. Where the Laramie Poudre Tunnel bores its way through the Laramie Mountains is a picnic area. The highway comes to Poudre Falls just before the Big South Campground. From the campground, Big South Trail stays within the wilderness edge of the Comanche Peaks Wilderness until it reaches Peterson Lake just outside the wilderness. You can also get to Peterson Lake through Box Canyon on a side road off of State Route 14 about six miles south of Big South Campground.

By staying on State Route 14 past the Box Canyon Road, the highway goes along the western side of Joe Wright Reservoir and climbs to Cameron Pass where it leaves the Roosevelt National Forest. That part of State Route 14 from Box Canyon Road to Cameron Pass goes through a narrow corridor between the Rawah Wilderness to the west and the Neota Wilderness to the east. The Neota Wilderness is primarily a high-mountain area of dense forests and flat-topped ridges, ranging from 10,000 to 11,800 feet in elevation. Nearly 4,000 acres of this wilderness is in alpine tundra above timberline. The wilderness is fairly inaccessible to hikers, with only the Neota Creek Trail at the southern tip of the wilderness maintained. To reach this trail, drive the dirt road from Peterson Lake to Long Draw Campground at the upper end of Long Draw Reservoir. If you stay to the north shore of the reservoir, you may drive to the Grand View Campground and then up to La Poudre Pass where the road ends. At the end of the road, Neota Creek Trail heads west into the wilderness and the La Poudre Pass Trail enters Rocky Mountain National Park.

One other option lets you explore a little more of the Roosevelt National

Forest. Between Big South Campground and the Box Canyon Road, County Road 103 branches off of State Route 14 and heads due north, following the eastern boundary of the Rawah Wilderness. The wilderness encompasses the southern end of the Medicine Bow Mountains. Immediately upon turning onto County Road 103, Chambers Lake is all that separates the gravel road from the wilderness. The southern half of the Rawah Wilderness is pocked with many lakes, and most of these can be reached through a series of interconnecting hiking trails. Many of the lakes are surrounded by forests of conifers. The most popular trail in the Rawah Wilderness is Blue Lake Trail, which is entered from the west end of Chambers Lake and goes to Blue Lake and beyond. Two miles southwest of Blue Lake is Clark Peak, the highest in the wilderness at 12,951 feet.

County Road 103 follows the Laramie River all the way to Wyoming. About 13 miles north of the upper end of Chambers Lake is a side road to Brown Peak Campground and a trailhead at the edge of the wilderness. The trail follows McIntyre Creek until it comes to Houssner Creek. If you take the trail along Houssner Creek to the northwest, you come to picturesque Shipman Peak. You can ultimately climb to the Medicine Bow Trail at Ute Pass and then stay on the crest of the range to 10,827-foot Shipman Peak.

Long Lake Fens

Near the dwindling mining town of Ward, Colorado, a well-maintained Forest Service road climbs steadily westward toward the Continental Divide, the boundary between the Atlantic and Pacific watersheds. A backward glance to the east reveals the Great Plains, here only 20 miles from the Divide. After passing Brainard Lake, a popular recreation area and campground, the road ends at the Long Lake parking lot. Long Lake lies a little beyond, within the Indian Peaks Wilderness, a 43,000-acre region administered by the Arapaho and Roosevelt National Forests. The wilderness gets its name from several mountains more than 12,000 feet tall, including Arikaree, Navajo, Apache, Shoshoni, and Pawnee.

The trail to Long Lake begins in a small patch of subalpine forest. Engelmann spruce is the dominant tree at this elevation, 10,521 feet, whereas thickets of the wiry-stemmed, pale-green-leaved bilberry, a relative of the blueberry, occupy much of the understory. Myriad shade-loving mountain wildflowers—including a blue Jacob's ladder, a white lousewort, a yellow arnica, and twisted stalk (a cream-colored, lilylike flower)—brighten other parts of the forest floor.

The trail soon emerges into a huge opening containing Long Lake, with a backdrop of towering mountains more than 13,000 feet above sea level. The high peaks within view include Mount Navajo to the west on the Continental Divide and Mount Audubon to the northwest. As it approaches Long Lake, the trail divides. One branch continues westward along the north side of the lake, the other leads southward across a bridge over bubbling South St. Vrain Creek. Hikers can walk around the lake either way or turn off to more remote destinations. One route leads across the Continental Divide at Pawnee Pass, whereas another goes to Isabelle Glacier, a nearby reminder of the Ice Age glaciers that sculpted the region.

Immediately east of the bridge, in a shallow depression carved out long ago by glaciers, lies a five-acre meadowlike area kept moist by the seepage of groundwater fed by the abundant snowmelt. This type of wet meadow is known as a fen because the water has become alkaline (a result of flowing through limestone rock). A narrow band of the shrubby flat-leaved willow, growing in foot-deep standing water, forms the fen's tangled border. Within, spongy, moundlike hummocks of assorted flowering plants overlie a gradually built-up foundation of sphagnum. Shallow rivulets of cold water separate the hummocks into tiny islets.

Near the fen's edge grow a tiny purple gentian, a magenta-flowered shooting star, and the remarkable elephant's head, whose purplish flower is an excellent replica of an elephant's head, complete with trunk. The fen's interior contains an astounding collection of grasses, sedges, and rushes, many of them rare, all of them adapted to a cold, wet habitat and short growing season of no more than three-and-a-half months. The most common grass is the tufted, bluish-tinted hair grass. An occasional white, fluffy head of narrow-leaved cotton grass (actually a sedge) stands a foot above the other vegetation. Various sedges, none more than six inches tall, make up most of the rest of the vegetation.

The sedges are perennials whose underground rhizomes store an abundance of carbohydrates to fuel the plants during the short growing season. The rhizomes also stabilize the plants in their soggy habitat. In six black-headed species, dark scales surround the plants' tiny, petalless flowers, giving the flower clusters a black appearance and absorbing heat. In this fen even the flower heads of the somewhat taller Nebraska sedge appear black, although at lower elevations they are more typically brown.

A second fen lies west of the trail, in a narrow strip along the south edge of Long Lake. A fringe of subalpine forest along its southern border provides shade for a number of species that do not survive out in the open. Among them are another black-headed sedge, white marsh marigold, and Parry pink primrose. Also distinguishing this fen are isolated patches of fine, yellowish gravel covered by about one inch of clear water. Although most plants find this wet, rocky ground inhospitable, two diminutive rushes grow here, each barely four inches high.

Routt National Forest

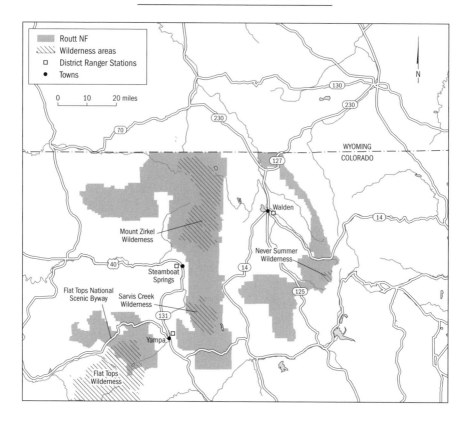

SIZE AND LOCATION: Approximately 1,130,000 acres in north-central Colorado, on either side of the Continental Divide, extending northward to the Wyoming border. Major access routes are U.S. Highway 40 and State Routes 13, 14, 131, and 134. District Ranger Stations: Steamboat Springs, Walden, and Yampa. Forest Supervisor's Office: 2468 Jackson Street, Laramie, WY 82070, www.fs.fed.us/r2/mbr.

SPECIAL FACILITIES: Boat launch areas; winter sports area.

SPECIAL ATTRACTIONS: Mud Slide Area; Forest Blowdown Area; Fish Creek Falls.

WILDERNESS AREAS: Mount Zirkle (160,568 acres); Sarvis Creek (47,140 acres); Flat Tops (135,035 acres, partly in the White River National Forest); Never Summer (20,692 acres, partly in the Roosevelt National Forest).

The popular resort town of Steamboat Springs is surrounded by the Routt National Forest, with the Park Range dominating to the east and the Flat

Tops Plateau to the southwest. The Elkhead Mountains lie northwest of Steamboat Springs. The most familiar landmark in the national forest is Rabbit Ears Peak, whose crumbly twin peaks give this formation its name. Rabbit Ears Peak has been a landmark for travelers for centuries. The peak can be reached by a two-mile trail from Dumont Lake, just north of U.S. Highway 40. Dumont Lake also features a lakeside campground. You may also take an unimproved dirt road north from Dumont Lake to a two-mile trail that makes its way through old-growth Engelmann spruces, subalpine firs, and boggy areas to Lost Lake. On the way to Dumont Lake from Steamboat Springs is a great observation point at the Ferndale Picnic Area; beyond that are the Meadow and Walton Creek campgrounds. U.S. Highway 40 crosses Rabbit Ears Pass three miles south of the peak.

A lesser known but equally impressive landmark is Hahn Peak north of Steamboat Springs. This perfectly shaped volcanic dome can be seen from miles away as it is approached on the road to Steamboat Springs State Park. A nice trail to the summit of Hahn Peak runs from just east of Columbine, and several roads from Columbine branch into the most extreme northern part of the national forest.Only at Hahn's Peak Lake does the Forest Service have facilities, including a campground and a boat launch area.

Roads to the east of Columbine head toward the Mount Zirkel Wilderness. A trail from Diamond Park climbs into the wilderness and to the Continental Divide. From there the trail comes to the Encampment Meadow Trail, which goes north to Wyoming. Mount Zirkel Wilderness is long and relatively narrow, stretching for 30 miles north to south and nearly 15 miles east to west. The wilderness includes the Park Range, with Mount Zirkel the highest peak at 12,180 feet. The Continental Divide cuts the wilderness nearly in half. More than 75 lakes dot the wilderness.

Slavonia Trailhead is perhaps the most popular entry point into the Mount Zirkel Wilderness. The trailhead can be reached by following the side road from Clark that follows the Elk River to the Seedhouse Campground and then to Slavonia. A loop trail from Slavonia circles northward along Gilpin Creek to Gilpin Lake, swings through the Gold Creek Meadows, and returns to Slavonia along Gold Creek, passing pretty Gold Creek Lake along the way. Both lakes have small brook trout. You may also follow the Gold Creek Trail eastward over Ute Pass and to the east side of the wilderness. About 1.5 miles from Slavonia on the Gilpin Creek Trail is a trail north to sparkling Mica Lake with Big Agnes Mountain providing a scenic backdrop.

On October 25, 1997, a windstorm with velocities reaching 120 miles per hour whipped through a part of the national forest, demolishing four million trees in a swath five miles wide and 30 miles long. Much of the devastating blowdown is in the Mount Zirkel Wilderness, but a segment of it may be observed by taking the road from Seedhouse Campground to Three Island Trailhead. The blowdown is south of the road. Two miles from the Three Island Trailhead is Three Island Lake. The lake is particularly scenic with its three islands. From the lake the trail climbs to the Continental Divide where it joins the Wyoming Trail. The Wyoming Trail follows the irregular crest of the Continental Divide.

A popular drive is from Steamboat Springs to Buffalo Pass and the Summit Lake Campground on the Continental Divide. Summit Lake is at the south end of the Mount Zirkel Wilderness, and the Wyoming Trail can be joined here. Six miles north on the Wyoming Trail is a side trail to lovely Luna Lake with a sandy beach but very cold water. In another mile along the Wyoming Trail is a trail to Lake of the Crags. The views across to Mount Ethel and Lost Ranger Peak are outstanding. South from Buffalo Pass is a road that dead-ends at Granite Campground and the Fish Creek Reservoir where there is a boat ramp. The road east of Buffalo Pass to Grizzly Campground is too rough for most passenger vehicles.

Three miles east of Steamboat Springs is thunderous Fish Creek Falls, one of the most scenic attractions in the national forest. The Main and North Forks of Fish Creek begin on the Continental Divide and converge to form the 283-foot falls. A one-eighth-mile handicap-accessible trail from a parking area allows the visitor to have several good views of the falls. From the falls is a 6.7-mile trail to the Wyoming Trail on the Continental Divide.

From Missouri Street in Steamboat Springs, County Road 36 heads north to Strawberry Park Hot Springs, a popular area with therapeutic springs.

To access the east side of the Mount Zirkel Wilderness, approaches off of State Route 14 are the best. County Road 24 goes to Grizzly Creek Campground and then northward to Teal Lake where there is a campground and a boat ramp. County Road 6W makes a long, circuitous route to within three miles of the Wyoming border before a branch road to the southwest goes to Big Creek Lake Campground and the Seven Lakes Trailhead. A good trail here passes between the two largest lakes to Big Creek Falls at the edge of the wilderness. Big Creek Lake also has an interpretive trail. Look for the beautiful wild aster in summer (pl. 11).

A section of the Routt National Forest can be reached by following State Route 14 southeast from Walden. Although State Route 14 never actually enters the national forest, a side road from Gould does. You know you have entered the national forest when you come to the Aspen Campground. The side road then leads to the Pines Campground and continues to a trailhead at the end of the road. This trail follows South Fork into the Never Summer Wilderness and in six miles comes to Baker Pass. The wilderness south of the pass is in the Roosevelt National Forest.

The Never Summer Wilderness is well named, for it always feels like winter with more than 60 percent of the land alpine tundra and boulder fields above the timberline. When you do find forests at lower elevations, they consist of Engelmann spruce, subalpine fir, and lodgepole pine.

From the junction of Three-way 13 miles north of Walden, State Route 127 passes between two small units of the Routt National Forest before entering Wyoming. East of State Route 127 is a road along Pinkham Creek, whereas west of State Route 127 is a scenic road along Camp Creek.

Kremmling is situated between two large parts of the national forest at the south end of the forest. Rabbit Ears Range is northeast of Kremmling and, although it covers many square miles, the area has little road access. A few hiking trails cross this part of the forest, including one that follows Haystack Creek over Troublesome Pass.

West of Kremmling on U.S. Highway 40 and east of Yampa on State Route 131 is a segment of the Routt National Forest that encompasses the Sarvis Creek Wilderness. This wilderness includes the southern end of the Park Range, and the mountains are low enough that few alpine tundra areas exist. Most of the wilderness is forested with lodgepole pine, Engelmann spruce, subalpine fir, and aspen. Two trails, one near the middle of the wilderness and one near the south end, cross the wilderness.

South of the Sarvis Creek Wilderness is a sizeable area of the Routt National Forest with roads throughout. Morrison Creek Road ascends to Lynx Pass where there is a campground and tranquil Lagunita Lake before coming to State Route 134. State Route 134 climbs again past Blacktail Creek Campground to Gore Pass where there is another campground. A phenomenon known as The Slide is along Muddy Creek about six miles north of Tonopah. This is a huge mud slide with silt being continuously deposited into Muddy Creek.

The town of Hayden lies midway between Steamboat Springs and Craig on U.S. Highway 40. County Road 80 north out of Hayden enters the Routt National Forest in about 18 miles. After passing over Quaker Mountain, the road comes to California Park, a large, open, grassy area where deer, antelope, and elk often may be seen. The park was a favorite hunting ground for Theodore Roosevelt.

From Craig, State Route 13 runs northward to the Wyoming border. In about 12 miles, three roads go into the extreme western edge of the Routt National Forest. The more southerly route, County Road 27, climbs onto McInturf Mesa and into the Elkhead Mountains. Sawmill Campground is along the way, as is the Lost Peak Trailhead just north of the Lost Creek Guard Station. If you continue on this road, you come to a short side road to Bears Ears Peaks where there are magnificent views from the summit. The middle road, County Road 11, goes to the Freeman Reservoir where there are a campground and hiking trails. Farther north, County Road 38 reenters the Routt National Forest after about eight miles and comes to a trailhead north of Sand Point. The trail from here is very scenic, passing Mount Welba, Mount Oliphant, and Buck Point, with views of Black Mountain.

Southwest of Steamboat Springs and west of State Route 131 is the great Flat Tops Plateau whose sheared-off mountain peaks still rise to 12,000 feet. Fast-flowing rivers and streams have carved deep canyons through the sandstone plateau. Much of the Flat Tops Plateau is in the Flat Tops Wilderness, and only the northern part of the wilderness is in the Routt National Forest, the larger southern part being in the White River National Forest. The Flat Tops National Scenic Byway is a wonderful way to observe the Flat Tops. The scenic byway is 82 miles long, from Yampa to Meeker, with much of the route as far west as Buford across national forest land.

About seven miles west of Yampa, the scenic byway climbs out of a broad valley and follows Spronks Creek to the entrance of the Routt National Forest. A side road takes you to Chapman Reservoir and Campground. The highway drops through a narrow canyon, and another side road goes to Sheriff Reservoir where there is a campground as well as a trail into the wilderness. The road then climbs gently to Dunckley Pass where wildflowers abound during summer. From this 9,763-foot pass you can see Orno Peak (12,133 feet), Pyramid Peak (11,532 feet), and the Little Flat Tops to the south. From Dunckley Pass, the scenic byway descends into a broad river valley to the Pyramid Guard Station that dates back to 1934. The road climbs past Vaughn Lake and Campground to Ripple Creek Pass at 10,343 feet where there is a super view across the White River Valley. The scenic byway on the other side of the pass is in the White River National Forest.

Steamboat Ski Area is a winter sports mecca on the mountain just east of Steamboat Springs.

San Isabel National Forest

SIZE AND LOCATION: Approximately 1,110,000 acres in the Rocky Mountains, extending from Leadville in the north to Trinidad in the south. Major access routes are Interstate 25, U.S. Highways 24, 50, and 285, and State Routes 12 (The Legends Highway), 69, 82, 91, 96, and 165 (Greenhorn Highway). District Ranger Stations: Cañon City, Leadville, and Salida. Forest Supervisor's Office:1920 Valley Drive, Pueblo, CO 81008, www.fs.fed.us/r2/psicc.

SPECIAL FACILITIES: Boat launch areas; winter sports areas; ATV areas.

SPECIAL ATTRACTIONS: Top of the World National Scenic Byway; The Legends Scenic Byway; Twin Lakes Recreation Area; Turquoise Lake Recreation Area; Lake Isabel Recreation Area; Spanish Peaks.

WILDERNESS AREAS: Collegiate Peaks (116,938 acres, partly in the Gunnison and White River National Forests); Holy Cross (122,797 acres, partly in the White River National Forest); Mount Massive (30,540 acres, a small part in the east managed by the Leadville National Fish Hatchery); Spanish Peaks (18,000 acres); Greenhorn Mountain (22,040 acres); Buffalo Peaks (43,410 acres); Sangre de Cristo (226,420 acres, partly in the Rio Grande National Forest).

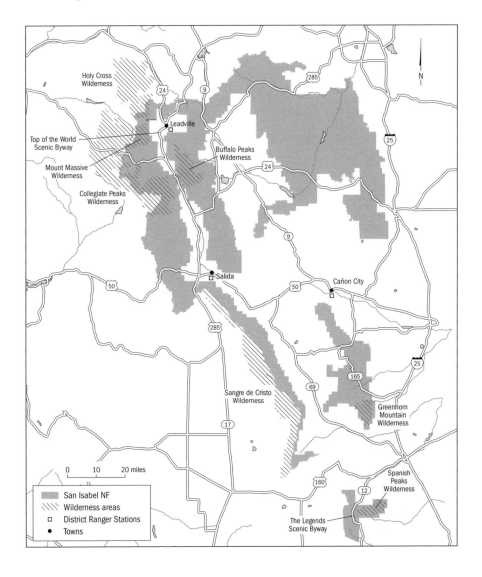

The highest peak in Colorado, Mount Elbert, at 14,433 feet, is in the San Isabel National Forest, but there are 20 peaks in the forest with an elevation of 14,000 feet or more. Dotted throughout the forest are beautiful mountain lakes and streams.

The San Isabel National Forest comprises three noncontiguous units. The northernmost unit, and by far the largest in extent, includes the Sawatch Range, the highest mountain range in Colorado, and the Sangre de Cristo Range. South of Cañon City is a unit of the national forest that encompasses the Wet Mountains. The southernmost unit, located west of Walsenburg and Trinidad, includes the famous Spanish Peaks, an extension of the Sangre de Cristo Range.

Because the Continental Divide runs the length of the San Isabel, and with several 14,000-foot peaks in the forest not even on the Continental Divide, there are numerous mountain passes. Each of these passes was a way to cross the Rocky Mountains, either as a travel route for the populace or as a means of reaching mines and hauling ore to its destination. Some of these passes today are on major paved highways, others are crossed only by roads suitable for four-wheel-drive vehicles, whereas still others have been virtually abandoned or reduced to hiking trails.

To explore the northern part of the San Isabel National Forest, you may wish to begin in the two-mile-high city of Leadville. After enjoying the historic buildings and sites of this mining town, take any road and it will lead you into some part of the national forest. Heading northward on U.S. Highway 24, you come to Tennessee Pass in about 11 miles and the monument to the U.S. Army's Tenth Mountain Division. During World War II, this was the training area for soldiers who would be fighting in mountainous terrain. Tennessee Pass crosses the Continental Divide separating the Tennessee Fork of the Arkansas River and the Eagle River watersheds. To the east of the pass is Ski Cooper, one of the forest's most popular winter sports areas. As you descend the north side of the pass, you are in the White River National Forest.

If you drive northeast of Leadville on State Route 91 for about 12 miles, you reach Fremont Pass, another pass on the Continental Divide. North of the pass you see massive mining activities from the Climax Molybdenum Mine.

Westward out of Leadville a paved forest road quickly comes to beautiful Turquoise Lake, a large impoundment that extends for four miles from the east to the west and is nearly 1.5 miles wide at its eastern end. The forest road completely encircles the lake and offers numerous photographic opportunities at Valley View, Mosquito View, and Shimmering Point. Along the forest road are several Forest Service campgrounds, most of them clustered at the eastern end of the recreation area. At Molly Brown Campground, there is an interpretive nature trail into the surrounding woodland. You may observe Sugarloaf Dam at a parking area. At the extreme western end of the road, where it makes a sharp turn back to the east, there is a trailhead for hikers to enter the Holy Cross Wilderness. Only a small part of this wilderness is in the San Isabel National Forest. One trail in the San Isabel follows Lake Fork to Timberline Lake; another longer one skirts the eastern edge of the wilderness past Galena Lake and Bear Lake.

Near the southwest corner of Turquoise Lake, a gravel road branches westward off the Turquoise Loop Road. In four miles this road comes to a trailhead at the upper end of the Mount Massive Wilderness. A trail from here westward is a one-miler to picturesque Windsor Lake. Another trail to the south goes for several miles, past a number of lakes and into the Leadville National Fish Hatchery. For the more adventurous hiker or for persons with a four-wheel-drive vehicle, take the rugged jeep trail north from the trailhead, curving and winding and climbing to Hagerman Pass on the Continental Divide. This historic pass is above two tunnels that were bored through the mountain by Colorado Midland Railroad magnate J. J. Hagerman for railroads from the mines between Leadville and Aspen and a third tunnel constructed by the U.S. Army Corps of Engineers for their Fryingpan-Arkansas Water Diversion project. The location of the first tunnel beneath the pass reputedly came to J. J. Hagerman in a dream. A hike to the entrance of the tunnels is possible. If you continue across the pass, either on foot or via vehicle, you cross into the White River National Forest. Hagerman Pass is one of the most historic in the forest.

Leaving Leadville on U.S. Highway 24 heading south, you pass a string of houses that make up Stringtown and then a tiny railroad stop called Malta. From Malta take the paved road due west of U.S. Highway 24. In about one mile, a gravel road southward follows Halfmoon Creek. This road ends at the Elbert Creek Campground where the hiking trail into the Mount Massive Wilderness goes to the 14,421-foot summit of mighty Mount Massive. On a picture-perfect day, anyone in good physical condition should be able to hike to the summit and back in a day. Mountain climbing experience is not necessary to conquer this mountain. More than 70,000 acres extend above the timberline on Mount Massive, and a remnant glacier is present year-round. The hiking trail goes through lodgepole pine forests before climbing through spruces and firs and then into the alpine tundra. Several small glacial lakes may be seen along the trail. Total elevation gained during the hike to the summit is 4,367 feet.

Continuing southward on U.S. Highway 24, less than one mile after crossing the Arkansas River bridge, turn right onto a gravel road and continue about five miles to a small lake known as Mount Elbert Forebay. At the lake's dam, the road becomes paved and parallels the southern boundary of the lake to the trailhead for Mount Elbert. Mount Elbert is the highest mountain in Colorado and the second highest in the lower 48 United States after California's Mount Whitney. The 14,433-foot summit may be hiked by the inexperienced mountain climber in a half-day. One mile south of the trailhead, the gravel road joins State Route 82 that stays near the northern edge of the huge Twin Lakes Reservoir. A left turn onto State Route 82 returns you to U.S. Highway 24; a right turn is the Top of the World National Scenic Byway that climbs to rugged Independence Pass on the border between the San Isabel and White River National Forests. Along the stretch of the highway from Twin Lakes to Independence Pass, there are features such as the old Vesuvius and Lost Chance mines, picturesque Snyder Falls, interesting rock formations such as Monitor Rock, and hiking trails into Echo Canyon, along Black Cloud Creek, and a longer one that eventually goes to the base of Ouray Peak and Grizzly Peak, both in the Collegiate Peaks Wilderness. This last trail begins from the site of the mining town of Everett.

If you are looking for a wild adventure, return to U.S. Highway 24 and retrace your route north of the bridge over the Arkansas River. A half-mile north is a gravel road east of the highway that proceeds to the south and east. In three miles the road comes to the Mount Massive Lakes before entering the San Isabel National Forest. The road now becomes very rough and narrow as it follows Union Creek for nearly nine miles, eventually climbing to Weston Pass in the Mosquito Range. The road skirts the northern end of the Buffalo Peaks Wilderness. This wilderness has one of the largest herds of bighorn sheep in Colorado. Weston Pass is not on the Continental Divide, but is on the boundary between the San Isabel and Pike National Forests. Access to the hiking trails in the wilderness is at the south end of the wilderness or from the Weston Pass Campground in the Pike National Forest three miles south of the pass itself.

Five miles south of the intersection of State Route 82 and U.S. Highway 24 is another gravel road, this one proceeding west and passing the ghost towns of Vicksburg, Rockdale, and Winfield. From the cemetery at Vicksburg, a hiking trail leads southward into the Collegiate Peaks Wilderness. It follows Missouri Creek up Missouri Gulch, climbing rapidly to Elkhead Pass, at 13,200 feet. Surveyors in 1955 found the skull of an elk at the pass. It is nearly five miles back to Vicksburg, but if you have the time and the stamina, the trail continues on for several miles to U.S. Highway 24. On the way it passes Bedrock Falls, located midway between Mount Oxford to the north and Mount Harvard to the south.

Eventually U.S. Highway 24 comes to Buena Vista, which can be your base for other exciting ventures into the San Isabel National Forest. If you take the paved road north out of Buena Vista between U.S. Highway 24 and the Arkansas River, you soon cross the river and enter the forest at a rock formation known as Elephant Rock. In six miles, with the dirt road becoming rougher and more difficult the farther you go, you come to the trailhead at the south end of the Buffalo Peak Wilderness. The trail, which starts in the San Isabel and ends in the Pike, stays west of the rounded domelike West Buffalo and East Buffalo Peaks. This trail goes through forests of Douglas fir, Engelmann spruce, lodgepole pine, limber pine, and aspens, with large wildflower meadows interspersed. Buffalo Meadows, about five miles north of the trailhead, is particularly attractive. Dry, south-facing mountain slopes have small stands of bristlecone pine.

You won't want to miss the road that leads west from the center of Buena Vista. The road is paved all the way to Cottonwood Pass in the Sawatch Range on the border with the Gunnison National Forest. The road over the pass was originally used by miners to get to the Tincup mines from Buena Vista and Leadville. The pass is at an elevation of 12,126 feet. The road follows Cottonwood Creek all the way to the pass, through pine and fir forests and alpine tundra, and parallels the southern border of the Collegiate Peaks Wilderness. A short hiking trail at the summit leads above the pass.

Two miles south of Buena Vista, U.S. Highway 24 intersects with U.S. Highway 285. At this intersection, U.S. Highway 24 abruptly turns eastward, and it and U.S. Highway 285 stay together for about 15 miles to Antero Junction. The road to Antero Junction goes through the San Isabel National Forest and the remains of past mining activities. Just before reaching Antero Junction, the highway climbs to Trout Creek Pass, one of the oldest as well as lowest and easiest of the crossings from the South Platte drainage to the Arkansas drainage. It was an old Indian trail when Europeans arrived in 1779. As U.S. Highways 24 and 285 to Trout Pass turn north and start their ascent to Trout Creek Pass, a dirt road leads to the south. In a couple of miles this road brings you to remarkable granite rock formations known as The Castle and Castle Rock. These are popular for rock climbers. You may see bighorn sheep and peregrine falcons in the area.

U.S. Highway 285 continues south after its junction with U.S. Highway 24. In five miles a paved road heads west toward the Sawatch Range. After about seven miles on this side road, the San Isabel is entered with the white Chalk Cliffs to the north and the Boot Leg Campground to the south. A little farther along the road, from the Chalk Lake Campground, there is a short trail to scenic Agnes Vaille Falls. Mountain peaks are visible everywhere on either side of the paved road, the most formidable being Mount Princeton (14,197 feet) to the north and Mount Antero (14,269 feet), Boulder Mountain (13,528 feet), Mount Mamma (13,553 feet), and Grizzly Mountain (13,708 feet) to the south. Past the Chalk Lake Campground, the road changes to gravel as far as the ghost town of St. Elmo. History buffs must not miss the old cemetery northeast of the townsite. From St. Elmo you have two choices. For those with a four-wheel-drive vehicle, take the rather poor road that heads west and climbs to Tincup Pass. The road continues over the pass and into the Gunnison National Forest and eventually the historic town of Tincup. This was the most direct route from Tincup to St. Elmo. Standard vehicles can make it on the dirt road southward from St. Elmo to the ghost towns of Romley and Hancock, the latter having had a population of about 200 in 1881. From Hancock you can hike or four-wheel-drive three miles west to Williams Pass or two miles south to Hancock Pass and Hancock Lake. Williams Pass is near the abandoned Alpine Tunnel, which is an 1,800-foot-long bore that was an engineering wonder when it was constructed during the early 1880s. Just below Williams Pass is a beautiful meadow. From Hancock Lake a National Recreation Trail leads to Monarch Pass, first going over Chalk Creek Pass and following the saddle between Monumental Peak to the west and Mount Aetna to the east. For a distance, the trail follows the Middle Fork of the Arkansas River. The trail then swings abruptly westward past Bass Lake Reservoir, climbing to Bald Mountain on the Continental Divide. Following the crest of the Continental Divide, the trail reaches Monarch Pass in about five miles.

The easier way to reach Monarch Pass is on U.S. Highway 50 from Poncha Springs. The trip to the pass is about 17 miles, but the road is good and the scenery unbeatable. The Monarch Pass you cross today on U.S. Highway 50 was completed in 1939; prior to that, the crossing was one mile to the south. The original crossing is now called Old Monarch Pass and can be reached by trail or road off of U.S. Highway 50.

U.S. Highway 285 continues southward from Poncha Springs to Mears Junction before it begins its climb over Poncha Pass, which is just outside the national forest. You still have plenty to do, however, if you take the gravel road southward from Mears Junction. A fork to the left brings you to the site of Shirley, another vanished mining town. The townsite is crossed by the Rainbow National Recreation Trail. If you are in the mood for long-distance hiking, you may take the trail either east or west. If you elect to go east, you will circle around the northeastern border of the Sangre de Cristo Wilderness. If you hike west, you will follow Silver Creek until the trail enters the Rio Grande National Forest. Instead of taking the left fork to Shirley, the right fork brings you to O'Haver Campground and Lake, the latter excellent for fishing. The gravel road continues over Marshall Pass (10,842 feet) and into the Gunnison National Forest. A trip in a passenger vehicle affords sightseers a great drive. The road is fairly gentle and passes through a long subalpine meadow before reaching the pass. The road from Mears Junction to Marshall Pass was originally a railroad line and then a toll road operated by Otto Mears. Railroads continued to use the pass until 1953. Two miles north of Marshall Pass is Mount Ouray, named for the famous Ute chief. On the east side of the mountain is a large glacial cirque that resembles a chair. Legend has it that the spirit of Chief Ouray resides here.

The Sangre de Cristo Wilderness is long and narrow, oriented in a northwest to southeast direction. Its northern end is five miles south of Salida and its southern end is near Blanca, a distance of approximately 70 miles. The western part of the wilderness is in the Rio Grande National Forest.

A gravel road from Coaldale follows Hayden Creek to Hayden Creek Campground at the edge of the wilderness. From the campground, a rough and narrow jeep road rapidly ascends 2,780 feet in 3.5 miles to Hayden Pass, the northernmost pass in the Sangre de Cristo Range.

Because U.S. Highway 30 turns eastward out of Coaldale and leaves the national forest as it heads for Cañon City, you must take State Route 69 south from Cotopaxi to explore the San Isabel side of the Sangre de Cristo Wilderness. However, because this side consists of extremely rough terrain with massive blocks and sheer vertical cliffs that are impenetrable, much of the wilderness is not accessible.

A road westward out of Westcliffe on State Route 69 comes to the national forest border in about seven miles. From here, a jeep road to Hermit Lake then follows fast-flowing Middle Fork of Taylor Creek to Hermit Pass. Although steep, the road is beautiful as it alternates between pine forests and wildflower meadows. Just after passing timberline is one of the prettiest alpine vistas in Colorado, with lovely Horseshoe Lake before you. Eastward from the pass are the Wet Mountains; to the south are the Crestone Peaks and Crestone Needle; to the west are the serrated 13,000-foot peaks of the Sangre de Cristo Range.

Venable Falls can be reached via trail from the Alvarado Campground about three miles south of the Hermit Lake Road. An accessible falls is along Dry Creek a few miles to the south.

The remarkably triangular-shaped Crestone Needle with its sheer face is in the wilderness, but it is difficult to reach, requiring a four-wheel-drive vehicle to negotiate the terrain between Humboldt Peak and Marble Mountain and stamina to hike from the end of the road for a little more than a mile to the Needle. The Needle is next to mighty Crestone Peak (14,294 feet), and these mountains were the last 14,000-footers to be conquered by mountain climbers.

Three other entrances to the eastern side of the Sangre de Cristo Mountains are feasible, but all require some extra effort. Four-and-a-half miles south of Westcliffe, take the paved Colfax Lane Road off of State Route 69. After 5.5 miles, the road comes to a T; take the left fork, now gravel, and continue to the national forest boundary in about four miles. A jeep road can be used as a hiking trail that climbs to Music Pass. The distinctive sound of "singing sands" that sometimes is heard at the summit accounts for the name of Music Pass. The sound is caused by strong wind blowing through loose sand resulting in internal avalanches within the sand.

A dirt road off of State Route 69 rises to Medano Pass. Long used by Indians, the trail over Medano Pass was reputedly used in 1807 by Capt. Zebulon Pike who was searching for the headwaters of the Arkansas River. From the pass is a jeep road that has been permitted to penetrate between two units of the wilderness. If you drive or hike this rugged road, you come to the Great Sand Dunes National Monument where the road continues into the monument. Or you may take a paved road from Gardner on State Route 69 and head for Mosca Pass. For nearly 18 miles, this road through non–national forest land becomes gravel and then dirt. When it reaches the San Isabel National Forest, it becomes even rougher and narrower for the last half-mile to Mosca Pass, which is at the edge of the wilderness. From the pass it is a 3.5-mile hike to the Great Sand Dunes National Monument and State Route 150. The trail comes in nearly at the monument's visitor center.

The Wet Mountains constitute a disjunct unit of the San Isabel National Forest south of Cañon City. The northern part of this unit has several points of interest with rather easy access. A paved road due south out of Cañon City comes to the forest boundary in fewer than four miles where there is a trail-head for a five-mile trail to Tanner Peak. The road, now gravel, continues along Oak Creek to the Oak Creek Campground at the edge of forest property. Eleven miles south of Florence, State Route 67 junctions with State Route 96, which comes in from Pueblo. Head southwest on State Route 96. This road follows Hardscrabble Creek and passes through five miles of the national forest between two of the taller peaks in the area—Adobe Peak (10,183 feet) to the north and Rudolph Mountain (10,334 feet) to the south. At the forest's edge, take State Route 165, the Greenhorn Highway, south. You may drive through much of this part of the national forest. Ophir Creek and Davenport are nice campgrounds off of State Route 165 before coming to the Lake Isabel Recreation Area. The lake is surrounded by the scenic Wet Mountains, and a good hiking trail to the west follows the St. Charles River.

A short distance beyond the national forest boundary, State Route 165 comes to the Cuerna Verde Park settlement. From here a trail leads into the Greenhorn Mountain Wilderness; the trail climbs to 12,220-foot North Peak. As you hike this trail, Greenhorn Mountain (12,347 feet) looms to your left. You hike through dry oak woods in the lower areas to ponderosa pine forests midway up the trail to spruce and fir forests higher up. Both North Peak and Greenhorn Mountain are capped by alpine tundra. The Greenhorn Mountain Wilderness is unusual in that it contains no mountain lakes, although Apache Falls on the east side of the wilderness is an attractive diversion.

A smaller isolated unit of the San Isabel National Forest lies west of Walsenburg and Trinidad and is dominated by the towering Spanish Peaks. State Route 12, known as The Legends Highway, begins in Trinidad and eventually climbs into the national forest, finally coming to Cucharas Pass. This highway stays in the national forest for several miles before dropping into the plains and the small town of La Veta.

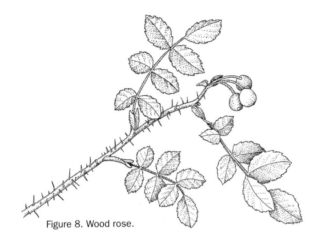

Figure 8. Wood rose.

At 13,626 feet and 12,683 feet, respectively, West Spanish Peak and East Spanish Peak are dramatically 7,000 feet above the plains. The Spanish appellation is derived from the first exploration by white men when an expedition was led by Governor Diego de Vargas of Spanish New Mexico in 1694. The peaks are massive igneous and sedimentary rocks but are of geological interest because of great dikes that radiate from the mountains like spokes of a wheel. The peaks and their associated dikes can be easily observed and hiked to from the Apishapa Scenic Highway. Eighteen thousand acres of land, including Spanish Peaks, were designated a wilderness in 2000.

The Apishapa Scenic Highway is a dirt road that begins at Cucharas Pass. The road climbs to Cordova Pass where there is the John B. Farley Memorial Trail. The first flower to bloom along the trail in spring is the pasqueflower with six lovely purple petals that form a cup-shaped flower (pl. 12). This is followed in turn by blue columbine, western bluebells, wild geranium, small-flowered shooting star, false hellebore, and white valerian. Two shrubby species, wood rose and shrubby cinquefoil, add color when they are in flower (fig 8).

Observation points dot the scenic drive. At one place, if you look to the south, you can view the headwaters of Purgatoire River. The river is reputedly named for a group of Spanish conquistadors who died in an Indian ambush near the river. Because they had not been given the last rites of the Catholic Church, the river was named Purgatoire. Local residents often refer to the river as the Picketwire. At another place, if you look to the west, you see the Culebra Range, a southern extension of the Sangre de Cristo Mountains, dominated by 14,067-foot Culebra Peak (not in the national forest). Soon, a great view of West Spanish Peak is before you. A three-mile-long trail to the timberline on West Spanish Peak takes off from the scenic highway. The road becomes very curvy and comes to the impressive Apishapa Arch, a rock masonry arch built through a dike in 1940 to allow for the road to pass through. A short trail along a ridge to the south gives you views of several dikes as well as an assemblage of bristlecone pines, limber pines, ponderosa pines, Engelmann spruces, Douglas firs, white firs, and alpine firs.

Near Cucharas Pass on State Route 12 is a road to the Cuchara Recreation Area and two great fishing lakes—Bear and Blue—and another road to the Cuchara Valley Ski Area.

Twin Peaks

The snowcapped peaks of the Sangre de Cristo Mountains rise 40 miles west of the southern Colorado town of Trinidad. Much of the intervening terrain ranges from flat to rolling, with few mountains. But two peaks, with elevations of 13,626 and 12,683 feet, stand out about 6,000 feet above the surrounding plateau. Known as the Spanish Peaks, they have served as landmarks to Native Americans, Spanish explorers, European settlers, and today's inhabitants and travelers in the area. Lending them distinction are a great number of dikes—freestanding walls of volcanic rock—ranging from one to 100 feet thick and about 100 feet tall, which radiate from the peaks like the spokes of a wheel. A dike is formed when molten rock is injected into a fissure and then hardens. If the enveloping rock, because of its composition, erodes more quickly, the dike is left standing alone. Because of these volcanic features, the Spanish Peaks have been designated a National Natural Landmark. One plant that grows only on these dikes is rock celery, a member of the carrot and parsley family. Its tiny yellow flowers appear in umbrella-like clusters, and its leaves smell like celery when crushed.

The peaks may be visited by setting out from Trinidad on State Route 12, which meanders westward and then northward for some 70 miles. On its westward leg, the road parallels the Purgatoire River, famous for its trout.

Soon it passes the nearly abandoned coal mining town of Cokedale, which had its heyday in the 1920s. It then begins its ascent into the mountains. About 30 miles due west of Trinidad, the road comes to a sandstone formation, 260 feet thick and more than 150 feet tall, known as the Stonewall; a small town of the same name is nestled at its base.

Ten miles short of the heart of the Sangre de Cristo range, State Route 12 turns northward, passing through a montane forest of pine, spruce, and aspen. After a dozen miles, the road enters the San Isabel National Forest, climbing through Cucharas Pass at an elevation of 9,941 feet. Here the forest suddenly opens up into a meadow, a good place to stop and hike around. Typical of the meadows in the southern Colorado Rockies, it has moist and springy soil and myriad wildflowers, most of which bloom during summer and early fall.

Although State Route 12 continues north from Cucharas Pass, it is about 15 miles to a junction with U.S. Highway 160, and the Spanish Peaks are reached by turning east onto Forest Road 46. In five miles, this twisting gravel and sometimes dirt road ascends another 1,300 feet to reach 11,248-foot Cordova Pass. At this higher elevation, the montane forest is replaced by a subalpine zone. Here almost all of the trees are conifers, and the understory of flowering plants is not as varied and dense. At the rustic campground at Cordova Pass, visitors can venture around for a look at the forest vegetation.

From the pass, a good trail winds up West Spanish Peak, the nearer of the two. The trail climbs through subalpine forests to the treeline at 12,000 feet and into an arctic-alpine zone of alpine meadows with lichen-covered rocks and tundra vegetation. The severe conditions on the summit prevent the growth of most woody plants, but several wildflowers grow in protected rock crevices. Most rewarding to encounter is old-man-of-the-mountains, a dwarf plant with large, daisylike heads. The species is confined to the southern mountains of Colorado. The hike up West Spanish Peak requires an extra day's time, but on a one-day drive, visitors can continue along Forest Road 46, which eventually reaches Interstate 25, 21 miles north of Trinidad. The road passes through Apishapa Arch, an opening in one of the volcanic dikes, carved by members of the Civilian Conservation Corps in 1940.

Wet meadows host a profusion of yellow flowers, including the leafy arnica, hairy groundsel, golden banner, yellow rattle, owl's-clover, large-leaved avens, and at least two kinds of evening primrose. One particularly distinctive species is called silverweed, with the underside of its leaves covered in silver-colored hairs.

Purple flowers in the meadow include the leafy Jacob's ladder (of the phlox family), purple avens, a small-flowered gentian, and elephant's head—

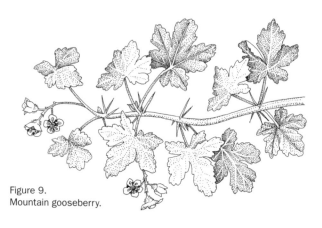

Figure 9.
Mountain gooseberry.

whose flowers come complete with a miniature "trunk." Meadow rue and northern bedstraw have clusters of small, white flowers, whereas white checker mallow has much larger but fewer flowers.

Corn lily, or false hellebore, which usually grows along rivulets in the meadow, gets its name from its large, strongly veined leaves that resemble green corn husks. Although its greenish white flowers are small, their great numbers give the plant a spectacular appearance when in bloom. Also growing in and along the rivulets are blue iris, white bistort (a type of smartweed), hemlock parsley, a couple of species of buttercups, and monument plant (pl. 13). The flowers of a monument plant each have four purple-speckled petals with a line of stiff hairs down the middle. The hairs guide the pollinating insect to the base of the petal, where there is a purplish nectary covered by a flap of tissue.

Montane forest consists principally of lodgepole pine, Douglas fir, ponderosa pine, Engelmann spruce, and quaking aspen. Wildflowers that may be visible from the road include Rocky Mountain bluebells, monkshood, golden smoke, New Mexican groundsel, white lousewort, Easter daisy, and the lovely mariposa lily, whose flowers have three large white petals. Blue- and white-flowered asters line the road throughout fall and until the end of the growing season. Other species can be encountered by penetrating deeper into the forest. The dense undergrowth beneath the evergreen trees holds shrubby mountain blueberry, false Solomon's-seal (pl. 14), pink-headed wild clover, and pearly everlasting.

Subalpine forest trees—Engelmann spruce, subalpine fir, and limber pine—are accompanied by such shrubs as the red-fruited and prickly mountain gooseberry (fig. 9) and buffalo berry, whose leaves are covered with rusty scales. Wildflowers include a wild larkspur, a wild geranium, a bottle gentian (whose purple petals do not spread open but form a bottle-shaped flower), and a deep purple thistle that is seldom found outside the Spanish Peaks area.

Arctic alpine wildflowers growing only in the southern mountains of Colorado are old-man-of-the-mountains and a red-flowered Indian paintbrush. More widespread species include a couple of white-headed daisy fleabanes, a blue bellflower, purple-fringed phacelia, golden saxifrage, moss pink, and alpine species of spring beauty, primrose, avens, and cinquefoil.

San Juan National Forest

SIZE AND LOCATION: Approximately 1,880,000 acres on the western side of the Continental Divide in southwestern Colorado. Major access roads are U.S. Highways 84, 160 (Navajo Trail), 550 (San Juan Skyway), and 666 and State Routes 145 (San Juan Skyway) and 151. District Ranger Stations: Bayfield, Dolores, and Pagosa Springs. San Juan Public Lands Center: 15 Burnett Court, Durango, CO 81301, www.fs.fed.us/r2/sanjuan.

SPECIAL FACILITIES: Winter sports area; boat launch areas.

SPECIAL ATTRACTIONS: San Juan Skyway; Durango and Silverton Narrow Gauge Railroad; Chimney Rock Archaeological Area; McPhee Reservoir; Vallecito Reservoir; Lemon Reservoir.

WILDERNESS AREAS: Lizard Head (41,193 acres, partly in the Uncompahgre National Forest); South San Juan (158,790 acres, partly in the Rio Grande National Forest); Weminuche (488,200 acres, partly in the Rio Grande National Forest).

The San Juan Mountains comprise the longest continuous mountain range in the United States and are among the highest, with many peaks more than 13,000 feet. Many of the San Juans are in the San Juan National Forest where the average elevation is around 10,400 feet, and the Continental Divide passes through them. The San Juan National Forest is one of the prettiest, with uncountable canyons, numerous streams, an abundance of high mountain lakes, cascading waterfalls, archaeological remains, dense forests, alpine tundra, and wildflower-laden meadows, but much of it is seldom seen because of some very remote areas, many of which are in the forest's three wilderness areas.

One unique way to get a glimpse of the San Juan National Forest is to take the nine-hour round trip on the Durango and Silverton Narrow Gauge Railroad that has been in existence since 1882. The route of the train is along the Animas River and enters the national forest a few miles north of Durango.

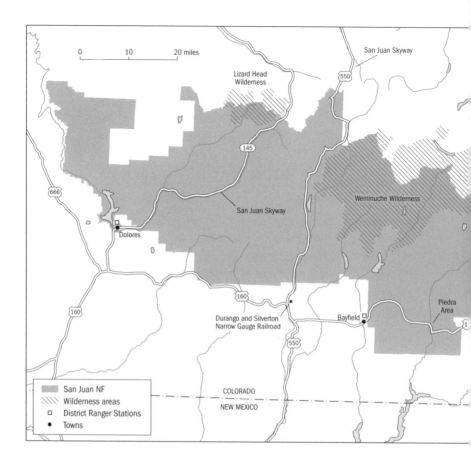

Where the river suddenly bends to the northeast, the railroad follows it through a narrow corridor between two segments of the Weminuche Wilderness. The railbed, blasted from solid rock, is about 300 feet above the river. From the railroad, you may see to the west the West Needle Mountains in the smaller segment of the wilderness, and to the east, the incomparable Needle Mountains in the major part of the wilderness. Occupying nearly a half-million acres, the Weminuche is Colorado's largest wilderness area, extending for nearly 60 miles west to east and into the Rio Grande National Forest.

When the narrow gauge comes to Tiff Spur, it has climbed 1,200 feet from Durango. It then goes through Needle Creek Canyon and stops at the wayside of Needleton to take on water. Those desiring to hike into the Weminuche may disembark the train here. This is the quickest way to get to the Chicago Basin that is nestled among the Needle Mountains. These are the most rugged mountains to be seen anywhere in the country, and they offer a challenge to rock climbers. To enjoy the magnificence of Mount Eolus, North Eolus, Windom Peak, all of them 14,000-footers, as well as Twin Thumbs,

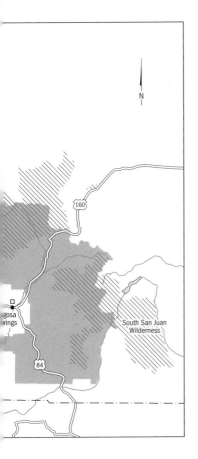

Jupiter Mountain, and peaks numbered one through 16, you don't need to climb them. Just being among them and looking up at them is awe inspiring. After reaching the Chicago Basin, the main trail curves to the south, climbs over Columbine Pass to Columbine Lake, and then splits. The south fork ascends again east of Bullion Mountain to Trimble Pass, passes the old Pittsburg Mine, and onto Silver Mesa. The fork to the southeast enters the Vallecito Basin and follows Johnson Creek to the east.

From Needleton, the train crosses Ruby Creek, Noname Creek, and Tenmile Creek before passing along the west side of 13,065-foot Mount Garfield. The railroad then goes through Elk Park, a wider area between the two segments of wilderness, before crossing Molas Creek and leaving the national forest. If you elect not to take the train, you can hike to Needleton from a trail that begins across from the Purgatory Ski Area Road on U.S. Highway 550.

You may also drive from Durango to Silverton on U.S. Highway 550, which is part of the San Juan Skyway. The skyway is a 236-mile loop that goes into a part of the Uncompahgre National Forest as well as the San Juan. The east side of the skyway goes from Durango to Red Mountain Pass in the San Juan. The first 26 miles of the skyway north of Durango are in the Animas River Valley, which is not in the national forest, but the forest land is close by on either side. Not until the skyway makes a sharp turn across Cascade Creek does it enter the national forest. The national forest has several access areas from this stretch of the highway, however. From Trimble and the Trimble Hot Springs about five miles north of Durango roads wander through the national forest between U.S. Highway 550 and the Weminuche Wilderness. Although these roads are often scenic, crossing several creeks, there are no developed Forest Service sites. Two miles north of Trimble is Hermosa where the steam train makes a stop next to an old wooden water tank, in use since 1882. The high ridge to the east is Missionary Ridge, named for a company of soldiers stationed nearby following the Civil War when most of these soldiers had fought a battle in Missionary Ridge, Tennessee. The area burned in 2002. A hiking trail leaves from the East Animas Forest Service Crew Station north of Durango and runs across the ridge to several fishing creeks. A paved road northwest from Hermosa initially follows Big Creek to the Lower Hermosa Trailhead, where you have several trails from which to choose.

Farther north on U.S. Highway 550 at the Chris Park Group Campground is a sign that describes the old stage and wagon road that connected the Animas Valley with Silverton prior to 1881. The Haviland Lake Campground on the east side of this pretty lake is one mile to the north. In a few miles, a side road to the west twists and climbs to the Durango Mountain Resort Ski Area, a very popular winter sports site. In summer, Purgatory has live theater and concerts. The forest road continues westward past the ski area to Sig Creek Campground and the Upper Hermosa Trailhead. From this trailhead at the end of the gravel road, a rough jeep road turns north and follows Hermosa Creek past Graysill Mine to Bolam Pass. From the pass is a spectacular view to the north, including Lizard Head Peak and the San Miguel Mountains.

As you look northward from Purgatory, Engineer Mountain looms ahead, its red rocks easily seen on a clear day. U.S. Highway 550 enters the San Juan National Forest 1.5 miles north of the Purgatory Road and begins to climb again to 10,640-foot Coal Bank Pass. From the pullout at the pass are vast views east into the Weminuche Wilderness and west to Engineer Mountain. You may hike to Engineer Mountain from here. U.S. 550 then descends the northern side of Coal Bank Pass and crosses Lime Creek before climbing again to Molas Pass. Molas Pass sits above picturesque glacial lakes such as Little Molas Lake and Andrews Lake. Molas Lake, larger than Little Molas and Andrews, has concessions operated by the town of Silverton. The Animals River Gorge lies below the pass as you look eastward, whereas scenic 13,368-foot Sultan Mountain is due north about four miles. Beyond Molas Pass, the skyway leaves the national forest temporarily as it drops into Silverton.

From the south end of Silverton, U.S. Highway 550 climbs westward back into the San Juan National Forest and parallels Mineral Creek for a while. The 24-mile stretch of the San Juan Skyway from Silverton to Ouray is the famous Million Dollar Highway. The highway may or may not have cost a million dollars, but the views from it are worth that much. The highway goes along the face of a mountain with several drop-offs of 1,000 feet or more. After about 11 miles past Silverton, the Million Dollar Highway climbs to Red Mountain Pass, the highest elevation on the highway and the site of an old mining town. Beyond the pass, the highway is in the Uncompahgre National Forest.

County Road 204 is another route you may take northward out of Durango to other scenic and exciting parts of the national forest. Within five miles is Junction Creek Campground. The Animas Overlook Trail begins here. After switchbacks to a saddle, there is a nice overlook into Animas City.

The road then twists its way to Animas Overlook where there are not only magnificent views but also a handicap-accessible interpretive trail. The road then winds past Sliderock Mountain and Monument Hill where there are abandoned mine sites. Just before the road comes to an end, a hiking trail to the west soon ascends to Kennebec Pass where the views of the mountains to the north are truly spectacular. The trail across this pass, known as the Highland Loop Trail, is said to date back to prehistoric time. You may hike the trail for scores of miles until it eventually turns eastward all the way to Molas Pass.

U.S. Highway 160, called the Navajo Trail, proceeds westward from Durango through Mancos to Cortez. At one point, about seven miles before reaching Mancos, the highway touches a corner of the San Juan National Forest where the Target Tree Campground has an interpretive sign explaining how the Ute Indians used ponderosa pine in their daily lives.

Once you reach Mancos, take State Route 184 north out of town for about two-tenths mile and turn right onto paved County Road 42. This road becomes a gravel road before it enters the national forest just beyond Jackson Gulch Reservoir. At the Transfer Campground, the easy Big Al Trail goes to a platform where you have an outstanding view of West Mancos Canyon below and the La Plata Mountains to the southeast. The road beyond the Transfer Campground passes Jersey Jim Lookout and eventually comes to Haycamp Mesa. The road then stays on the mesa before dropping down to Corner Canyon where it leaves the national forest and rejoins State Route 184.

From Cortez, State Route 145 northward is the western side of the San Juan Skyway loop. This highway, which eventually enters the Uncompahgre National Forest and the town of Telluride, bisects a large part of the San Juan National Forest and provides access to the west side of the national forest. The highway enters the national forest at the southern tip of large McPhee Reservoir. A paved side road follows rather closely to the southern end of the reservoir where there are boat launch areas and the McPhee Campground. From nearby Ridge Point Overlook, you can see Mesa Verde National Park to the south and much of the narrow length of McPhee Reservoir.

To explore the extreme western part of the San Juan National Forest, take State Route 184 from the south end of McPhee Reservoir for a few miles west until the highway comes to U.S. Highway 666 near the small community of Cahone. From Cahone take a gravel road back to the east and cross historic Bradfield Bridge over the upper end of the Dolores River. From the nearby Bradfield Campground are three unpaved roads, one that heads southeast, one due east, and one northeast. The southeast road follows along the north side of Dolores River through a scenic canyon. The Cabin Canyon and Ferrin Canyon Campgrounds are along the way. At the Ferrin Canyon Campground, near where the road crosses to the south side of Dolores River, is the site of the Lone Dome Ranger Station, the first station in what was the old Montezuma National Forest in 1912. An interpretive sign describes the landmark. The road continues past Ferrin Canyon to the McPhee Reservoir Dam where it ends. Another interpretive sign at the dam site describes this Dolores River water project. The dam also has picnic facilities.

The middle road from the Bradfield Campground is the Ormiston Point Stock Driveway along a ridge known as Ormiston Point. This road goes between Narraguinnep Canyon to the north and the Cabin Canyon to the south. The road curves around the small Ferrin Reservoir and then follows above Ferrin Canyon before it ends above the Dolores River.

The road to the northeast from the Bradfield Campground travels the ridge above the remote and wild and woolly Narraguinnep Canyon and its Research Natural Area. The natural area has been set aside to preserve a magnificent undisturbed example of interior canyon and mesa country of southwestern Colorado. The depth of the canyon is approximately 1,300 feet below the rim, which is at an elevation of 8,000 feet. Because the slopes into the canyon are steep and rocky, access to the natural area is extremely difficult. Ponderosa pine is prevalent on the rim of the canyon, particularly at the north end, but it occurs in all parts of the natural area. The larger trees are in excess of 100 feet tall. Three kinds of junipers usually grow in the rockier terrain: alligator juniper, one-seeded juniper, and Utah juniper are usually multitrunked and gnarly. The junipers are usually associated with piñon pine. A very common shrub in the natural area is mountain mahogany, which is a good source of food for mule deer. When the gravel road reaches the small Corner Reservoir, it has reached the northern end of the natural area. From Corner Reservoir, a multitude of gravel, dirt, and jeep roads run northward to the remainder of the San Juan National Forest, but there are no Forest Service amenities in this large area of canyons, mesas, and small lakes.

Returning to Dolores, the San Juan Skyway heads east out of town and then follows the West Dolores River through the national forest. Two miles before Stoner, at the junction of Stoner Creek and West Dolores River, a good graded road provides access to all points of interest in the northern part of the San Juan National Forest. This road follows West Dolores River all the way to the southern end of Lizard Head Wilderness. The river continues into the wilderness with the road turning southeast along Meadow Creek to join State Route 145 at the Cayton Campground. This long stretch of gravel road along West Dolores River passes three campgrounds and two trailheads along the way. As the road progresses toward the Lizard Head Wilderness, you may want to make a couple of short stops just before reaching the Dunton Guard Station. About one mile up Geyser Creek is Geyser Springs, the only true geyser in Colorado. Two miles farther up the road, take a very short trail to a nice waterfall on Eagle Creek. The Navajo Lake Trailhead is located where the West Dolores River enters the Lizard Head Wilderness. After crossing the footbridge over the river, the trail follows the river into the San Miguel Mountains. The trail winds through meadows and forests, past a waterfall, and then drops into Navajo Basin where lovely Navajo Lake is located. The trail then continues around 14,017-foot Wilson Peak and into the Uncompahgre National Forest.

Returning to State Route 145, the San Juan Skyway continues eastward past the West Dolores River side road. In about 11 miles you come to the trailhead for Bear Creek Trail to the south and, a mile farther on, the Priest Gulch Trailhead. The skyway climbs steadily toward Lizard Head Pass. Where the highway crosses Scotch Creek, a Forest Service interpretive sign describes the Scotch Creek Wagon Road that is now a jeep road. At one time it led all the way to the Animas Valley. In a mile or two are historic charcoal ovens along the west side of the skyway, just before Rico. Several jeep trails run out of Rico to various mines of the past.

Beyond Rico is the Cayton Campground and the final climb to Lizard Head Pass. The pass, in an open grassy meadow, is just outside the Lizard Head Wilderness. Lizard Head is a prominent narrow peak, fewer than four miles from the pass. Although Lizard Head is at an elevation of 13,113 feet, it is overshadowed by the Wilson Group of peaks behind it. The Wilson Group dominates the eastern half of the wilderness, with towering Mount Wilson (14,246 feet), El Diente Peak (14,159 feet), Wilson Peak (14,017 feet), and Gladstone Peak (13,913 feet), the last two in the Uncompahgre National Forest. The western end of the wilderness has the Dolores Group of peaks: Dolores Peak (13,290 feet), Middle Peak (13,261 feet), and Dunn Peak (12,595 feet). Lizard Head Trail begins at the pass and climbs quickly to the wilderness, passing Black Head on its way to Lizard Head. From Lizard Head, which is so crumbly that it is not wise to attempt to climb it, the trail enters the Uncompahgre National Forest. Beyond Lizard Head Pass, the San Juan Skyway enters the Uncompahgre National Forest on its way to Telluride.

Most of the eastern half of the San Juan National Forest lies between U.S. Highway 550, which is the eastern loop of the San Juan Skyway between Durango and Silverton and U.S. Highway 160, the highway that crosses Wolf Creek Pass. A significant part of the national forest lies east of U.S. Highway 160.

The entire northern part of this side of the San Juan National Forest is in the Weminuche Wilderness, although the extreme eastern side is in the South San Juan Wilderness. Few roads of any kind access the northern reaches of the Weminuche.

You may wish to begin your exploration of the eastern half of the forest from Durango. From the center of town, take County Road 240 that eventually follows Florida River, reaching the San Juan National Forest boundary at the southern tip of Lemon Reservoir. The Forest Service maintains the Middle Creek Campground and the Upper Lemon fishing access area on the eastern side of the reservoir. If you continue northward on the east side of the reservoir, there are two more campgrounds and a trail to scenic Lost Lake.

By continuing on County Road 240 east of Lemon Reservoir, there is a paved side road north to large Vallecito Reservoir where there are several Forest Service campgrounds and hiking trails, one for 1.2 miles to lovely Lake Eileen and longer ones into the Weminuche Wilderness. A road completely encircles the Vallecito Reservoir. From the northeast corner of the reservoir, Middle Mountain Road is a four-wheel-drive road full of switchbacks as far as it goes up between two lobes of the wilderness. At the road's end, you may hike three miles to the remains of Tuckerville, a short-lived mining town.

From Bayfield, U.S. Highway 160 makes its way eastward through the national forest, past numerous canyons, peaks, and abandoned mines. At the small community of Piedra, the highway is near the south end of the Piedra Area, a wild, roadless region that is not yet a designated wilderness. Trails into the Piedra Area are off of First Fork Road north of town.

Six miles east of Piedra, State Route 151 branches to the south and descends through piñon pines and junipers as it curves around the south end of Chimney Rock and into the Chimney Rock Archaeological Area. The San Juan Mountains Association operates a small visitor center at the turnoff to Chimney Rock on State Route 151. Visitors must sign up for guided tours here. The twin pinnacles of Chimney Rock and Companion Rock that thrust 1,200 feet into the air are the centerpiece of an ancient Chaco ruins. Now a national historic site preserving these 1,000-year-old remains, the site contains the Great House, Ridge House, Grand Kiva, and about 200 other structures in a six-square-mile area. Great Kiva Trail Loop is handicap accessible from the Great Kiva to the Pit House and around several other structures. The Pueblo Trail is one mile long and more rugged as it climbs 200 feet to the Great Pueblo. An interpretive program is offered several times daily. The area is accessible by guided tour only in summer and closed to all entry in winter.

Two miles before Pagosa Springs, County Road 600, the Piedra Road, heads northward and eventually twists around the upper end of the Piedra Area. With a good forest map, you can make your way to within a half-mile of the Weminuche Wilderness. From the end of the road is a mile-long hiking trail to Piedra Falls. Northward from the end of the road are two geological phenomena: The Notch and The Keyhole. The main road continues northward past large Williams Creek Reservoir to the southern boundary of the wilderness. A trail from the end of this road enters the wilderness and joins the Continental Divide National Recreation Trail just north of the Williams Lakes. Northeast of Williams Creek Reservoir is the Williams Creek Research Natural Area, which contains large undisturbed stands of white fir. Other plants you may encounter are Douglas fir, Engelmann spruce, Rocky Mountain maple, pachistima, thimbleberry, creeping Oregon grape, mountain snowberry, Fendler's meadow rue, and several species of wild violets. The natural area is on a bench between the reservoir and the steep cliffs of an adjacent mountain. The Continental Divide National Recreation Trail runs east and west from here, forming the border between the San Juan and Rio Grande National Forests.

At Pagosa Springs, U.S. Highway 160 heads northward and climbs to famous Wolf Creek Pass at 10,857 feet on the Continental Divide. On the approach to the pass, a side road eastward to East Fork Campground continues on to pretty Silver Falls. Just before U.S. Highway 160 begins its tortuous ascent to Wolf Creek Pass, there is a short interpretive trail to 110-foot Treasure Falls. Back on the main highway, at the first severe switchback, a pullout permits a spectacular view into the San Juan River Valley.

U.S. Highway 84 proceeds southward from Pagosa Springs. To the east of this highway is a view into the South San Juan Wilderness, the western half of which is in the San Juan National Forest. Because of steep slopes, deep glacier-carved ravines, and spirelike pinnacles, the wilderness is very remote and access is limited. From Silver Falls, mentioned earlier, there is a four-mile gravel road along Quartz Creek to a trail into this wilderness. About 12 miles south of Pagosa Springs, the Blanco Basin Road goes east toward the wilderness, eventually ending at a trail that follows Rio Blanco Creek into the wilderness.

A little farther south on U.S. Highway 160, there is a two-mile-long road to the Blanco River Campground. Nearby is the Murray Place, an early homestead in the forest.

Uncompahgre National Forest

SIZE AND LOCATION: Approximately 950,000 acres in south-central Colorado. Major access routes are U.S. Highway 550 (San Juan Skyway) and State Routes 141 and 145 (Unaweep-Tabeguache Byway). District Ranger Stations: Montrose and Norwood. Forest Supervisor's Office: 2250 Highway 50, Delta, CO 81416, www.fs.fed.us/r2/gmug.

SPECIAL FACILITIES: Horse trails; boat launch areas.

SPECIAL ATTRACTIONS: Dry Mesa Dinosaur Quarry; Roubideau Canyon Area; Tabeguache Canyon Area; Unaweep-Tabeguache Byway.

WILDERNESS AREAS: Mount Sneffels (16,565 acres); Uncompahgre (102,721 acres); Lizard Head (41,189 acres, partly in the San Juan National Forest).

The Uncompahgre National Forest consists of two distinct units, each very different but both beautiful in their own way. The western unit encompasses most of the Uncompahgre Plateau, a mesa at extremely high elevation penetrated by rugged canyons. The eastern unit includes the northern mountains of the San Juan Range, with jagged peaks, glacial lakes, narrow canyons,

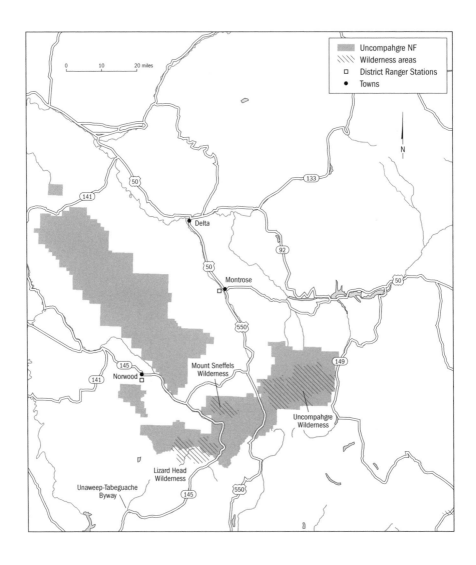

and swift, clear streams. These mountains between Ouray and Telluride are often referred to as the Switzerland of America.

The Uncompahgre Plateau runs diagonally northwest to southeast, roughly between State Route 145, the Unaweep-Tabeguache Byway, to State Route 62, the highway from Ridgway to Placerville. At Placerville, State Route 145 goes through Norwood and meets State Route 141 near Naturita. From Placerville to Gateway, the highway parallels the southwestern edge of the plateau. The Divide Road bisects the plateau for much of its length, and driving this road is a good way to see many of the plateau's features.

Beginning from the north of the plateau, leave State Route 141 on the Divide Road about 20 miles southwest of the junction of State Route 141 and U.S. Highway 50. After going through scenic Jacks Canyon, the Divide Road enters the Uncompahgre National Forest in about eight miles. The Divide Road winds for scores of miles, exiting the national forest a few miles south of Placerville. East of here, the Uncompahgre Plateau is on Bureau of Land Management and private land.

The Divide Road crosses more creeks and side canyons than you can keep up with, and there are hiking trails along most of these creeks and canyons, as well as four-wheel-drive roads. The entire route across the Uncompahgre Plateau is between approximately 8,400 and 9,700 feet, gradually climbing as it heads eastward. The Forest Service maintains one picnic area and three campgrounds along the route. Near the Columbine Campground at Columbine Pass, about midway along the Divide Road, is the intersection with the Delta-Nucla Road. If you choose to take the Delta-Nucla Road northward, you climb onto 25 Mesa and follow Monitor Creek out of the national forest. Along Escalante Creek is a Research Natural Area. This small riparian area contains an undisturbed stand of blue spruce. The blue spruce is restricted to the creek bottom but disappears above the riparian area. Trees associated with the blue spruce are Engelmann spruce, Douglas fir, ponderosa pine, and aspen. Wildflowers in the area include Indian paintbrush, oxeye daisy, meadow rue, marsh marigold, and yarrow. Should you take the Delta-Nucla Road southward, you circle around the east side of Tabeguache Canyon, drive through Coal Canyon below Pinto Mesa, and come to the small community of Nucla. A side road over Pinto Mesa takes you to the Indian Trail that has been in use for hundreds of years. The trail provides views of and access to the Tabeguache Canyon area, a tremendously scenic roadless region containing a very deep canyon where layer upon layer of sedimentary rock has been exposed by the cutting action of Tabeguache Creek. Layers of Chinle, Wingate, and Kayenta sandstone are particularly colorful. The Wingate Cliffs have been formed from sand dunes deposited eons ago. Several caves are in the canyon. The Indian Trail follows Forty-seven Creek at the western edge of the area and circles south of Round Mountain to Starvation Point on the north side of the canyon.

Beyond Columbine Pass, the Divide Road passes tiny Hinkley Pond and the natural Ouray Spring. Transfer Road branches to the north near the Antone Springs Campground and goes along the east side of the Roubideau Canyon area. This roadless area features 20-mile-long Roubideau Canyon, the steep sandstone walls of which are strikingly beautiful. Roubideau Creek at the bottom of the canyon is lined with willows and cottonwoods, and nice riparian communities are scattered along the creek. Several hiking trails cross the canyon, each requiring a steep descent to the creek and a steep ascent on the other side. Three miles up Transfer Road is Roubideau Trail, which is 5.5 miles long from the trailhead on the east side to the trailhead on East Bull Creek Road on the west side. Nine miles farther north, the Transfer Road crosses the midsection of the canyon, beginning at Oak Hill on the east and ending on the west at Payne Mesa Road. This is a 7.5-mile trail that circles around Davis Point. At the upper end of the canyon within the Uncompahgre National Forest is the Traver Trail. It begins from a trailhead on Roatcap Road, which branched west off of Transfer Road. Roubideau Canyon continues a few miles farther to the north on Bureau of Land Management land. The lower elevations within the canyon have a desertlike appearance with sagebrush and bitterbrush common. As one ascends the canyon sides, one passes through a piñon pine–juniper woods with scattered specimens of Gambel's oak to ponderosa pine forests to Engelmann spruce and subalpine fir stands on top of the canyon walls.

The most remarkable feature of the Uncompahgre Plateau is the Dry Mesa Dinosaur Quarry along the East Fork of the Escalante River. It is the largest dinosaur quarry in the United States, and some of the most remarkable skeletal remains of dinosaurs and other creatures have been discovered here. Since the quarry was discovered in 1971, 17 different genera of dinosaurs have been unearthed, as well as fossil remains of crocodiles, birds, fish, mammals, and pterosaurs. Some of these dinosaurs are very rare, including *Torvosaurus tanneri*. Dating back 148 million years, this dinosaur, when standing, is 30 feet long and 12 feet tall. The only reconstructed specimen known is on display at the Earth Science Museum on the campus of Brigham Young University. Tours of the Dry Mesa Dinosaur Quarry are available during summer months for a fee, and those wishing to work with a paleontologist can sign up for half-day or all-day digs. For tour times and dig reservations, call (970) 874-6638. The quarry can be reached by taking Forest Road 501 westward out of Delta on a continuation of 5th Street and driving this road southwestward across Sawmill Mesa. After about 22 miles, much of it on Bureau of Land Management land, the road enters the Uncompahgre National Forest and, after five miles, comes to a junction with a dirt road heading north. Take this road for three miles, then a side road to the left for 2.5 miles. The quarry is in front of you.

The eastern part of the Uncompahgre National Forest has some of the most spectacular mountain scenery in the country. Although old abandoned mine sites have scarred the flanks of many of the mountains, the remains add a touch of historical charm to the area.

The northwest corner of this part of the Uncompahgre contains Mount Sneffels, at 14,150 feet the second highest peak in the national forest, located at the far end of the Mount Sneffels Wilderness. To the west of this huge granite mountain are steep-sloped mountains with jagged peaks. Because of their granitic nature, these mountains are dangerous to climb. The best hiking trail in the wilderness is the Blue Lake Trail that circles Mount Sneffels and passes between the azure Blue Lakes. The northern trailhead is at the end of East Fork Dollar Creek Road that comes down from the north five miles east of Ridgway on State Route 62. The southern trail into the Mount Sneffels Wilderness is at the end of Yankee Basin Road, and this trail climbs to Blue Lake Pass. The Blue Lakes are at an altitude of 13,000 feet. If you stop at Yankee Basin from late July to late August, one of the very best alpine wildflower displays to be seen anywhere greets you (pl. 15). In fact, most of the wilderness is above timberline, and alpine flowers can be enjoyed at many places. Just below timberline are fine stands of Engelmann spruce and subalpine fir.

Ouray and Telluride are situated in the most magnificent mountain settings, and the areas around the towns and between them are sure to excite any visitor. Ouray may be reached from the south by following U.S. Highway 550 through the San Juan National Forest to Red Mountain Pass and then through the Uncompahgre National Forest to town. This stretch of highway has been designated as the San Juan Skyway, and it is also known as the Million Dollar Highway. The skyway from Red Mountain Pass descends down the flanks of the mountain, passes Crystal Lake, and finally parallels the Uncompahgre River for three miles until it drops into Ouray. Originally established as a mining town, Ouray is now geared for tourists.

After touring Ouray but before leaving for the wild areas that surround the town, take the short road at the south end of town to the Amphitheatre Campground. From here there is an interpretive trail to an overlook of Cascade Falls.

The easiest way to get to Telluride, on the other hand, is to take State Route 145 through the San Juan National Forest to Lizard Head Pass and then continue on in the Uncompahgre National Forest to town. This is the west side of the San Juan Skyway loop. Like Ouray, Telluride is now a thriving tourist town, although even more isolated than Ouray. This has been intensive mining country, and the area particularly between Ouray and Telluride has supported hundreds of gold and silver mines. The weathered structures of many of these mines is a vivid reminder of the past. To get to the mines, crude roads had to be constructed, and the area between the two towns is crisscrossed by dozens of roads. Some of these have been reduced to hiking trails, and others probably should have been. Still, many miles of road abound where jeep enthusiasts have a heyday. Unless you are an experienced jeep driver in the mountains, do not attempt most of these roads. Jeep rental offices are available in Ouray and Telluride, and if you want a nearly heart-stopping experience, take a tour with one of the operators.

The most notorious of these hairy-scary roads is the Black Bear Road, immortalized in a humorous song of the same name by C. W. McCall. (The music for this song was written by Chip Davis, the leader of Mannheim Steamroller.) This road begins on the south side of Red Mountain Pass in the San Juan National Forest and makes its way via unbelievable narrow switchbacks to the old Black Bear Mine located near the top of Telluride Peak at 12,000 feet. If you look at the face of Telluride Peak from Imogene Pass (see p. 123), you see many zigzag lines going back and forth up the mountainside. This is Black Bear Road. These zigzag switchbacks are flanked on either side by Ingram Falls and Bridal Veil Falls, two shining long and narrow waterfalls. Bridal Veil Falls is Colorado's highest at more than 350 feet. As you pass abandoned mines along the Black Bear Road, you see many of the weathered mining structures. If you plan to explore around any of them, remember that these are old mines, and deep shafts are here and there, so you must be extremely cautious. Several switchbacks bring you to the base of Bridal Veil Falls. You may hike to the top of the falls by following the closed jeep road along Bridal Veil Creek. From the base of the falls, the drive into Telluride can be negotiated by ordinary passenger cars.

Another jeep road crosses Ophir Pass and the mining district between Ouray and Telluride. This road also begins in the San Juan National Forest, this time about 5.5 miles south of Red Mountain Pass. The road follows Middle Mineral Creek to the rocky Ophir Pass. The pass is crowded between Lookout Peak (13,661 feet) to the north and South Lookout Peak (13,357 feet) to the south. A huge valley spreads out before you as you look to the west, with the jagged peaks of the Ophir Needles in the distance. At one time, this road went through to the community of Ophir, but the road is closed to vehicular traffic before reaching another jeep trail to the Carbonero Mine. Ophir may be reached, instead, by driving State Route 145 down the north side from Lizard Head Pass. Between the pass and Ophir is the picturesque Trout Lake, the Priest Lakes, and the Matterhorn Campground. Just beyond the north end of Trout Lake is Trestle Road, a good dirt road from State

Route 145 that follows the east side of Trout Lake. In one mile, turn left onto Hidden Lake Road. In about another 2.5 miles, at one of the switchbacks, is a trail south to Lake Hope. Although this trail is 2.5 miles long, it is gentle enough for families to hike.

From Ophir, State Route 145 continues to Telluride. Two miles north of Ophir, however, another exciting road runs to the east that passenger cars should be able to handle as far as the mining town of Alta, or the little that is left of it. From here you may hike or drive a four-wheel-drive vehicle over a short road to Alta Lakes or take a seven-mile jeep road north to State Route 145, about one mile west of Telluride.

In Ouray, another good day trip is to drive the paved road for six miles to Camp Bird, the site of the old Camp Bird Mine. The highway climbs along scenic Canyon Creek where you can observe various creeks and gulches on either side. For most people, this six-mile drive is sufficient excitement, but for the more adventurous, there are two jeep roads from Camp Bird. The right fork follows Sneffels Creek past the townsite of Sneffels and into Yankee Boy Basin and the trailhead for hikes into the Mount Sneffels Wilderness. The left fork follows Imogene Creek into Imogene Basin and eventually climbs to Imogene Pass via a narrow shelf road at 13,114 feet. The descent down the south side from the pass is on switchbacks to a tranquil opening called Savage Basin. From here the road is on another shelf past Pandora to Telluride.

To the west of Ophir and State Route 145 is the Lizard Head Wildernesss whose northern third is in the Uncompahgre National Forest. The eastern side of the wilderness in the Uncompahgre includes Wilson Peak (14,017 feet) and Sunshine Peak (12,930 feet). A hiking trail crosses the wilderness from Lizard Head Pass and circles Bilk Basin to the northern end of the wilderness. The western side of the wilderness is accessible via trails from Woods Lake where there are picnic tables and the fishing is good. Woods Lake can be reached by taking the Fall Creek Road southward from State Route 145 between Sawpit and Placerville. Two sections of the Uncompahgre National Forest are west and northwest of the Lizard Head Wilderness. Immediately west of the wilderness is Lone Cone, reached by a very crooked dirt road from Woods Lake. Northwest of Lone Cone is an isolated 40-square-mile part of the Uncompahgre National Forest with Naturita Creek flowing through the entire section.

The Uncompahgre National Forest also extends for 18 miles north of Ouray and 20 miles west. More than half of this large forested area is in the Uncompahgre Wilderness. The wilderness contains fascinating rock castles and several towering mountains, including the national forest's highest, Uncompahgre Peak, at 14,309 feet. Nearly as imposing are Wetterhorn Peak

(14,015 feet) and six other mountains in excess of 13,000 feet. The wilderness is crossed by many streams and is so rugged that access is limited. A branch of the Big Blue Trail goes to Uncompahgre Peak, and the West Fork Pack Trail crosses to the west of Coxcomb Peak and into Wetterhorn Basin. Cow Creek Canyon at the western side of the wilderness is deep but inaccessible because of the dense forests.

U.S. Highway 550 proceeds northward out of Ouray and soon leaves the Uncompahgre National Forest. At Ridgway, there is a road that reenters the national forest just north of the Uncompahgre Wilderness and follows Spruce Ridge to Owl Creek Pass. From the pass, the impressive Chimney Rock can be seen one mile to the south within the wilderness boundary. West of Chimney Rock is Courthouse Rock, another isolated mountain with a jagged top. After driving over Owl Creek Pass, the road swings northward along the Cimarron River, flanked by Turret Ridge to the east and Cimarron Ridge to the west. The road then makes a sharp bend around large Silver Jack Reservoir, which has a campground at the north end and an interpretive trail at the campground. Within the next two miles are two more campgrounds before the road leaves the Uncompahgre National Forest.

State Route 159 northward out of Lake City goes along the eastern edge of Uncompahgre Wilderness. Two miles beyond the northeast corner of the wilderness, the Alpine Road leads west off of State Route 149 and follows Soldier Creek across Alpine Plateau to the Blue Campground and the north entrance to the Big Blue Trail.

White River National Forest

SIZE AND LOCATION: Approximately 2,300,000 acres in northwestern and north-central Colorado, in the Rocky Mountains. Major access routes are Interstate 70, U.S. Highway 24, and State Routes 13, 82, 131, and 133. District Ranger Stations: Aspen, Carbondale, Eagle, Meeker, Minturn, Rifle, and Silverthorne. Forest Supervisor's Office: 900 Grand Avenue, Glenwood Springs, CO 81601, www.fs.fed.us/r2/whiteriver.

SPECIAL FACILITIES: Winter sports areas; boat launch areas; ATV areas.

SPECIAL ATTRACTIONS: West Elk Loop and Scenic Byway; Flat Tops Scenic Byway; Wild and Scenic Rivers: South Fork of the White River, Deep Creek, Cross Creek.

WILDERNESS AREAS: Hunter Fryingpan (81,866 acres); Collegiate Peaks (166,938 acres, partly in the Gunnison and San Isabel National Forests);

Maroon Bells–Snowmass (181,117 acres, partly in the Gunnison National Forest); Holy Cross (122,797 acres, partly in the San Isabel National Forest); Eagles Nest (132,906 acres, partly in the Arapaho-Roosevelt National Forest); Ptarmigan Peak (12,594 acres); Raggeds (64,992 acres, partly in the Gunnison National Forest); Flat Tops (196,344 acres, partly in the Routt National Forest).

The Rocky Mountains dominate the White River National Forest, with many peaks in excess of 14,000 feet. The incomparable Maroon Bells is the most photographed region in any national forest and is the centerpiece of the Maroon Bells–Snowmass Wilderness. This wilderness, whose southern portion is in the Gunnison National Forest, has six of the 14,000-foot peaks, nine passes above 12,000 feet, and more than 100 miles of hiking trails. Because of the great number of visitors to the area during summer, the 10.5-mile road from Aspen to Maroon Lake at the edge of the wilderness is closed to private vehicles (except for campers and the physically disabled) from 8 A.M. to 5 P.M. from June through Labor Day. Visitors to Maroon Lake must take a shuttle from Aspen Highlands ski area; it leaves from the Rubey Park Transit Center every 20 minutes, and there is a charge for the shuttle. From Maroon Lake several fine trails cross into the wilderness. One goes 1.5 miles to Crater Lake and then follows West Maroon Creek for five miles, eventually climbing over West Maroon Pass. It skirts the east side of 14,014-foot North Maroon Peak and 14,156-foot Maroon Peak. Another trail goes from Crater Lake up Minnehaha Gulch and then over Buckskin Pass, eventually coming to Snowmass Lake and Snowmass Peak.

Another popular trailhead is at the former Janeway Campground situated along scenic Crystal River about three miles north of Redstone along State Route 133. The trail enters the western side of the wilderness and follows Avalanche Creek for many miles, ultimately going to Avalanche Lake in the heart of the wilderness. Castle Peak, the highest in the wilderness at 14,265 feet, is located on the border between White River and Gunnison National Forests. You can get to within one mile of Castle Peak by following the old four-wheel-drive road south of the mining town of Ashcroft for about six miles. This is a highly scenic and very rough road that follows Castle Creek until it turns abruptly north to the old Montezuma Mine. Another popular trail begins at the end of a spur road that branches off the Castle Creek Road about five miles south of Aspen. This trail parallels Conundrum Creek for nearly 11 miles, eventually climbing over Triangle Pass before entering the Gunnison National Forest. On the way it passes tranquil Silver Dollar Pond and is sandwiched between Kiefe Peak and Cathedral Peak. A pioneer grave sits along the trail, about 1.5 miles west of 14,022-foot Conundrum Peak. At any place where the trail goes above timberline into alpine tundra habitats, keep a sharp lookout for rare wildflowers such as Kotzebue's grass-of-Parnassus (fig. 10), rockcress draba, alpine meadowrue, and pygmy alpine buttercup, which grows only a half-inch tall.

Southeast of Aspen is the Collegiate Peaks Wilderness, although most of this wilderness is in the Gunnison and San Isabel National Forests. The White River National Forest part of the wilderness contains pretty Truro Lake at the base of Truro Peak. These gems of the forest can be reached by following the trail along Tabor Creek from a trailhead off the Lincoln Creek Road three miles southeast of the Lincoln Gulch Campground, but take a map with you because the trails are sometimes difficult to follow.

Immediately north of the Collegiate Peaks Wilderness and separated from it by State Route 82 is the Hunter Fryingpan Wilderness. State Route 82 connects Leadville with Aspen, entering the White River National Forest at Independence Pass. This is one of the most rugged passes crossed by a paved road. A small bog near the pass is home to the very rare altai cotton grass. West of the pass are the weathered remains of the old Independence mining town.

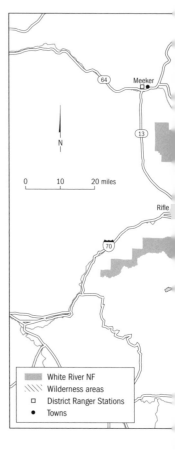

The highway then continues westward. At Lost Man Campground, a rugged nine-mile trail passes Lost Man Reservoir and follows Lost Man Creek into the Hunter Fryingpan Wilderness, eventually climbing over South Fork Pass. Just before timberline is a nice forest of Engelmann spruce and alpine fir. Farther down the mountain are forests of lodgepole pine. A little farther along State Route 82 are The Grottos and Devils Punchbowl, both points worth seeing. Fishing is a major activity in the Hunter Fryingpan Wilderness with Hunter Creek and Fryingpan River and their tributaries teeming with trout. The craggy Williams Mountains run through the center of the wilderness.

One particularly exciting drive is the forest road that leads due east from the small community of Basalt on State Route 82, 18 miles southeast of Glenwood Springs. From Basalt, the forest road follows the Fryingpan River for about 13 miles to the Rocky Fork Picnic Area at the west end of the long and narrow Ruedi Reservoir. Along the way to the reservoir, the road crosses several picturesque mountain streams and scenic gulches such as Seven Castles

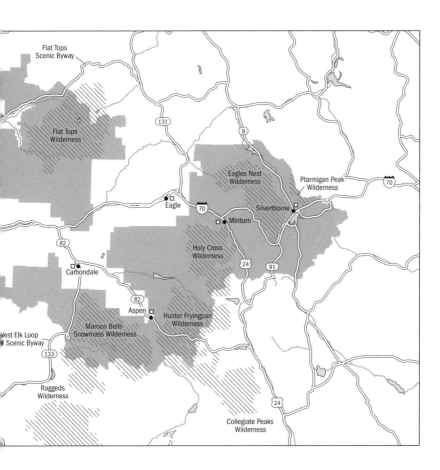

Creek, Willow Creek, Frenchman Creek, Hoover's Bend Gulch, Burke Gulch, and Snyder Gulch. Staying near the northern border of the Ruedi Reservoir, the forest road passes several Forest Service campgrounds and circles the northern side of Gyp Hill before reaching the tiny community of Meredith at the east end of the reservoir. Deerhammer Campground with a boat launch area is located here. In two miles, the road comes to the few houses that comprise Thomasville. In another mile, a dirt road branches northward from the paved road. The paved road continues for several miles after the Chapman Campground, providing marvelous views of the Sawatch Mountain Range to the east in the Holy Cross Wilderness and the Williams Mountain Range to the southeast in the Hunter Fryingpan Wilderness. From the end of the paved road, there are several gravel and dirt roads, many of them suitable only for four-wheel-drive vehicles. One of these eventually climbs over Hagerman Pass and into the San Isabel National Forest just outside the northern boundary of the Mount Massive Wilderness. The dirt road northward from Thomasville, passable in a passenger car, is an exciting route with narrow switchbacks. The road more or less follows Lime Creek to a tranquil meadow known as Lime Park. It then swings northward to Crooked Creek Reservoir before snaking its way over Crooked Creek Pass. From here, the road traces the course of West Brush Creek for about 20 miles to Eagle on Interstate 70. Two miles before the road leaves the White River National Forest, an improved road branches to the east and follows East Brush Creek. South of the old mining community of Fulford are numerous natural springs and Fulford Cave. The Yeoman Park and Fulford Cave campgrounds are nearby. You should allow a full day for this trip so you can enjoy the scenic vistas, admire small waterfalls, and hike a trail or two. Under one of the waterfalls you may see the pretty but rare nodding, yellow-flowered Barneby's columbine.

U.S. Highway 24 leads southward from Interstate 70 about five miles west of Vail. After going through the town of Minturn, the highway becomes very scenic. In 2.5 miles, as the highway begins climbing via a switchback, there is a highway sign designating Tigiwon Road to the right. After 6.5 miles on the rather primitive Tigiwon Road, there is a lodge and, shortly beyond, the Tigiwon Campground. The log lodge, built in the 1930s, was originally a shelter for people traveling to the Mount of the Holy Cross. The mountain received its name from two intersecting narrow ravines that appear in the shape of a cross when they are filled with snow. Because of its crosslike appearance, the mountain has been the object of various religious pilgrimages since photographer William Henry Jackson saw it in 1873. So many people hiked to see the mountain that a stone shelter, still standing, was built as a refuge on Notch Mountain east of the Mount of the Holy Cross. After two more miles, the road ends at the Half Moon Trailhead, where trails lead westward into the vast Holy Cross Wilderness and Notch Mountain. Mount of the Holy Cross, at 14,005 feet, looms over the wilderness, although 25 other peaks attain an elevation of at least 13,000 feet. Between the peaks are gorgeous glacier-carved U-shaped valleys. Alpine vegetation occurs on all the peaks, and black bears, bobcats, and lynx are present.

After returning to U.S. Highway 24, the road drops down to Eagle Creek. Just before the bridge, the road east to Redcliff is a must. After passing the little mining community of Redcliff, the dirt road follows Turkey Creek, climbing through forests of aspens and pines to Shrine Pass for one of the finest views of Mount of the Holy Cross. Beyond Shrine Pass, the road is more improved until it reaches Interstate 70 at Vail Pass. For the more adventurous, there are several four-wheel-drive roads off of the Shrine Pass Road, including one to Benson's Cabin and another that parallels Lime Creek.

If you return to U.S. Highway 24 and continue southward, the highway is

Figure 10. Grass-of-Parnassus.

known as the Tenth Division Memorial Highway and it passes a large flat valley that is the site of Camp Hale. Now a historic site with a campground, Camp Hale was a 1,210-acre training site during World War II for the U.S. Army's Tenth Mountain Division where 17,000 soldiers trained for mountain warfare. About 10 miles west of Camp Hale, just outside the eastern boundary of the Holy Cross Wilderness, is the site of Holy Cross City, an old mining town, and Homestake Reservoir. From the north end of the reservoir are trails into the wilderness. Where U.S. Highway 24 leaves the White River National Forest and enters the San Isabel National Forest it crosses Tennessee Pass with an elevation of 10,424 feet. A monument at the pass honors the Tenth Mountain Division.

Six miles west of Avon, Interstate 70 leaves the national forest for a while, entering it again as it begins its descent into the remarkable Glenwood Canyon. From an exit off Interstate 70 within the canyon is a marvelous short but steep trail to Hanging Lake and Bridal Veil Falls.

Another fascinating part of the White River National Forest lies south of Glenwood Springs. From Carbondale, State Route 133 stays alongside sparkling Crystal River as it borders the western side of the Maroon Bells–Snowmass Wilderness. Mighty Mount Sopris looms east of the highway, and directly across State Route 133 from the mountain is Assignation Ridge, a wild 12,000-acre area being considered for wilderness status. Historic Redstone Castle lies to the east of the highway and is adjacent to the privately owned Redstone Inn. Across from Redstone are the remains of coke ovens from scores of years ago. Two miles south of the coke ovens is picturesque Haynes Creek Falls.

Before State Route 133 crosses McClure Pass into the Gunnison National

Forest, take the improved road eastward to Marble and its famous marble quarries that were active many years ago. Huge slabs of marble may be seen strewn across the countryside. A 100-ton slab of marble from these mines was formed into the Tomb of the Unknown Soldier. Before continuing beyond the village of Marble, check to see if the road is safe for passenger vehicles.

For spelunkers, Hubbard Cave is a must. To reach it, drive State Route 82 south out of Glenwood Springs for about two miles, then turn left onto Red Canyon Road and follow this for 2.5 miles. Take another left past a gravel pit for two more miles. The road gets rougher and rougher, and a four-wheel-drive vehicle is recommended. After climbing a narrow road crowded by Gambel's oaks for another 4.5 miles, you reach the parking area for Hubbard Cave. Then hike about one-fourth mile to the cave. Hubbard Cave has 3,000 feet of passages through limestone and dolomite. A large room called the Grape Room contains many gypsum flowers, gypsum blisters, selenite needles, and angel's hair. All of these formations in the cave are protected by law. A very short distance near Hubbard Cave is an ice cave.

A substantial amount of the White River National Forest lies north of Interstate 70, including the huge Flat Tops Wilderness and the western side of Eagle Nest Wilderness. One of the most exciting adventures in the national forest is to drive the road northward from the Dotsero exit on Interstate 70. After about 15 twisty miles, the road comes to a breathtaking overlook into Deep Creek. At the bottom of the canyon you may get glimpes of the Deep Creek National Wild and Scenic River. A mile beyond the Deep Creek Overlook is Coffee Pot Springs, with a pleasant campground nestled among the trees. The narrow, crooked road then winds its way in piñon pine and juniper forests past Broken Rib Spring and Grizzly Creek Falls (a mile west of the road). After patches of aspens, the road comes to the meadowlike Crane Park and eventually Heart Lake and Deep Lake at the southern edge of the White River Plateau and the Flat Tops Wilderness. By now the forests are dominated by spruces and firs. The Flat Tops are high mountains with broad, flat tops.

Although more of the Flat Tops Wilderness is in the White River National Forest, with a smaller part in the Routt, the best way to observe the wilderness is to take the Flat Tops National Scenic Byway that begins from Yampa on State Route 131 and winds through the Routt National Forest to Ripple Creek Pass where it enters the White River National Forest. Do not miss the Ripple Creek Overlook for a nice overview of the area. Trails enter the wilderness from Lily Pond and Mirror Lake. Near Mirror Lake is a side road that dead-ends at Trappers Lake at the foot of the spectacular Chinese Wall.

The Gore Range is the major mountain mass in the Eagle Nest Wilderness. The range rises abruptly to 13,000 feet above the Blue River valley to the east. The western side of the wilderness is in the White River National Forest and can be entered at several places off of Interstate 70 east of Vail. Two miles south of the East Vail exit is the Gore Creek Campground where one trail ends at Deluge Lake and a longer one climbs alongside Gore Creek to Red Buffalo Pass and into the Arapaho National Forest. Trails from the frontage road along Interstate 70 follow Bighorn Creek, Pitkin Creek, and Booth Creek into the wilderness, the latter passing a lovely falls before terminating at Booth Lake.

For a short day trip, take the Piney Lake and Red Sandstone Road northward from Vail. This dirt road is scenic with numerous red rocks visible. After 10 miles, park at the Piney River Ranch and take the trail along Piney River. You may be rewarded with colorful wildflowers during spring and summer and nice leaf coloration in mid-September. Longer trails from the Piney River Ranch penetrate the northwest corner of the Eagle Nest Wilderness.

From the village of Buford near the forest's northwestern corner is a gravel road that travels for 44 miles southward through the White River National Forest to New Castle on Interstate 70. After two miles of meadows just south of Buford, the road makes its way for four more miles until it enters the national forest. As the elevation increases, the vegetation changes from aspens and pines to spruces and firs. To the west are Lake Beaver and Lake Gilley; to the east is Burro Mountain. A side road to the top of Burro Mountain provides a marvelous view of the Flat Tops Wilderness. After alternating between mountain meadows and stands of spruces and firs, the main road comes to refreshing Hines Spring. From the spring is a side road to the east that goes to Meadow Lake, where there is a campground, or to Cliff Lake, finally ending at the South Fork Campground and a trailhead at the western edge of the Flat Tops Wilderness. At Cliff Lake a trail follows the South Fork of the White River through blue spruces, aspens, and gnarly oaks. The main gravel road goes south from Hines Spring to Triangle Park. From Triangle Park you have two choices. One is a rough road along Little Box Canyon to the Three Oaks Campground and eventually to the town of Rifle. The other alternates between forests and meadows as it winds its way to New Castle.

Winter sports activities of all kinds may be enjoyed throughout the forest, particularly in the areas around Aspen, Avon, Breckenridge, Frisco, and Vail.

Flat Tops

In northwestern Colorado, within the Rocky Mountain range, is a mass of blunt-topped mountains known as the Flat Tops. Rising roughly 9,000 to 12,000 feet above sea level, they have been carved out at the eastern edge of a large plateau by the action of wind, water, and Ice Age glaciers. The patterns of erosion are striking, in part because underlying layers of sedimentary rocks (sandstone, limestone, shale) are capped by a harder layer of basalt, a volcanic rock. Some of the mountains have amphitheater-like shapes that almost appear to be the work of giants.

Flowing westward across the plateau from the Flat Tops is the White River, a tributary of the Colorado. It passes through rolling grasslands and, in less arid areas (generally on north- or east-facing slopes and along streams), through stands of forest. Above the timberline, which lies at about 10,200 feet, the habitat is alpine tundra. From there down to about 8,200 feet is a subalpine forest with subalpine fir, mountain ash, and other species. Below this level grows montane forest, where the subalpine fir and mountain ash are virtually absent. A spruce bark beetle epidemic in the middle of the twentieth century left behind many standing dead trees in the Flat Tops area, so it is often referred to as the Ghost Forest.

The Flat Tops fall within the Routt and the White River National Forests, where a nearly 370-square-mile section has been designated the Flat Tops Wilderness. This rugged zone is off-limits to motorized vehicles and equipment, but a scenic, mostly gravel County Road 8 meanders across the plateau for 82 miles, connecting Yampa and Phippsburg in the east to Meeker in the west. Taken westward, the highway first climbs to Dunckley Pass and, after following a roller-coaster course, tops out at 10,343 feet at Ripple Creek Pass. It then descends toward the North Fork of the White River. The main road continues downriver, but a turnoff can be followed upriver for about eight miles to Trappers Lake, near which are a campground and the Trappers Lake Lodge and Resort.

Trappers Lake, which lies within the wilderness area, was the first U.S. National Forest property to be treated as a wilderness, long before Congress passed the Wilderness Act of 1964. In 1919 the U.S. Forest Service hired a young landscape architect, Arthur H. Carhart, and told him to do a survey for a road around the lake and several homesites on the lakeshore. Carhart completed the task but urged his supervisor, Carl Stahl, to save the lake for wilderness recreation. Stahl backed the recommendation, and in 1920 Trappers Lake was declared an area to be kept free of roads and development. Carhart's vision remains a landmark in the history of wilderness preservation: "There is a limit to the number of lands of shoreline on the lakes; there is a limit to the number of lakes in existence; there is a limit to the mountainous areas of the world," he wrote in a memorandum. "There are portions of natural scenic beauty which are God-made, and . . . which of a right should be the property of all people."

Trappers Lake is nearly 1.5 miles long and a little more than a half-mile across at its widest point. Providing a formidable background on its northeast side is the so-called Chinese Wall, a sheer face that rises 1,650 feet. This basalt formation extends for nearly five miles in an irregular semicircle. Because of its sheerness and direct exposure to the afternoon sun, the side facing the lake is bare of vegetation except for some rock-hugging lichens.

The view of the Chinese Wall from below is rewarding enough, but adventuresome hikers may take various trails that provide a close-up experience. (Horses also may be hired at the lodge.) My wife Beverly and I took the Stillwater Trail, which heads eastward from Trappers Lake, passes Little Trappers Lake, and, after a sharp bend, comes face to face with the wall. At this point the trail climbs over the formation's south end, rising from 10,000 feet to 11,000 feet in little more than a mile. After another mile, we turned onto the Chinese Wall Trail, which winds northward near the top of the mountain.

Two miles along this route, we came to a branch trail leading eastward to the Devils Causeway. This, we soon discovered, is a craggy ridge only four feet wide, with precipitous 1,500-foot drop-offs on both sides. The passage looked far too scary, so we retraced our steps and continued northward. After a gentle descent, we reached the Lost Lakes Trail and headed back south to complete our round trip.

Grassy meadows often appear on flat terrain and on west- or south-facing slopes. Slender wheatgrass, nodding brome, fringed brome, oat grass, red fescue, junegrass, and mountain muhly are the most abundant species in dry areas. Most of these flower beginning in July and continue into fall. Where a moister habitat is created by seeps, the grasses include redtop and meadow foxtail. Above 10,000 feet, Thurber's fescue and alpine timothy are common.

Roads that penetrate the Flat Tops have disturbed the native vegetation, permitting the invasion of European species such as orchard grass and timothy, along with opportunistic wildflowers such as milfoil (yarrow), oxeye daisy, fleabane, Indian paintbrush, and thistles.

Alpine tundra vegetation consists of dwarf willow trees about one foot tall, an abundance of sedges, and various deep-rooted, mat-forming wildflowers. Among these are nailwort, slender mountain sandwort, rock jasmine, and alpine primrose.

Subalpine forest has subalpine fir, Engelmann spruce, lodgepole pine, and mountain ash. Myrtle-leaved blueberry is the most abundant shrub, although a species of wild rose and Colorado currant are also common. Wildflowers include larkspurs, Colorado blue columbine (the state flower), several species of beardtongue, sweet cicely, Jacob's ladder, and several kinds of arnica, the flower heads of which resemble those of daisies, except that they are entirely yellow or orange.

Montane forest has lodgepole pine and Engelmann spruce, with Douglas fir often found on north slopes and ponderosa pine on south slopes. Other trees include aspen and Rocky Mountain maple. The mixture of shrubs and wildflowers is similar to that in the subalpine zone, but wax currant grows here instead of Colorado currant.

Wetland habitat appears along the streams and around the ponds and lakes that dot the plateau. These areas are often lined with Geyer's willow, Wolf's willow, and narrowleaf cottonwood. Red elderberry, snowberry, and a species of prickly currant are common in the shrub zone. Among the wildflowers are a wild geranium, an aster with smooth stems and leaves, leafy arnica, yellow monkey flower, meadow rue, and bluebells. The scouring rush, a spore-bearing plant with jointed, leafless stems, is also common.

NATIONAL FORESTS IN LOUISIANA

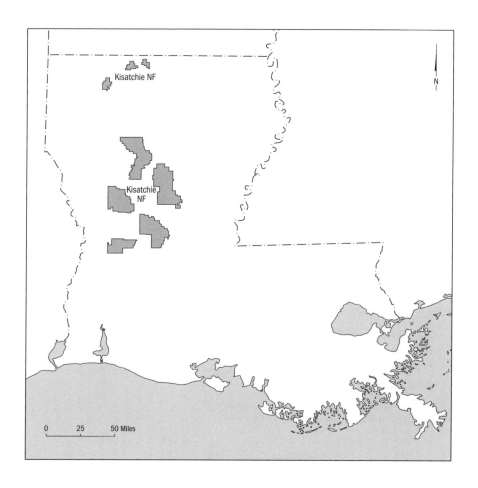

The Kisatchie National Forest is the only national forest in the state of Louisiana. It is in Region 8 of the U.S. Forest Service.

Kisatchie National Forest

SIZE AND LOCATION: 604,000 acres in north, west, and central Louisiana. Major access roads are Interstates 20 and 49, U.S. Highways 71, 79, 84, and 165, and State Routes 8, 9, 10, 28, 112, 117, 121, 123, 126, 156, 159, 463, 472, 488, 500, and 520. District Ranger Stations: Bentley, Boyce, Homer, Natchitoches, and Winnfield. Forest Supervisor's Office: 2500 Shreveport Highway, Pineville, LA 71360, www.fs.fed.us/r8/kisatchie.

SPECIAL FACILITIES: Boat launch areas; swimming beaches; ATV trails.

SPECIAL ATTRACTIONS: Longleaf Vista Recreation Area; Saline Bayou National Wild and Scenic River; Longleaf Trail National Scenic Byway; Longleaf Scenic Area.

WILDERNESS AREA: Kisatchie Hills (8,679 acres).

The Kisatchie National Forest is scattered in eight discrete parcels in northern, western, and central Louisiana. Although the forest consists of some swampland, there is rugged terrain with steep slopes, rocky outcrops, and flat-topped mesas more reminiscent of the western United States. Many lakes, some natural, others artificial, are dotted across the national forest and provide ample opportunity for water-based activities. Miles of hiking trails cross streams and swamps and climb over rolling hills. The forest has a surprising diversity of natural habitats, some of them not normally associated with Louisiana.

Because of the mesalike buttes and various rock formations, the Kisatchie Hills are the most prominent land features in the national forest and quite unlike the concept that many people have of Louisiana. The hills are in the western section of the national forest in the vicinity of Fort Polk Military Reservation. The most undisturbed part of this hilly region has been designated the Kisatchie Hills Wilderness. The often rocky hills in the wilderness are sometimes up to 300 feet above a vast wetland known as Bayou Cypre. Whereas the swamp is inhabited by egrets, herons, alligators, bald cypresses, and tupelo gums (fig. 11), the hills are forested with pines and hardwood species and occasional prairielike plants in forest openings. The 12-mile-long Backbone Hiking Trail stays on the ridges above the wetland.

One of the best ways to get an overview of the Kisatchie Hills is to drive the Longleaf Trail National Scenic Byway. This highly scenic road begins just outside the eastern edge of the wilderness where there is an overlook into the swamp at Bayou Pierre Vista. A second impressive viewpoint at Bayou Cypre

Vista is 1.5 miles farther west. The scenic byway then follows the southern and western borders of the wilderness. A short side road to Longleaf Vista not only brings you to another marvelous view into the Kisatchie Hills but to a 1.5-mile loop interpretive nature trail. This rather easy trail is a great way to get a close-up experience of the rocks, vegetation, and topography of the Kisatchie Hills. Because sandstone outcroppings are not common in Louisiana, where they do occur usually means that plants that live in sandstone crevices and on exposed blufftops may be found. Species unique to this habitat in Louisiana are woolly lip fern, fragile fern, and small flower of an hour, the latter with succulent cylindrical leaves and pink flowers that are open for only one hour during the day. Many of the outcrops along the trail

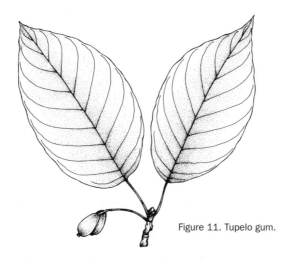

Figure 11. Tupelo gum.

are flat-topped and stand up to 30 feet above the surrounding area. The forest along the trail is dominated by longleaf pine. These forests, known as longleaf pine savannas, often have a parklike appearance because the plants beneath the pines are primarily grasses.

If you are up to a longer hike, the 12.5-mile Caroline Dorman Trail may be what you are looking for. This trail begins 1.5 miles west of Longleaf Vista and heads away from the wilderness, eventually reaching the Kisatchie Bayou Campground in the midst of a mixed pine and hardwood forest at the edge of a bayou. Dominant trees around the campground are loblolly pine, American beech, white oak, and the evergreen southern magnolia. Instead of a grassy understory, this woods has a dense growth of shrubs and vines. You may also drive to the Kisatchie Bayou Campground from a side road off of the Longleaf Trail National Scenic Byway.

When the scenic byway comes to Middle Branch, there is a boggy habitat on the side of a hill known as a seepage bog. Here are found growing among sphagnum a species of pitcher plant, a sundew, a butterwort (all three of these latter species are carnivorous), meadow beauty, narrow-leaved sunflower, willow-leaved goldenrod, a creeping ground pine, and several kinds of sedges, rushes, and yellow-eyed grasses. A shrub community in the bog consists of wax myrtle, raisin tree, black chokeberry, dahoon holly, and poison sumac. Trees that shade the bog include longleaf pine, sweetbay magnolia, and red bay. Huge royal ferns add to the beauty of the bog. The scenic byway continues west past Lotus Campground, crosses the Kisatchie Bayou, and ends at the Dogwood Campground on State Route 117.

A few miles southwest of Kisatchie Bayou Campground is the small but attractive Kisatchie Falls, lined on either side by a dense thicket of brookside alder. On the terrace above the stream are silverbell trees whose silver white bell-shaped flowers are a delight in late April.

About 20 to 25 miles south of the Kisatchie Hills district is a unit of the national forest that features the Longleaf Scenic Area and the Fullerton Lake Recreation Area. The Longleaf Scenic Area has stands of longleaf pine. Depending on the soil depth and the amount of moisture in the soil, shortleaf pine, slash pine, and loblolly pine may be associates of longleaf pine. Big Branch Trail encircles the scenic area.

Fullerton Lake is about seven miles east of the Longleaf Scenic Area and consists of a campground, picnic area, swimming beach, boat launch area, and the Whiskey Chitto Trail. This five-mile trail winds through stands of longleaf pine, follows Whiskey Chitto Creek for a short distance, and then joins Big Branch Trail near the Longleaf Scenic Area.

A large district of the Kisatchie National Forest is a few miles southwest of Alexandria and highlights several interesting areas in the national forest. Kincaid Reservoir and Valentine Lake both have campgrounds, swimming areas, boat launch areas, and picnic facilities. Castor Creek Scenic Area preserves a handsome stand of tall trees and is crossed by the Wild Azalea Hiking Trail. This trail is most colorful from mid-March to mid-April. It begins in the town of Woodworth and ends 31 miles later at Valentine Lake. At the eastern edge of the national forest and just north of Oden Lake are 37 acres of huge bald cypress and tupelo gum trees in the Bayou Boeuf Research Natural Area. These mammoth trees are along the bayou in the floodplain of the Red River.

The Winn Ranger District west of the town of Winnfield contains the Saline Bayou National Wild and Scenic River. Most of the 21 miles of the river offer an exciting canoeing experience as the canoe passes beneath Spanish moss-laden bald cypress trees. Expect to see alligators somewhere along the route. The vast amount of sand along the river, some of it encrusted with salt, is conducive to the growth of plants that can tolerate sandy or salty con-

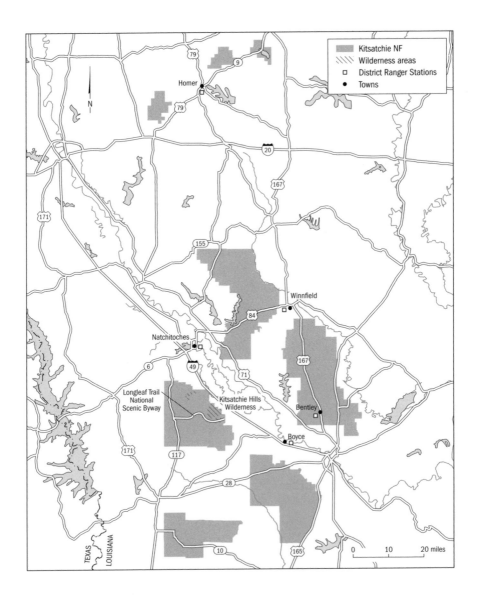

ditions. A salt flat just south of the community of Goldonna is habitat for salt-tolerant species such as creeping alternanthera, camphor weed, prairie cord grass, and several kinds of spike rushes, flat sedges, and bulrushes. Nearby areas of deep sand harbor a number of rare species, such as cupleaf beardtongue, many-flowered wild buckwheat, longleaf wild buckwheat, large clammyweed, southern jointweed, Mohlenbrock's umbrella sedge, smooth twistflower, and Riddell's spikemoss. Near the northern end of the wild and scenic river is an occurrence of the beautiful and rare Louisiana bluestar.

A narrow area about five miles long and up to one mile wide occurs about two to three miles east of Saline Lake. The soil in this area consists of calcareous marly clay, which is responsible for two unusual vegetational communities, a calcareous prairie and a calcareous forest. Kieffer Prairie consists of open areas within a forest where prairie-type species grow. Typical prairie plants found in the calcareous prairies are big bluestem, little bluestem, bushy bluestem, Indian grass, compass plant, pale coneflower, white and purple prairie clovers, yellow puff, bundleflower, poppy-mallow, blue sage, a species of Zornia, and several kinds of blazing stars, goldenrods, and asters. Louisiana rarities in the prairie openings may include purple bluets, slender heliotrope, prairie parsley, prairie acacia, and a ground plum.

In areas where the calcareous marly clays hold enough moisture for the growth of trees, forested communities occur. Dominant trees are post oak, Shumard oak, white oak, American beech, nutmeg hickory, white ash, and shortleaf pine. A zone of shrubs and small trees include flowering dogwood, redbud, red buckeye, fringe tree, rusty nannyberry, arrowwood viburnum, and Carolina buckthorn. Trees with very white trunks are chalk maples. In fall, the very uncommon ear-leaved goldenrod flowers.

A short distance southeast of the calcareous forests and prairies is the Gum Springs Recreation Area where there is a campground and an equestrian trail. A short loop Dogwood Trail may be hiked along the south end of U.S. Highway 84 about 1.5 miles south of Saline Lake.

Between Winnfield and Catahoula is the rather large Catahoula District of the Kisatchie National Forest. This is an area of forests and small streams, crisscrossed by a maze of roads. One of the roads is the Dogwood Auto Tour, which is spectacular when the flowering dogwoods bloom in April. This is a circular drive a few miles northwest of Catahoula. Small Stuart Lake with a campground is toward the southern end of the district. The Glenn Emery Hiking Trail is here.

Three small disjunct units of the Kisatchie National Forest lie north of Interstate 20, on either side of the town of Homer. Large Corney Lake is the central feature of the northeastern unit, with a campground and boat launch area near the southern end of the lake. A small unit of the national forest south of Colquitt has no recreational facilities, although the Forest Service has provided two hunter camps. The Caney District begins two miles north of Minden and contains two Caney Lakes and the interesting Sugar Cane Trail. This 7.5-mile trail loops around Upper Caney Lake through forests of pine and hardwood trees on rolling terrain. The remnants of an old sugar mill and its associated earthern terrace system may be seen.

Kisatchie Hills

For nearly 100 miles, from Harrisonburg, in northern Louisiana, west to the Texas border, a series of 300-foot-high ridges, hills, and flat-topped mesas mark what was once a continuous ridge. Throughout this length, the seaward slopes of the hills are gentle, but to the north, the terrain drops nearly 180 feet to valleys drained by meandering streams. Modern geologists call such ridges or hills—those with one gentle slope and one steep slope—cuestas, using the Spanish word for "slopes." Part of this terrain runs through the Kisatchie National Forest, where it is known as the Kisatchie Wold.

The wold had its origin tens of millions of years ago when rivers carried soil from lands to the north of Louisiana southward, depositing successive layers of sediment that dipped gently beneath the sea. These deposits, hundreds to thousands of feet thick, became clays, shales, sandstones, and other types of sedimentary rocks. Through this deposition, as well as through uplifts, the land that is now northern Louisiana slowly emerged from the sea. The softer sedimentary rocks eroded to form valleys, whereas the harder ones, such as sandstone, remained as ridges. The Kisatchie Wold is one such ridge.

Many of the ridgetops and the upper reaches of the slopes are dominated by a mixture of longleaf, shortleaf, and loblolly pines. Beneath them is an abundance of wax myrtle and yaupon holly. Here and there, where conditions are more moist, longleaf pine is the only tree, accompanied by an understory of bluestem, panicum, three-awn, and several other grasses. These parklike areas are referred to as longleaf pine savannas.

On most of the lower slopes, moisture becomes too abundant for the longleaf pine, which has narrow requirements. Here, loblolly pine becomes the dominant conifer, growing alongside a mixture of oaks, hickories, and other hardwood trees. On the valley floors, red oak, white oak, and mockernut hickory combine with southern magnolia, sweet bay magnolia, red bay, and sweet gum to form a dense canopy. Violets and trilliums brighten these woods in spring.

The most striking elements of the Kisatchie Wold are the occasional flat-topped mesas, or buttes, that resemble in miniature the buttes of the western states. One vantage point from which to appreciate them is the Longleaf Vista Overlook. The surface of the buttes is exposed sandstone. The trees adapted to this dry habitat, such as sandjack oak and blackjack oak, grow short and gnarled. Wild azaleas and huckleberries, both members of the heath family, flourish beneath the trees. Perhaps more than any other family of flowering plants, heaths have become well adapted to living in acidic soils such as sandstone.

About 4.5 miles west of Longleaf Vista Overlook, toward Middle Branch Creek, the longleaf pine forest is punctuated by a series of low sandstone outcrops that drip with water. These ledges, at most only six feet tall, feed a hillside seep bog, a habitat unusual for this part of the world. Botanists Edwin Bridges and Steve Orzell speculate that water in the uplands, draining downward through the soil, reaches impermeable sandstone bedrock, then flows through cracks in the bedrock until the rock layer is exposed on the slope. Because the seepage from the sandstone ledges is so low, it spreads out over a large area and does not form a gully such as a stream might produce.

This hillside seep bog covers only about two acres and is relatively open, with only a few trees and shrubs. Sedges are the most abundant plants. Sphagnum, or peat moss, a plant usually associated with bogs, is present but not conspicuous.

Growing in the bog and on the dripping wet ledges is the pale pitcher plant, whose slender pitchers are yellow green and marked with an intricate pattern of red veins. The deep red color is also found on the inner surface of the lid that arches over the pitcher. The pitcher plant is referred to as a carnivorous plant because it is able to trap insects and other small animals and then digest their bodies with enzymes. The nutrients so obtained are utilized by the pitcher plant, although recent studies seem to indicate that the plant can survive equally well without this supplement to its diet.

Scattered within the bog are other carnivorous species, including the thread-leaved sundew, bladderworts, and a blue-flowered butterwort. In a yearlong study, Barbara and Michael MacRoberts, botanists from Shreveport, found six carnivorous species, along with 100 other kinds of flowering plants.

Since 1986 the Forest Service has burned the bog on a regular basis during winter to prevent encroachment by the surrounding plant species. In the past, natural fires in the drier longleaf pine savannas up-slope probably burned into this bog. Fires kill shrubs and tree seedlings without affecting herbs, whose dormant buds or seeds lie within the boggy soils.

Without fire, the bog would fill in with a junglelike growth of woody vegetation, ultimately destroying the fragile bog species that depend upon an open habitat. Such a jungle apparently has developed down-slope from the bog, where a rather soggy soil, perhaps once supporting a bog, is now covered by a dense growth of sweetbay magnolia, red bay, myrtle, yaupon holly, alder, possum haw viburnum, and red chokeberry, all overrun by an entanglement of laurel-leaved greenbrier.

NATIONAL FORESTS IN
MINNESOTA

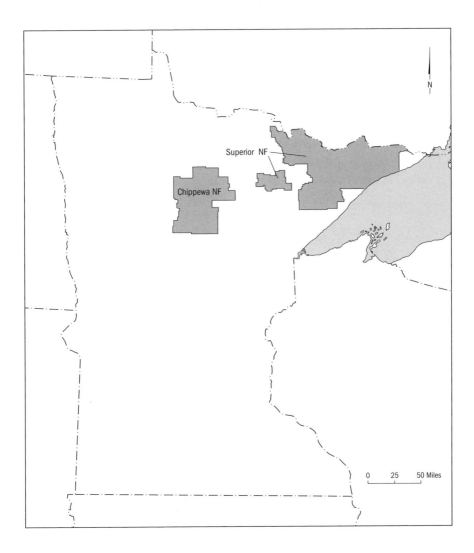

The two national forests in Minnesota have fine examples of boreal forests. Numerous lakes, streams, bogs, and marshes make these forests prime destinations for outdoors persons. The forests are in Region 9, the Eastern Region of the U.S. Forest Service. Regional forest office: 310 W. Wisconsin Street, Milwaukee, WI 53202.

Chippewa National Forest

SIZE AND LOCATION: 666,522 acres in central Minnesota. Major access routes are U.S. Highways 2 and 71 and State Routes 6, 9, 29, 34, 35, 38, 39 (The Scenic Highway), and 46. District Ranger Stations: Blackduck, Deer River, and Walker. Forest Supervisor's Office: 200 Ash Avenue NW, Cass Lake, MN 56633, www.fs.fed.us/r9/chippewa.

SPECIAL FACILITIES: Mi-Ge-Zi Bike Trail; boat ramps; swimming areas; cross-country ski areas; canoe streams.

SPECIAL ATTRACTIONS: Avenue of Pines National Scenic Byway; The Scenic Highway; Edge of the Wilderness National Scenic Byway; Woodtick Trail Auto Tour; Heartland Tour; Camp Rabideau.

WILDERNESS AREAS: None.

The Chippewa National Forest is in the heart of the Great North Woods of the United States. Minnesota is the Land of Lakes, and the Chippewa has more than 700 lakes, 940 miles of rivers and streams, and countless marshes, bogs, fens, and forested wetlands. More than two-thirds of the national forest are either water or wetlands.

Several automobile routes that have been designated as scenic byways provide access to most parts of the national forest. A byway known simply as The Scenic Highway (State Route 39) is on the western side of the Chippewa. Beginning in the north at the community of Blackduck, the road extends for 28 miles to Cass Lake. This is remote territory, and most of the route takes you through forested areas. You see limited signs of human habitation along the highway, and those that are there are usually concealed from view.

Five miles from Blackduck is the site of Camp Rabideau, once a busy place that served as a Civilian Conservation Corps (CCC) camp in the 1930s. The 13 weathered green buildings of the camp that remain are located on 112 acres beneath tall trees. At one time the camp had 25 buildings that housed up to 200 people at any given time. These inhabitants spent much of their time working in the Chippewa National Forest. A one-mile-long interpretive trail runs through the camp.

Just north of Camp Rabideau and visible from it is Benjamin Lake, where there is a hiking trail beneath basswood (fig. 12), sugar maple, and paper birch. South of Camp Rabideau is a loop road, Anderson Lake Road, off The Scenic Highway to Anderson Lake and Webster Lake, two popular lakes for fishing and camping. A six-mile hiking trail links the two lakes. Bur oaks, red

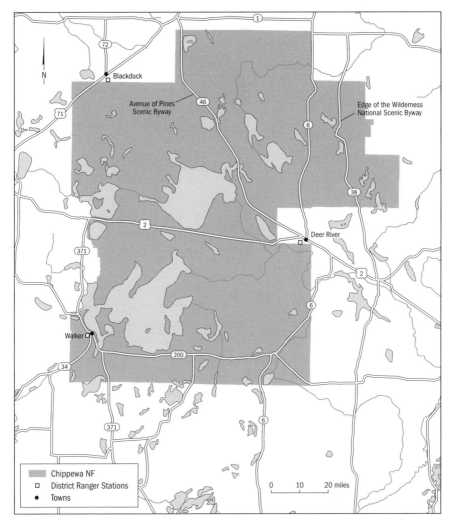

Chippewa NF
□ District Ranger Stations
● Towns

0   10   20 miles

pines, aspens, paper birches, and red maples are common. A trail with a boardwalk across a bog encircles Webster Lake.

Before The Scenic Highway reaches the community of Pennington is one of the most pristine natural areas in Minnesota, the Pennington Bog State Natural Area. Growing beneath a canopy of white cedar, balsam fir, and black spruce is an incredible diversity of bog-loving species. Among the wildflowers are a dozen kinds of wild orchids, some of them rare. The area is not managed by the Chippewa National Forest, and anyone wishing to visit it needs permission from the Minnesota Department of Natural Resources State Natural Areas Program at telephone number (651) 296-2835. However, similar boggy areas exist throughout the national forest that you may wish to ex-

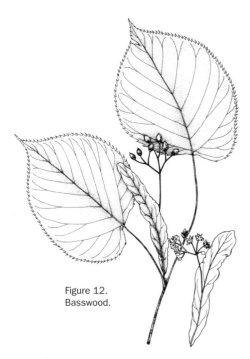

Figure 12.
Basswood.

plore. Most of these areas have a dense cover of sphagnum across the forest floor beneath the white cedars and black spruces. One must be careful while exploring these bogs because of the numerous holes among the tree roots that are concealed by the heavy growth of mosses. One Forest Service area in the vicinity of Pennington Bog is dominated by white cedars, with occasional larches, balsam firs, and black spruces present. Bog birch is a common shrub in this bog. Pitcher plants, twin-flower, and dwarf raspberry commonly grow in the deep mats of sphagnum, but it is the orchid population that is most impressive. Botanists have discovered 11 kinds of native orchids in this bog, some large and showy, some inconspicuous.

Two miles south of Pennington, the route crosses the Mississippi River approximately 70 miles from its source at Lake Itasca State Park. The Forest Service maintains recreation areas along Cass Lake at Knutson Dam and Norway Beach. There is a Forest Service visitor center and a white-sand swimming beach at Norway Beach. The visitor center lodge was constructed with red pine logs by the CCC during the 1930s.

The Scenic Highway ends at U.S. Highway 2, with the latter going westward between Cass Lake and Pike Bay before reaching the town of Cass Lake. Just before U.S. Highway 2 reaches Pike Bay, there is a forest road to the south that goes to the Big Pine Forest Trail in about two miles. This one-fourth-mile trail is beneath huge red pines that are more than 300 years old. The stumps and fallen logs you see are the result of a violent storm in July 1940. Several attractive wildflowers may be seen from the trail, including wild ginger, bunchberry, starflower, and shinleaf pyrola. In wetter areas are blue iris, marsh marigold, and red osier, a dogwood with bright red twigs (fig. 13). The Mi-Ge-Zi Bike Trail encircles Pike Bay and leads visitors to the town of Cass Lake and to Norway Beach.

When in the town of Cass Lake, be sure to stop at the forest supervisor's office housed in a historic two-story log building. The building was completed in 1936 by the CCC, constructing it from red pine logs that came from

Star Island and Lake 13 areas. The 8,500-square-foot building is trimmed with birch, oak, maple, and white pine. The red pine logs were notched and grooved, a process requiring no chinking. The floor beams are held in place by heavy wooden pegs. The fireplace and chimney are 50 feet high and run through the center of the building. Made of glacial boulders found in the area, the fireplace is 14 feet by 14 feet at its base, tapering to 10 feet by 10 feet at the top. It consists of 265 tons of rock. The railing of the staircase is made of maple limbs that were frost damaged, giving them a gnarly appearance. The split-log steps were hand hewn with a broad ax and planed until smooth. The building is on the National Register of Historic Sites.

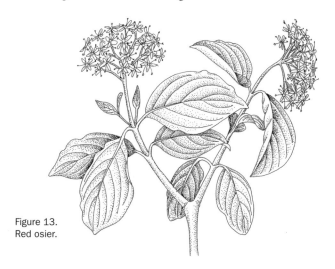

Figure 13.
Red osier.

Star Island is a short boat ride from the Norway Beach area on Cass Lake. Campers may stay at one of two backcountry campsites on the island. Star Island is unique in that is has its own lake within it known as Lake Windigo.

The Heartland Trail is another scenic paved bike route from Cass Lake to Park Rapids, a distance of 45 miles. About seven miles south of Cass Lake, the Oak Point Road goes to Oak Point Peninsula, a narrow mass of land with Steamboat Bay on one side and Sunken Bay on the other. The point has a campground. South of Walker are several Forest Service campgrounds, boat launch areas, and the Shingobee Ski Area.

State Route 371 heads into the extreme southern part of the Chippewa National Forest. Two routes cross this part of the Forest. State Route 200 is the main highway to Remer. For half its length it is near the southern end of huge Leech Lake. County Road 13, a side road to the north, divides with the left fork ending at Pine Point and the right fork at Stony Point, both narrow land masses that protrude into Leech Lake. Much of Pine Point is designated as the Pine Point Research Natural Area, a 1,230-acre tract with fine stands of jack, red, and white pines. More than half of the area is red pine. The uplands have small patches of virgin pines that are 150 years old, mixed with paper birch, northern red oak, and quaking aspen. The western side of the natural area is rolling hills, whereas the eastern part is nearly level to gently rolling with kettle holes filled with bogs. Henry Schoolcraft's historic expedition in 1832 camped on Pine Point. Schoolcraft was one of the first to explore the upper midwest during the early 1800s. The smaller Stony Point Research Natural Area supports both wet and mesic hardwood forests. Most of the trees at Stony Point are American elm, slippery elm, sugar maple, basswood, quaking aspen, blue-leaf willow, northern red oak, and bur oak, with a limited amount of paper birch. Smooth sumac and elderberry are common shrubs.

South of the junction of State Routes 371 and 200 is a primitive road across the national forest. Known as the Woodtick Auto Trail, this is a 14-mile route along an old railroad grade that was used to haul logs in the 1890s. The auto trail has 14 stops along its route. The North Country National Hiking Trail crosses the Woodtick Auto Tour. The tour passes sparkling Diamond Lake and the Webb Lake Swamp. This is a good place to see a typical north-country wetland and its characteristic plants and animals. The road passes a beaver lodge, a fine fishing lake called Moccasin Lake, and the Goose Lake Trail system that connects to Barnum Lake on the national forest boundary. The Woodtick Auto Tour ends when it reaches State Route 5.

The Avenue of Pines National Scenic Byway (State Route 46) is a 39-mile route running from north to south through the center of the Chippewa National Forest. The scenic byway enters the national forest from the north, two miles south of the community of Northome. Island Lake is a large fishing lake with Elmwood Island near its center. The island is reached by boat from a ramp on the north side of the lake. Once on the island, you may hike the half-mile trail. After passing several lakes to the east of the highway, the scenic byway comes to a narrow corridor between Squaw Lake to the east and Round Lake to the west. Squaw Lake, like many of the lakes in the area, is nearly covered by dense stands of native wild rice that has been harvested for hundreds of years. Round Lake has a boat ramp.

The scenic byway passes through stands of large red pines for which the route takes its name. The Continental Divide Picnic Area is located in the region where waters to the north flow into Lake Superior, and waters to the south eventually reach the Gulf of Mexico. Just south of the Continental Divide is the Cut Foot Experimental Forest where the forest is punctuated by several small lakes and penetrated by a number of back roads. Within the experimental forest is Sunken Lake, surrounded by swamps and wooded ridges. The lake was formed by a glacial moraine that prevented natural drainage of the lake. After heavy spring rains in 1915, a cut in the moraine where logs had been skidded gave way, dropping the lake 15 feet below its original level.

Sunken Lake and the area surrounding it is a designated 50-acre natural area and contains a red pine forest, a successional forest dominated by aspens and birches, a black spruce and larch bog forest, a shallow marsh, a wet meadow, and an open bog. One of the roads, Farley Hill Road, in the experimental forest climbs to the Farley Hill Lookout Tower, which is no longer used. The Cut Foot Sioux National Recreation Trail can be joined from the lookout. You may follow this trail eastward, crossing the scenic byway and winding your way to the Indian village of Inger on the banks of the Bowstring River. You may also drive to Inger on County Road 35 from the Cut Foot Sioux Visitor Center. The visitor center is a must stop on the scenic byway so you may learn about the history of the area and see the oldest remaining Forest Service ranger station in the eastern United States. Just south of the Cut Foot Sioux Visitor Center, the scenic byway comes close to the eastern edge of Cut Foot Sioux Lake. O-Ne-Gum-E Campground is situated beneath a plantation of red pines. At one point the lake is constricted between two land masses at a place called Williams Narrows, where there is a campground on the southern point of land. The 1748 battle between the Sioux and the Anishinabe (Chippewa) took place at the junction of Williams Narrows and Sugar Bush Point Roads. Nearby is the Battle Point Research Natural Area, a 329-acre protected forest dominated by sugar maple and basswood.

A short spur from County Road 9 goes to Winnie Dam on the Mississippi River and nearby Plug Hat Point, with an Army Corps of Engineers campground and a Forest Service picnic grounds in the area. A boat ramp at Plug Hat Point allows for easy access into gigantic Lake Winnibigoshish. Although the scenic byway continues to the community of Deer River, County Road 39 is an interesting alternate that goes along the west side of Ball Club Lake to the village of Ball Club. Along this stretch of highway you see big-tooth aspen, quaking aspen, paper birch, red pine, jack pine, red maple, and spruces, with dense growths of bracken fern on the forest floor.

The Edge of the Wilderness National Scenic Byway goes through the eastern part of the Chippewa National Forest between Bigfork and Grand Rapids, a distance of 22 miles. Much of the route parallels an old railroad grade that has been converted into a snowmobile trail. Shortly after entering the national forest, the scenic byway comes to a short side road, Jingo Lake Road, to Jingo Lake, site of a hunter-walking trail. Common along the trail are paper birch, red pine, red maple, and white spruce. Lady ferns are lush beneath the trees. The scenic byway passes several lakes with boat launch areas, and Clubhouse Lake also has a campground and a swimming beach. At the district ranger station in the village of Marcel, pick up brochures that describe the Edge of the Wilderness National Scenic Byway. During fall, the lovely colors of the deciduous trees blend beautifully with the greens of the conifers. The byway passes Pughole Lake, the Suomi Hills, Clubhouse Lake,

Rice River, and Jingo Lake. The Chippewa Adventure Trail explores the forest east of Marcel. The Suomi Hills is an adventurous mountain biking and cross-country ski area where basswood, northern red oak, sugar maple, red maple, big-tooth aspen, and paper birch prevail.

To the east of the Suomi Hills is the Trout Lake Management Area centered around picturesque Trout Lake. This large area actually contains 11 lakes with a combined 26 miles of shoreline. The terrain around the lakes is gently rolling uplands. Northern hardwood trees dominate the forest, with a minor amount of red and white pines. The scenic byway and national forest end at Pughole Lake.

To see the north-central part of the Chippewa National Forest, drive County Road 29 eastward from the village of Alwood. In a few miles you come to the Lost 40 Natural Area. This remarkable area has several acres of red pine forest that have never been cut, with another area dominated by large spruce and fir trees. In another area there is a white pine estimated to be more than 300 years old. One side of the natural area is a bog where black spruce and larch dominate, and a large alder marsh is also in the area. A half-mile trail penetrates the natural area. Other recreation sites along County Road 29 are at Dora Lake and Noma Lake.

Several canoe routes are available in the national forest where a different perspective of the region may be experienced. Easy canoe trails in the forest are the Turtle River, the Rice River, and the North Branch of the Turtle River. More difficult trails are in the Boy River and the historic Chippewa Headwaters and Pike Bay Loop. Bald eagles may be seen soaring over these rivers. Minnesota botanists Scott Milburn and Jason Husveth have discovered many specimens of the dwarf gnome fern in the national forest.

Superior National Forest

SIZE AND LOCATION: 2,172,500 acres in northeastern Minnesota, extending to the Canadian border. Major access routes are U.S. Highway 53 and State Routes 1, 61 (North Shore Scenic Drive), and 169 (Fernberg Road). District Ranger Stations: Aurora, Cook, Ely, Grand Marais, and Tofte. Forest Supervisor's Office: Duluth, MN, www.fs.fed.us/r9/superior.

SPECIAL FACILITIES: Canoe trails; winter sports areas; boat ramps.

SPECIAL ATTRACTIONS: North Shore Scenic Drive All American Road; Superior National Forest Scenic Byway (Forest Highway 11); Pictured Rocks.

WILDERNESS AREA: Boundary Waters Canoe Area Wilderness (809,750 acres).

The Superior National Forest is unlike any other national forest. It has more miles of connected water routes than it has automobile roads. The national forest has more than 2,000 lakes greater than 10 acres in size and 2,400 back-country campsites that can be reached only by water, about 200 outside the wilderness and 2,200 in the wilderness. The Superior National Forest is a mixed coniferous forest and aspen–birch forest, with white pine, red pine, jack pine, balsam fir, white spruce, and black spruce the dominant species for conifers. Here and there, however, are large stands of yellow birch, paper birch, white ash, sugar maple, and red maple. The entire forest is of the bo-

Figure 14. Bear with nautical instincts.

real type, and plants such as trailing arbutus, Labrador tea, shrubby laurel, wintergreen, leatherleaf, twinflower, and a variety of raspberries and blue-berries are testimony to this. Deer, black bears, moose, beavers, snowshoe hares, red squirrels, muskrats, skunks, woodchucks, and chipmunks are common animals in the Superior (fig. 14, pl. 16). Bird life draws many ornithologists to the area. Anglers come to try their luck in catching walleye, northern pike, lake, brown, rainbow, and brook trout, and panfish—crappies and sunfish.

The Boundary Waters Canoe Area Wilderness is the largest wilderness in the eastern United States, stretching for 150 miles along the Canadian border and up to 33 miles north to south. Within the wilderness are 1,500 miles of connected water routes. If you desire solitude in nature and like to canoe, this is the wilderness for you and the other 200,000 visitors to the wilderness each year. And yet, with all of these visitors, you can spend days without seeing any other humans, only bald eagles, moose, wild orchids, and maybe one of the 400 gray wolves that live here (pl. 17). All the streams and lakes are ideal for fishing. You need a permit to enter the wilderness, obtainable at

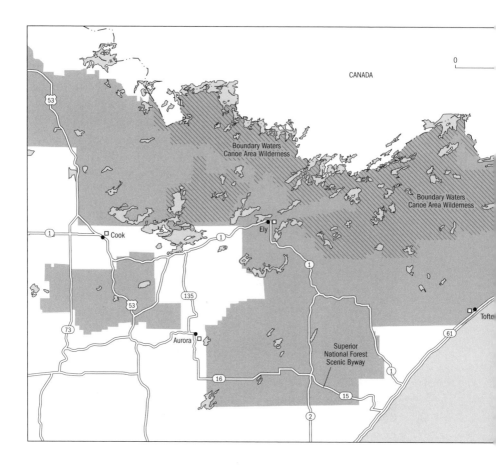

ranger stations or outfitters, and you must have the permit with you. If you plan to use a motorized watercraft, inquire at Forest Service offices on which lakes they are allowed. You may have cans or containers for fuel, insect repellent (which you definitely need in summer), medicine, and toiletries, but nonburnable beverage cans and bottles are not permitted.

Colorful rock formations with Indian hieroglyphics, known as Pictured Rocks, are another attraction in the Lake Superior area and can be seen in several places in the wilderness. When you see these in the remote wilderness, you get the feeling that you are the first person to see them. Fine Pictured Rocks are at the end of Kekekebic Lake, at Kattyman Lake, at Trease Lake, near the Canadian border above Wheelbarrow Falls, near the Fish Stake Narrows, on the western side of Crooked Lake, near the south end of Lac La Croix, and at the Little Vermilion Narrows. The only Pictured Rocks near enough to hike to are at Sea Gull Lake off of County Road 12 in a small northern protrusion of Minnesota.

Eagle Mountain is Minnesota's highest elevation at 2,301 feet. It is near

the southern boundary of the wilderness and you can hike to it on a four-mile trail from the Eagle Mountain parking area along Forest Road 170.

If you are not into canoeing, you can still enjoy some of the wilderness because it has 15 hiking trails that are only for hikers. Fewer than one-third of the Superior National Forest is in wilderness, and the remainder may be enjoyed by taking auto tours or color tours that have been designated. Brochures for these are available from the district ranger stations and several of the local communities' Chamber of Commerce offices. Dozens of forest roads also enter the backcountry of the national forest.

North Shore Scenic Drive (State Route 61), a premier scenic byway classed as an All American Road, follows the edge of Lake Superior for 58 miles between Schroeder and the Kadunce River State Wayside Park, which is about 11 miles northeast of Grand Marais. As you drive this scenic byway, you see waterfalls tumbling over ledges and into Lake Superior. Inland there are rolling, rocky hills. A hiking trail from Temperance River State Park enters the Superior National Forest, climbs the Sawtooth Mountains, and meanders past waterfalls, bogs, lakes, and streams for 125 miles. Just north of Tofte is a short, steep trail to the top of Carlton Peak where there are unforgettable views of Lake Superior and the Sawtooth Mountains. A little farther along the scenic byway are hiking trails to Leveaux Mountain, Moose Mountain, and Eagle Mountain. These are popular winter trails as well. Several side roads lead to many lakes within the Superior National Forest. Several vista points also are along the highway, one particular favorite is two miles east of Grand Marais and another is at Red Cliff.

One of the better forest roads is the Gunflint Trail (County Road 12), which leaves Grand Marais and penetrates into the heart of the national forest. County Road 8 off of the Gunflint Trail to the west goes along the north side of Devil Track Lake, which contains northern pike, perch, lake trout, and walleye. The lake also has a campground. To the north of Devil Track Lake is Two Island Lake and campground. Back on the Gunflint Trail, there are hiking trails and boat ramps at Kimball Lake and Northern Light Lake and a boat launch area at Swamper Lake. North of Swamper Lake, the Gunflint Trail follows a narrow corridor between two segments of the Boundary Waters Canoe Area Wilderness. Recreation areas are found at East Bearskin Lake, Flour Lake, Border Route, Fishhook Island, and at the appropriately named Trails End. Observation lookouts and scenic vistas are particularly striking at Honeymoon Bluff Overlook, which gives you a grand view of Hungry Jack Lake, Bogenho Lake Vista, Laurentian Divide Overlook, and Gunflint Overlook. A half-mile trail begins on the north side of Hungry Jack Lake and takes you to Caribou Rock. The trail permits the hiker to look down upon West Bearskin Lake. Forty miles up the Gunflint Trail from Grand Marais, the rugged Border Route Trail extends for 32 miles from Loon Lake 40 to Little John Lake. It crosses several rock ledges and offers spectacular views over Rose Lake and Mountain Lake.

The extreme eastern part of the Superior National Forest may be accessed by Forest Road 309 off of the Gunflint Trail and by County Road 16 above the small community of Hovland. Forest Road 309 and some of its branches nearly encircle large Greenwood Lake. County Road 16 passes near the lovely Portage Falls before ending at the extreme eastern edge of the Boundary Waters Canoe Area Wilderness.

The picturesque community of Silver Bay is situated on the banks of Lake Superior several miles southwest of Tofte. Silver Bay is the starting point for the Superior National Forest Scenic Byway, a 55-mile route that lets you get the feel of boreal forests between Silver Bay and Aurora. The scenic byway alternates between rich green coniferous forests and the bright coloration of aspens and birches. Bald eagles may usually be seen, and snowy owls appear regularly. Moose and timber wolves are here as well, but they are more elusive. After leaving Silver Bay, you enter the national forest on Forest Highway 11, the Superior National Forest Scenic Byway. The first major point of interest is the White Pine Picnic Area where a one-fourth-mile-long interpretive trail is beneath very tall 300-year-old white pines. The scenic byway continues west to Toimi, a once-thriving Finnish community. The Toimi School still stands, although it has not served as a school since 1942. Children at the school were permitted to converse only in Finnish during recesses.

The scenic byway crosses one sparkling creek after another. At Cadotte Lake, where fishing is good, there is a campground, swimming beach, and boat ramp. A forest side road, Otto Lake Road, to the southwest of Cadotte Lake goes near Otto Lake and Harris Lake, two of the most pristine bodies of water in the vicinity. The lakes can be reached by short portage trails, and another trail connects these two lakes that are about 2.5 miles apart. Both lakes have been stocked with walleye, but other catches may include white sucker, northern pike, yellow perch, pumpkinseed sunfish, and golden shiner. Three miles west of Cadotte Lake, the scenic byway crosses Jenkins Creek where there is a great hiking trail. Ruffed grouse are often seen along the trail.

The scenic byway makes an abrupt turn to the north and crosses Shiver Creek as it climbs to Skibo Vista where there used to be a lookout tower. Even without the tower, the view from the top of the hill is breathtaking. During the hawk migration season, hundreds of people view these marvelous raptors from Skibo Vista.

In a few miles, the scenic byway crosses St. Louis River, a great canoe stream with whitewater rapids. The river is lined by forested hills and bluffs. After crossing the St. Louis River, there are side roads, Stone Lake Road, to Stone Lake to the east and Norway Point to the south. Stone Lake is covered with wild rice, which is legal to harvest. A 1.5-mile trail from Stone Lake goes to large but secluded Big Lake, which has been stocked with walleye and northern pike. Norway Point is on a rocky outcrop above the St. Louis River. Red pines tower over the site. Bird Lake is adjacent to the scenic byway and is the origin of 10 miles of hiking trails in summer and cross-country ski trails in winter.

Less than one mile past the Norway Point turnoff, the scenic byway leaves the national forest and continues to the communities of Hoyt Lakes and Aurora. A short spur road crosses the Partridge River over a three-pin Timber Arch Bridge. The bridge is worth a visit.

Although not a designated scenic byway, State Route 1 is a superb road that crosses much of the nonwilderness portion of the Superior National Forest. If you start this drive from the south, you leave from Tettegouche State Park on the shore of Lake Superior. After several miles, the highway enters the national forest and eventually takes a northwesterly course, staying about 10 miles south of the Boundary Waters Canoe Area Wilderness. Little Isabella River, Lake Gegoka, McDougal Lakes, Campers Lake, another Harris Lake, and the South Kawishiwi River have Forest Service recreation sites or boat ramps. Plenty of side roads are available, including Wanless Road, which provides access to Divide Lake Recreation Area, Ninemile Recreation Area, and Wilson Lake, and Tomahawk Road, which ends at Isabella Lake on the southern border of the wilderness area. Beyond the South Kawishiwi River Recreation Area, State Route 1 heads north to Ely.

Ely offers several highway options. State Route 169 to the east, known locally as Fernberg Road, gives access to Fall Lake Recreation Area, several large lakes, including Moose Lake and Snowbank Lake, and numerous entry points to the Boundary Waters Canoe Area Wilderness.

North of Ely, County Road 116, the Echo Trail, passes Burntside Lake and Fenske Lake before becoming a light duty road that stays close to the southern boundary of the wilderness area. This road goes through many miles of coniferous forests, lakes, and bogs, to such popular areas as Meander Lake, the Sioux River, Lake Jeanette, and Echo Lake.

State Routes 1 and 169 are the same highway west of Ely until the road nears the Superior National Forest's southwestern area. Here, State Route 1 takes a northerly route through the national forest, with a side road to Pfeiffer Lake Recreation Area, and State Route 169 takes a southerly route to the Laurentian Divide Recreation Area. At the latter, you may hike to the top of Lookout Mountain. If you look eastward, you get a panoramic view of Laurentian Divide, the area where rivers and streams to the north flow into Hudson Bay, whereas those to the south flow to the Gulf of Mexico.

Near the western boundary of the Superior National Forest, from the village of Cusson on U.S. Highway 53, County Road 150 and Forest Road 203 take a circuitous route to Vermilion Falls where there is a fine waterfall and picnic area. The Vermilion River flows into Crane Lake through the Vermilion River Gorge, and the King William Narrows is at the northeastern corner of the lake. On the east side of the narrows are fine Pictured Rocks.

NATIONAL FORESTS IN MISSOURI

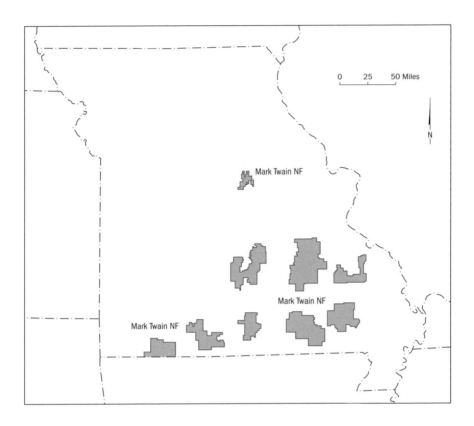

The Mark Twain National Forest is the only national forest in Missouri and includes what was at one time the Clark National Forest as well as the Mark Twain. It is in Region 9 of the U.S. Forest Service.

Mark Twain National Forest

SIZE AND LOCATION: 1.5 million acres across southern Missouri, entirely south of Interstate 70. Major access routes are Interstates 44 and 70, U.S. Highways 60, 63, 67, and 160, and State Routes 7, 8, 13, 14, 17, 19, 32, 39, 49, 49A, 72, 76, 86, 88, 95, 99, 112 (Sugar Camp National Scenic Byway), 142, and 181. District Ranger Stations: Ava, Doniphan–Eleven Point, Houston-Rolla, Poplar Bluff, Potosi, and Salem. Forest Supervisor's Office: 401 Fairgrounds Road, Rolla, MO 65401, www.fs.fed.us/r9/marktwain.

SPECIAL FACILITIES: Boat ramps; swimming areas.

SPECIAL ATTRACTIONS: Sugar Camp National Scenic Byway; Glade Top Scenic Byway; Eleven Point National Wild and Scenic River; Blue Buck Knob National Scenic Byway; Grasshopper Hollow; Cupola Pond; Hercules Glades.

WILDERNESS AREAS: Bell Mountain (8,977 acres); Devil's Backbone (6,595 acres); Hercules Glade (12,315 acres); Irish (16,277 acres); Paddy Creek (7,019 acres); Piney Creek (8,087 acres); Rock Pile Mountain (4,089 acres).

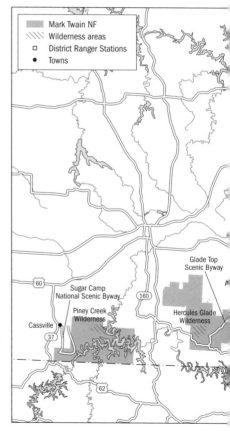

The Ozark Mountains are among the oldest in the United States, having been uplifted some 250 million years ago. The center of the uplift is the St. Francois Mountains in southeastern Missouri. Although the Ozark Mountains have great areas of exposed granite, they also have what is known as karst topography and a type of rock known as chert. Karst topography develops when water flows over limestone and dolomite rock, gradually dissolving and cracking the rock, resulting in the formation of caves, sinkholes, springs, arches, and natural bridges. All of these features are present in the Mark Twain National Forest. Chert is a type of quartz formed into tiny crystals and is acidic in nature. It often fills streams and accounts for the rocky bottoms so characteristic of Ozark streams.

The Mark Twain National Forest is in the southern half of the state, entirely south of Interstate 70. It extends from near the Illinois border on the east to near the Oklahoma border on the west and to the Arkansas state line to the south. The Mark Twain National Forest occupies much of this area, and all but one small district of the national forest is in the Ozark Mountains.

At one time the northern ranger districts of the Mark Twain National Forest were in the Clark National Forest, but the Clark and Mark Twain were combined into a single national forest. The Mark Twain National Forest consists of 6 ranger districts.

The most southeastern part of the forest lies immediately north of the town of Poplar Bluff. On the eastern edge of the region is Lake Wappapello, a popular Army Corps of Engineers recreation area that is surrounded by the Mark Twain National Forest. Much of this part of the national forest has been logged in the past so that the present forest is of relatively young age.

Markham Springs is three miles west of Williamsburg along the Black River. You can reach it via State Route 49A and Forest Road 421. The springs

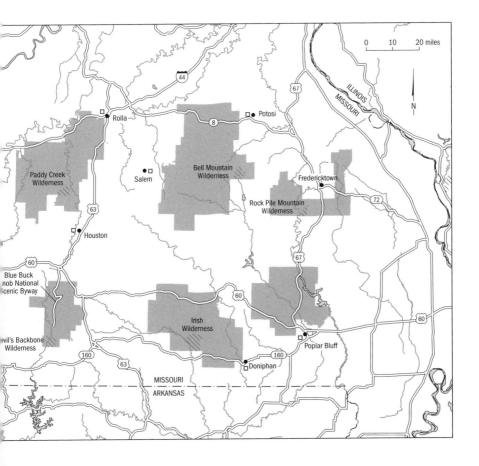

lie near the top of the Ozark Dome, which was formed by volcanic action more than one billion years ago. The springs have a flow of 3.5 million to 4.6 million gallons of water per day. The barrier-free Markham Springs hiking trail is a gentle gravel trail about 1,000 feet long that passes Bubbling Spring. Nearby Blue Hole and Brushy Creek flow underground. Prior to 1850, a mill was built here to grind grain, and a sawmill was adjacent. In 1874, Bill and Joe DeHaven built a two-story mill with a 14-foot water wheel. The mill was abandoned in 1907 and remained idle until the 1930s. In 1955, it was acquired by the U.S. Forest Service. The forest in this vicinity consists of sugar maple, Ohio buckeye, red buckeye, and box elder in moist habitats, and wild black cherry, white oak, and black gum in drier areas.

South of Markham Springs and State Route 49A is the Cane Ridge Management Area. The Cane Ridge Road across this area is a great way to become acquainted with this part of the Missouri Ozarks.

South of Wappapello Lake and one mile east of County Highway T is Mud Creek Natural Area. This 1,038-acre area preserves a wet bottomland forest along the creek and a dry oak–hickory forest on the slopes above. Along the creek is a good place to see Nuttall's oak, rare for this part of the country. The halberd-leaved tearthumb and purple fringeless orchid are also present.

Pineywoods Lake southwest of Elsinore and a half-mile south of U.S. Highway 60 is a popular fishing lake with a pier. A few miles west of Pineywoods Lake is the Doniphan–Eleven Point Ranger District that features spectacular Eleven Point National Scenic River. The river meanders through Ozark hill country from Thomasville near the southwest corner of the ranger district to Riverton at the southern edge, passing such fascinating places as Cane Bluff, Long Hollow, Boom Hole, Turner's Mill North, Turner's Mill South, Stinking Pond, Horseshoe Bend, Barn Hollow, Whites Creek, and Boze Mill. Archaeological evidence indicates that the river has been used by humans since around 10,000 B.C. Cane Bluff is a scenic dolomite cliff that towers 250 feet above the river. At this point, the river is lined with stands of giant cane, a type of bamboo, which forms extensive colonies known as cane brakes. Boom Hole is where an earthen breastwork with a wooden chute was built on a hill above a 200-foot-tall cliff to transport virgin shortleaf pine logs cut from the surrounding area. The logs plunged into the river and floated several miles downstream. Turner's Mill site was the focal point for the now defunct village of Surprise. The mill, which began operating around 1850, had a 26-foot steel wheel. The mill building and adjacent wooden flume are gone, but the giant wheel, a cement flume by the spring, and the chimney of Turner's house, still remain. Levi Boze built a mill at a spring farther downstream. The springwater powered the grist mill in the 1800s. The average flow of this spring is 12 million gallons per day. The old turbine used at the mill was made by James Leffel in Springfield, Ohio, in 1862.

South of Riverton about six miles, the Mark Twain National Forest has a small parcel of land along the Eleven Point River where the river crosses State Route 142. At this crossing is The Narrows, where four large springs flow more than 100 million gallons of water each day. The springs are Blue Spring, Morgan Spring, Sullivan Spring, and Jones Spring. Aquatic vegetation is abundant in and around the springs.

Some enjoyable adventures are to float lazily down the Eleven Point River through miles of natural beauty, or swim in the clear water, or fish from the banks. Just south of where State Route 49 crosses the Eleven Point River is Greer Springs, the second largest spring in Missouri with a daily flow of 220 million gallons.

North of Boom Hole is McCormick Lake, where bass fishing is usually good. A 3.7-mile hiking trail leads southward from the lake to the Eleven Point River, with an optional 2.1 mile loop into Duncan Hollow.

The Current River, another beautiful Ozark stream, flows near the eastern edge of the Doniphan–Eleven Point Ranger District. Floating is extremely popular in the Current River, as well. Near the Float Camp Picnic Area along the Current River is the pleasant 1.5-mile White Oak Forest Trail where you can see firsthand the typical forest vegetation in the area. The loop trail passes a wildlife pond.

The Doniphan–Eleven Point District contains several significant dedicated natural areas, many of them the result of karst topography, where rainwater filters down through limestone or dolomite to form large caverns or caves. When too much rock dissolves, the ceiling of the cave collapses, forming a sinkhole. Most sinkholes are dry, but a few of them are partially filled with water. Those that have water in them usually support rare plants, several of them primarily Coastal Plain species. Because of the unusual assemblage of plants, most of the sinkholes with water in them are protected as nature preserves, and they are often dissimilar in the makeup of their vegetation. The dominant tree in the Overcup Oak Sink Natural Area is overcup oak, named for its acorn that is almost completely enclosed by the cap; Red Maple Pond Natural Area is dominated by red maple; Tupelo Gum and Cupola Natural Areas are dominated by tupelo gums; Haney Pond Natural Area supports mostly sweet gums.

Overcup Oak Sink Natural Area is 6.5 miles by road northeast of the community of Winona. The sinkhole is filled with wetland vegetation surrounded by a dry, upland Ozark forest. Although large overcup oaks are usually confined to bottomland forests, the Overcup Oak Sink has a number of very large specimens. Growing in the sinkhole with the overcup oaks are buttonbush, pink St. John's-wort, sneezeweed, and trumpet creeper. The surrounding upland woods consists of white oak, black oak, scarlet oak, post oak, black hickory, and shortleaf pine. Shrubs and small trees in this community include serviceberry, farkleberry, cockspur thorn, slippery elm, flowering dogwood, blackberries, and wild grapes.

Red Maple Pond Natural Area is adjacent to County Highway C, about 12 miles north of the junction of County Highway C and U.S. Highway 160. Unlike most other sinkhole ponds in the Missouri Ozarks, Red Maple Pond is within the floodplain of an intermittent creek. Although this pond may dry out during late summer, the deep, mucky soils never become completely dry. Hummocks two feet high across the pond are formed by red maples with multiple trunks. The hummocks, which are raised, drier mounds of humus, support other plants such as buttonbush, brookside alder, pink St. John's-wort, wild azalea, and several species of sedge. Surrounding the hummocks are royal fern, manna grass, lizard's-tail, and orange-spotted touch-me-not.

Tupelo Gum Pond Natural Area and Cupola Pond Natural Area are both dominated by huge specimens of tupelo gums. The 32-acre Tupelo Gum Pond Natural Area is located about 14 miles southwest of Winona at the end of Forest Road 3239. Surrounded by a community of dry forest species, Tupelo Gum Pond is ringed by tupelo gums with huge, swollen bases. The central part of the pond supports spatterdock, cattails, and several sedges including the rare Canby's bulrush. The dry forest around the pond is dominated by white oak, black oak, scarlet oak, shortleaf pine, sassafras, and persimmon. Cupola Pond Natural Area, which is similar in appearance to Tupelo Gum Pond Natural Area, is located at the end of Forest Road 4823, a few miles east of County Highway J.

Three-and-a-half miles east of the Tupelo Gum Pond Natural Area is another significant natural area known as Brushy Pond. This 28-acre preserve has an abundance of rare sedges, including winged oval sedge, porcupine sedge, and cypress knee sedge, as well as an uncommon manna grass.

Marg Pond Natural Area is a sinkhole pond just north of U.S. Highway 60, five miles southeast of Winona. This shallow sinkhole is covered by a marsh community that includes four species of plants listed as rare for Missouri. Another significant sinkhole is Haney Pond Natural Area at the beginning of Forest Road 4823, just two miles southwest of Cupola Pond. The vegetation in Haney Pond is classified as a forested acid seep community, where the wet basin is fed by acidic groundwater that contains carbonic acid. The dominant tree in this pond is sweet gum, with buttonbush and rose mallow common beneath the trees. The natural area is crossed by the historic Bellevue Trail that was used by early explorers and settlers.

In addition to sinkhole ponds, there are several fens spread across the Mark Twain National Forest, and particularly the Doniphan–Eleven Point District. A fen is a natural community where soils are saturated from mineral-rich groundwater, creating rivulets and ooze areas. Wells Branch Fen consists of two acres of open land bordered by small trees and shrubs. Within the fen are several clumps (or tussocks) of sedges and rushes, white turtlehead, orange coneflower, false loosestrife, and yellow-eyed grass, along with such prairie species as Indian grass, big bluestem, little bluestem, and Riddell's goldenrod.

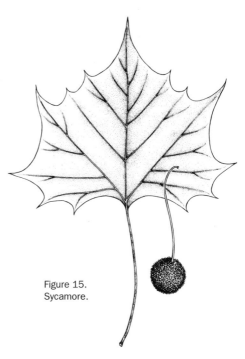

Figure 15.
Sycamore.

One of the more diverse natural areas in the district is Big Barren Creek Natural Area, where the principal features are an Ozark headwaters stream, a dry dolomite cliff, a narrow, rocky gorge known locally as a shut-in, a bottomland forest, a dry forest with cherty pebbles strewn across the forest floor, and a rocky, barren opening known as a dolomite glade. Big Barren Creek itself has shallow riffles, spring branches, and deep, permanent, spring-fed pools, with vegetation that includes watercress, spatterdock, golden ragwort, marsh coneflower, pale manna grass, water willow, and chairmaker's rush, the latter actually a sedge. The dry, dolomite cliff has occasional cavelike openings in the side, and plants such as alumroot and wild hydrangea cling to the sides of the cliff. Devil's Run is a tributary to Big Barren Creek, and this narrow stream is lined by vertical cliffs of chert, sandstone, and dolomite. The bottomland forest along Big Barren Creek is typical and includes bitternut hickory, white oak, red oak, sugar maple, sycamore (fig. 15), pawpaw, musclewood, and spicebush. The rocks scattered across the dry forest floor are chert particles, which are flintlike and weather very slowly. Dominant trees in the chert forest are white oak, scarlet oak, mockernut hickory, post oak, shortleaf pine, and black hickory. The dolomite glade is a rocky opening above Devil's Run shut-in and contains the Missouri coneflower and the wiry-stemmed prairie bluets, among a diversity of wildflowers. You can reach this fascinating area by taking County Highway C about 11 miles south of U.S. Highway 60, and then following County Road 167 for 2.3 miles.

Two-and-a-half miles southwest of Big Barren Creek Natural Area is Cowards Hollow Natural Area that is significant for its chert shut-in, vertical chert cliffs, a shelter cave, a waterfall, a fen community, and a seep-fed stream. The chert shut-in is a narrow steep-sided cleft through a chert cliff. A 10-foot waterfall marks the beginning of the shut-in. The fen is easily recognized by the presence of large clumps of royal fern and sensitive fern.

The largest dolomite glade in this part of the Missouri Ozarks is the 332-acre Bald Hill Glade Natural Area located southwest of Wolf Mountain and one-fourth mile west of Forest Road 3149. This large, rocky opening in the forest has a diverse flora of glade and prairie plants, including palmate-leaved coreopsis, Missouri coneflower, purple prairie clover, silky aster, false aloe, an uncommon morning glory, and the state's endangered umbrella plant, which is a type of wild buckwheat.

Tunnel Bluff Woods Natural Area is a 288-acre preserve that straddles the Current River and is managed by the U.S. Forest Service and the Ozark National Scenic Riverway. The woods along the Current River are a high quality bottomland forest on a raised floodplain terrace. Huge sycamore trees tower above dense stands of giant cane. East of the river is a natural tunnel remaining from a collapsed cave system.

Although not a dedicated natural area, Falling Spring is a scenic and nostalgic area in the Doniphan–Eleven Point District about 12 miles slightly southeast of Winona. At Falling Spring is a small mill reached by way of a boardwalk, a log cabin, an old cemetery, a pond, and a small spring that emerges from a rocky cliff behind the mill. The spring also has picnic facilities.

The Doniphan–Eleven Point District contains the Irish Wilderness, more than 16,000 acres of Ozark forested land dominated by several species of oaks and hickories, shortleaf pine, flowering dogwood, and sassafras. During the mid-1800s, Catholic priest John Hogan brought Irish immigrants to the area now occupied by the wilderness to escape animosity from people living in St. Louis. The Irish established a settlement, but during the Civil War, both Union and Confederate soldiers raided the community. At the war's end, nothing could be found of Father Hogan or any of the Irish settlers. The wilderness contains a number of sinkholes, natural springs, and Whites Creek Cave where crystalline formations may be seen. The Irish Wilderness is adjacent to the eastern side of Eleven Point River and extends to the east to County Highway J. Whites Creek Hiking Trail crosses the wilderness and forms a large loop.

Another part of the Mark Twain National Forest lies north of the Poplar Bluff and Doniphan–Eleven Point districts. East of the town of Fredericktown, the rocky terrain is in the foothills of the Ozark Mountains and is dominated by scarlet oak, shortleaf pine, and black hickory, with a colorful shrub layer that includes pink azaleas. West of Fredericktown are larger mountains—Blue, Grassy, Tims, Crane, Patterson, Keller, Smith, and Black. The St. Francis River bisects the area, with scenic tributaries such as Marble Creek, Rock Creek, and Marsh Creek. Granite outcrops occur throughout the district, and several rocky openings, or glades, are scattered here and there. The upland forests consist of shagbark hickory, black hickory, white oak, scarlet oak, and black oak above a shrub zone of fragrant sumac and low-bush blueberry. Common wildflowers are bee balm, skullcap, lobelia, and wild petunia. Among the glade plants are prairie dock, feverfew, rock spike-moss, prickly pear cactus, flower of an hour, and thread-leaved sundrops.

The Silver Mine Recreation Area is on the site of the Einstein Silver Mining Company, which began operation in 1877. In 1879, the St. Francis River was dammed to drive a turbine wheel, a smelting furnace was built, and machinery was brought in. A town of 900 residents grew up around the mine. After tons of lead, 3,000 ounces of silver, and traces of gold were extracted, the mine was shut down and the town disappeared. A hiking trail traverses the site, passing prospecting pits and the site of the old mine. An interesting array of natural balanced rocks may be seen along the trail. Seventy-two wooden steps lead down to an old dam across the St. Francis River. The Silver Mine Recreation Area is one mile south of State Route 72, about six miles west of Fredericktown.

Marble Creek Recreation Area is around the site of a grist mill that was in operation until 1935. The dam and building foundations are still present. The mill is adjacent to Marble Creek, which rushes for 20 miles through the rugged St. Francois Mountains. The creek is named for the colored, marble-like dolomite, known in the local building trade as Tom Sauk marble. The Marble Creek Recreation Area is along County Road E several miles southeast of Ironton. In the southwestern corner of the region is scenic Crane Lake. The lake was constructed many years ago when Crane Pond Creek was impounded at the upstream end of a shut-in. A five-mile double loop hiking trail snakes around the lake.

Rock Pile Mountain Wilderness is at the south end of the area. The wilderness is named for an ancient circle of granite rocks that were piled on top of the mountain by some early inhabitants. Many of the forested areas in the wilderness have never been cut, and handsome specimens of basswood, butternut, Kentucky coffee tree, black walnut, and sugar maple abound. The wilderness in the St. Francois Mountain Range has elevations between 520 and 1,305 feet. In addition to the rock pile, other attractions are steep limestone bluffs, rock formations, caves along the St. Francis River, narrow shut-ins, and granite glades.

The Salem and Potosi Ranger Districts are contiguous and lie west of Fredericktown and north of the Doniphan–Eleven Point District. The Bell Mountain Wilderness is one of the primary attractions in the Potosi District, capped by 1,702-foot Bell Mountain. The wilderness lies east of County Highway A a few miles northeast of the junction of County Highway A and State Route 49.

Bell Mountain is in the St. Francois Mountain Range and consists of steep-sided felsite and rhyolite outcroppings. Granite glades are scattered on the slopes of the mountain, providing openings in the predominant oak–hickory–shortleaf pine forests. Gnarly red cedars are sprinkled throughout the glades, and blackjack oak and winged elm often form borders of the glades. Shut-in Creek, which crosses the wilderness and has several shut-ins, is spring fed and flows year-round. Steep talus slopes along the creek make it very difficult to climb to the top of the mountain from the creek.

One good way to explore this section of the Mark Twain National Forest is to drive or hike the 24-mile Berryman Trail that winds through dry forests dominated by shortleaf pine, white oak, and black oak, and bottomland hardwood forests dominated by basswood, sugar maple, and honey locust. Occasional switchbacks make you feel as if you are in a higher mountain range. Along the trail are several springs, some with adjacent picnic tables, and a campground along Brazil Creek. The trail begins at an old Civilian Conservation Corps campsite 17 miles west of Potosi on State Route 8 and then one mile north on Forest Road 2266.

Worth visiting is Red Bluff, a towering cliff along Huzzah Creek, whose reddish color is due to the presence of iron compounds in the rock. This formation was deposited as sediment in an ancient sea about 500 million years ago. Nearby are a campground, picnic areas, and two hiking trails. Red Bluff is at the end of Forest Road 2011 north of County Highway V at the small community of Davisville.

The smaller Salem Ranger District has Grasshopper Hollow Natural Area, the Blair Raised Fen Natural Area, Sutton Bluff, Council Bluff Recreation Area, and Logger's Lake Recreation Area. Grasshopper Hollow Natural Area is the finest fen complex in Missouri, with four different kinds of fens—seep, deep muck, prairie, and forested. To reach the fen, follow State Route 72 eastward from Bunker until you come to Forest Road 860. Turn left onto the forest road and follow it for a half-mile.

The Blair Creek Raised Fen Natural Area is unique because it is the only raised fen in Missouri where a deep deposit of mucky organic soil has swollen, or raised, above the surrounding flat land. Many of the plants in this raised fen grow in dense mats, or tussocks. The species in the Blair Creek Raised Fen are somewhat different than those found in nonraised fens in

Missouri, and include umbrella sedge, stingless nettle, orange coneflower, tufted loosestrife, and marsh blue violet below a shrubby layer of heart-leaved willow and swamp dogwood. Birds often seen here include the common yellowthroat and swamp sparrow.

Sutton Bluff is a scenic cliff above a particularly picturesque part of the West Fork of the Black River. A small campground is adjacent.

Council Bluff Lake Recreation Area surrounds beautiful Council Bluff Lake. The recreation area has a swimming beach, boat ramp, and a campground. The lake is in the shadow of Johnson Mountain and is located about three miles north of the Bell Mountain Wilderness.

Logger's Lake Recreation Area contains a 25-acre impoundment constructed in the 1930s by the Civilian Conservation Corps. Located seven miles southwest of the village of Bunker off of Forest Road 2221, the lake is a great place to fish for largemouth bass and bluegill. The recreation area has a boat ramp, campground, and picnic area, and the mile-long Lake View Interpretive Trail encircles the lake.

At the Hazel Creek Campground are the ruins of a lead-smelting furnace. This area is along County Highway Z about 10 miles northeast of Viburnum.

West of Rolla is the Houston-Rolla Ranger District that features the Paddy Creek Wilderness, Kaintuck Hollow, Slaughter Sink, and two picturesque Ozark rivers—Gasconade and Big Piney.

Paddy Creek Wilderness is 7,019 acres of oak, hickory, and shortleaf pine forests above steep cliffs that hang above Paddy Creek. At the northeast corner of the wilderness is the Paddy Creek Recreation Area that consists of a campground and picnic facilities as well as the major trailhead into the wilderness. A short trail at the campground stays along Paddy Creek for a while before climbing steeply to the bluff top on the south side of the creek. After descending from the ridge, the trail crosses Paddy Creek on its way back to the campground.

One mile northeast of Paddy Creek Recreation Area is the Slabtown River Access where there is a campground, picnic area, boat ramp, and hiking trail along Slabtown River. The Slabtown Bluff Hiking Trail is two miles long and follows the river below picturesque bluffs.

Just outside the southwestern corner of Paddy Creek Wilderness is Roby Lake, a small body of water with adjacent picnic facilities. Big Piney Trail is a 17-mile trail across the wilderness, from the Paddy Creek Recreation Area to Roby Lake.

Although not a wilderness, Kaintuck Hollow is a wild area popular with hikers, mountain bikers, and horseback riders. Four trails, ranging in length from one to 15 miles, are in the hollow. Wilkins Spring, with a daily flow of three million gallons, is at the western side of Kaintuck Hollow, along County

Highway AA. Mill Creek, at the northern end of the hollow, is great for trout fishing. Both Wilkins Spring and Mill Creek have picnic tables.

Slaughter Sink is a large sinkhole in the middle of a heavily wooded forest. The vegetation in and around the sink is fascinating. The sink is immediately west of Interstate 44. Take the Jerome exit and follow the frontage road, Onyx Cave Road, on the west side of the interstate southward.

Another area worth visiting is Lane Spring, where 12 million gallons of water flow daily. Among the amenities are a sizable campground, picnic facilities, and a three-fourths-mile interpretive loop trail along the creek and bluff overlooking Little Piney Creek. The water is stocked with trout. You will see an abundance of watercress in the spring area.

The Gasconade River is a great one for a float trip. Anglers will be delighted by the variety of fish in the river.

The northernmost part of the Mark Twain National Forest extends nearly to Interstate 70. It consists of 21 noncontiguous parcels of land, mixed in with private property, between Fulton and Columbia and is the only district of the Mark Twain National Forest that is not in the Ozark Mountains. Most of the Forest Service land has been timbered, farmed, or stripped for coal in the past. The terrain is gently rolling, with forests and isolated patches of prairies. The Cedar Creek Hiking Trail consists of two loops and a 22-mile trail that alternates between patches of forest and gravel roads. The Devil's Backbone Hiking Trail is along Cedar Creek and begins at the Pine Ridge Campground and picnic area six miles east of Ashland on County Highway Y. This is a 20-mile loop primarily through oak–hickory forests.

The southwest area of Missouri is occupied by the Ava-Cassville-Willow Springs District. The area consists of low, round-topped mountains called knobs, clear, gravel-bottomed streams, ravines or hollows, dry ridges, and occasional springs. Floating is popular on the North Fork River, a picturesque clear stream fed by Blue Spring, Rainbow Spring, and other smaller natural springs. A short hiking trail leads to Blue Spring. The best accesses to the river are at Hebron and North Fork. The Blue Buck Knob National Scenic Byway bisects the district from north to south, at first following State Route 181 and then County Highway AP. The byway climbs to Blue Buck Knob where the historic lookout provides for unending panoramas of the Missouri Ozarks. The byway is good access to Noblett Lake and Blue Hole. Fishing is usually good at Noblett Lake and, in spring, the blossoms of redbud and flowering dogwood color the landscape. At Noblett Lake and Blue Hole are trailheads for the Ridge Runner National Recreation Trail. Near Noblett Lake this trail passes bubbling Hellroaring Springs, whereas at Blue Hole is the Missouri National Children's Forest. Blue Hole is named for the misty blue fog that often hangs over the valley.

The Missouri National Children's Forest came about after a devastating 1,300-acre forest fire in spring 1971 that destroyed all of the original beauty of the area. With funding from Hunt-Wesson Foods, the U.S. Forest Service rebuilt the forest, calling the project the "Investment in the Future." Children all around the United States participated in the program by saving labels from Hunt-Wesson food products. For each label, Hunt-Wesson contributed five cents toward the purchase of seedling trees. The Forest Service was responsible for planting the trees and building the nature trail. A capsule was placed in the Children's Forest holding each child's name who sent in a label and had a tree planted for him or her. Boy Scouts participated in the planting program. A total of 429,000 trees were planted. Young women 15 to 18 years old from the Youth Conservation Corps Camp in the Mark Twain National Forest made interpretive signs, helped construct the nature trail, cut weeds and brush, and constructed parking barriers.

Five miles west of the community of Willow Springs and just north of State Route 76 is the Indian Creek Research Natural Area. This area is significant in that it harbors a high concentration of rare plant species. This 385-acre area includes the streambed of Indian Creek, moist sandstone and dolomite cliffs, natural seeps, rugged upland ridges, and narrow ravines. The creek is swift-flowing with deep pools and numerous riffles. Sullivantia, a plant related to alumroot, barren strawberry, white death camas, and showy lady's-slipper are among the uncommon species. The ravine forests consist of sweet gum, sycamore, and American elm, whereas the upland forests are dominated by oaks, hickories, and shortleaf pine.

An area of narrow ridges and ravines separated by high, steep bluffs has been designated as the Devil's Backbone Wilderness. Located between County Highways CC and KK, the wilderness is extremely rugged with elevations between 680 feet along the North Fork of the White River and 1,200 feet on some of the ridges. Several trails crisscross the wilderness, bringing the hiker to Blue Spring, McGarr Spring, and Amber Spring. In mid-May, pink azaleas add color to the steep, wooded slopes, and small limestone glades are alive with colorful wildflowers during summer and fall.

The Mark Twain National Forest south of Ava receives numerous visitors, particularly during summer months, partly because of its proximity to Springfield and Branson. The crown jewel of the district is the Hercules Glades Wilderness with its colorful rocky glades, forested knobs, and open grasslands. The shallow, rocky soils are ideal for the optimum growth of glade and prairie plants. A unique feature of the wilderness is the presence of animals more usually found in the southwestern United States.

Glade Top Scenic Byway is one of the most scenic byways in the Mark Twain National Forest and gives the visitor a good orientation to glade country. The north end of Glade Top Scenic Byway is 10 miles south of the community of Ava on Forest Road 147 where it begins on top of Bristle Ridge. Immediately upon entering the national forest is Smoke Tree Scene where you should look at the scenery that unfolds all around you. To the north is a typical view of the Ozark Mountains. Across the hollow to the west is a barren ridge where yellowwood and smoke trees are scattered. The pendulous clusters of white flowers of yellowwood in early May are striking. Looking southward you see another hollow before the forested ridges stretch out into the distance. In about two miles the trail comes to Arkansas View and the Corbitt Potter Campground. The mountains that rise up to the south, some 40 miles away, are in Arkansas.

After a few more miles you come to the Caney Picnic Area and lookout tower. The rounded, nearly treeless hills in the near distance are called bald knobs and are where the notorious group known as Baldknobbers used to gather and plan its escapades. At the Willie Lee Group Campground, the Glade Top Scenic Byway forks. The left fork to the southeast follows Forest Road 147 to the nearly defunct village of Longrun. Just before reaching Longrun is a pullout for a panoramic view. The right fork leads southwest on Forest Road 149. A pullout for a spectacular view into Big Creek Basin is about five miles before the Glade Top Scenic Byway ends where Forest Road 149 and State Route 125 come together. If you take State Route 125 to the west, you come to Hercules Glades Wilderness in about three miles.

Five miles south of the community of Sparta, is the Chadwick ATV Area where more than 100 miles of trail-bike or motorcycle trails wind through densely forested hollows and down long ridge tops. The area has two campgrounds.

The southwesternmost part of the Mark Twain National Forest is near the community of Cassville. This region extends southward all the way to the Arkansas state line. Parts of the area surround huge Table Rock Lake and the extremely crooked White River and its tributaries, including Kings River.

Northwest of Table Rock Lake is Piney Creek Wilderness, a region of narrow ridges separated by steep, narrow hollows that are up to 400 feet below the ridge tops. Piney Creek flows across the wilderness, followed closely by the Piney Creek Trail. The creek is fed by numerous springs. Several rocky openings, or glades, are scattered in the wilderness. The surrounding upland forests consist primarily of black, red, white, post, and blackjack oaks and shortleaf pine.

Big Bay Campground at the edge of a neck of Table Rock Lake has a boat ramp and a short hiking trail.

Twelve miles south of Cassville, State Route 112 becomes the Sugar Camp National Scenic Byway when it enters the Mark Twain National Forest. After two miles, Forest Road 1035 branches off of State Route 112 and becomes the scenic byway. The drive is especially scenic in spring and fall as it passes the Sugar Camp Lookout Tower. The scenic byway then becomes Forest Road 197 to the community of Eagle Rock where it swings back to the north until it leaves the national forest at Natural Bridge about seven miles north of Eagle Rock.

Just south of the Sugar Camp Lookout is the Butler Hollow Glades Natural Area. The dolomite glades on steep slopes have few trees, with most of the vegetation made up of grasses and wildflowers. Glades occur in shallow soil on both sides of the hollow. Dominant grasses are little bluestem, Indian grass, and side-oats grama. One of the earliest wildflowers to bloom is wild hyacinth, followed by orange puccoon and Trelease's larkspur. Later, Bush's poppy mallow, Missouri evening primrose with saucer-sized yellow flowers, Indian paintbrush, shooting-star, pale purple coneflower, and compass plant bloom. A shrubby border of the glades often consists of red cedar, Ashe's juniper, aromatic sumac, winged elm, and gum bumelia. Among the forest trees in the hollow are yellowwood, black walnut, and basswood. Bird watchers thrill at the sight of Bachman's sparrow and the comical roadrunner. Tarantulas sometimes may be seen crawling over the rocks in the glades. Unfortunately, the area is not well maintained.

Grasshopper Hollow

In the first act of Edward Albee's *Who's Afraid of Virginia Woolf,* Martha describes husband George as "an old bog in the History Department. A bog. . . . A fen. . . . A [expletive deleted] swamp." Of these three terms, used interchangeably by Martha, "fen" perhaps stands out as a bit uncommon. All connote a wet, soggy habitat, with "swamp" implying a forested area with standing water and "bog" suggesting an open region quaking with layers of sphagnum. Like bogs, fens are mostly open areas, but they have an important distinction recognized by many biologists; the groundwater in bogs is acidic, whereas that in fens is basic, or calcareous.

In the early 1980s, Southern Illinois University botany student Steve Orzell discovered dozens of areas that qualified as fens in the St. Francois Mountains region of southeastern Missouri. The most fruitful spot was Grasshopper Hollow, a ravine some 85 miles southwest of St. Louis and two miles northwest of the tiny Ozark community of Reynolds. Here Orzell discovered a series of fens in a 300-acre area. Some of the land is in private hands and some was acquired by the Nature Conservancy, but much of it is administered by the Mark Twain National Forest.

Grasshopper Hollow owes its existence to Grasshopper Creek and its tributaries, which have carved deep defiles into the bedrock. Blocks and rock fragments are scattered throughout the area as a result of erosion high on the steep slopes. Water seeps from numerous places near the base of the slopes and, observes biologist Max Hutchison, collects to form small rivulets or spreads out into the valleys to form the fens. Because the bedrock is largely dolomitic limestone, the water that flows through the area picks up carbonate and becomes basic.

The fens in Grasshopper Hollow are of several kinds. The great majority—10 or 11 of them—are open, essentially treeless areas with moderate water that seeps over shoe tops. Three others are quite different. One contains so much water and soggy soil that anyone stepping into it sinks to the waist or deeper; Missouri botanist Paul Nelson describes such habitats as deep muck fens. Another, not as wet as most of the others, is dominated by typical prairie plants and has been called a prairie fen. Finally, there is a forested area that on appearances might logically be called a swamp but for which biologists prefer the term forested fen, on account of the calcareous water.

The more typical fens are found on somewhat sloping terrain, in the narrow valleys, or on the hills, where groundwater oozes up and fans out. The amount of standing water varies with the degree of slope, from as little as a trace to as much as 15 inches. Gravel and rock fragments are usually scattered liberally about among tussocks of sedges and grasses. Mixed in are several colorful summer- and fall-flowering wildflowers, such as orange coneflower, false dragonhead, swamp goldenrod, golden ragwort, purple fringeless orchid (pl. 18), and fen aster.

More forbidding and extremely difficult, if not impossible, to traverse is the deep muck fen, which lies on a flat valley terrace near the foot of a slope. The area is boglike and quaking, with standing water forming small, shallow ponds. The soupy, mucky soil is more than 40 inches deep, with no surface gravel and rock fragments. Grasses and sedges abound, along with a shrubby alder. One of the most graceful sedges, fringed sedge, stands as much as five feet tall, growing with the robust royal fern. Other wetland shrubs are swamp dogwood, silky willow, and buttonbush, whereas the wildflowers include the speckled wild sweet william.

Grasshopper Hollow's 17-acre prairie fen is considered by local botanists to be the most pristine such fen in the world. Prairie fens usually arise along valley terraces with a slight slope. They are easily traversed because there is little standing water and only a shallow layer of mucky soil. What distinguishes them is that the more typical fen vegetation is joined by tall prairie grasses and prairie wildflowers.

Orzell and Missouri botanist Donald Kurz have recorded 179 species of plants in the prairie fen in Grasshopper Hollow, more than 20 percent of which are characteristic of midwestern tall grass prairies. The principal tall grasses are big bluestem, Indian grass, and prairie cord grass. A sampling of the abundant prairie wildflowers includes New England aster, prairie dock, rosinweed, prairie sunflower (pl. 19), scarlet Indian paintbrush, butterfly weed, and Culver's root.

In the floodplain adjacent to Grasshopper Creek, where the soil is poorly drained and seasonally saturated, lies Grasshopper Hollow's forested fen. Green ash, red maple, and slippery elm grow up to 60 feet tall and so densely that they form a closed canopy. Such smaller woody species as blue beech, spicebush, and alder fill the midlayer of the forest. Sedges and grasses are common in the understory, but the conspicuous feature is the continuous cover of clumps of mosses that conceal the soil surface.

Fens have been recognized for many years in northern Europe, but North American botanists did not really begin to distinguish fens from bogs until the 1960s. Those first identified were in areas that, like northern Europe, had a history of Ice Age glaciation: Iowa, northern Illinois, northern Indiana, northern Ohio, and other northerly regions. Even though the eminent Missouri botanist Julian A. Steyermark had reported "calcareous wet meadows" from the heart of the Ozark Mountains in 1938, no one thought of these and other areas in southern Missouri as fens until Steve Orzell studied them in 1981. Only then did scientists realize that fens, among them those of Grasshopper Hollow, existed south of the line of maximum ice advance during the Ice Age.

Cupola Pond

Some 38 miles west of Poplar Bluff, on the generally dry and somewhat monotonous Ozark Plateau, a ridge suddenly drops 40 feet into a five-acre depression filled with water. This is Mark Twain National Forest's Cupola Pond, which was classified a Missouri Natural Area in 1981 and a Society of American Forests Natural Area in 1983 (pl. 20). These special designations acknowledge not only the rarity of a permanent source of water in the uplands but also the unexpected vegetation—a nearly pure stand of water tupelo trees. Water tupelos are normally found in lowland areas, growing in bald cypress swamps (the nearest to Cupola Pond being almost 40 miles away). As a result of the draining, logging, and cultivation of southern Missouri's bottomland, even such swampy habitats are no longer common.

Cupola Pond is a sinkhole, a depression caused by a dissolving of underlying rock, in this case limestone, followed by a collapse of the land surface. Paul Delcourt, Hazel Delcourt, and E. Newman Smith from the University of Tennessee have probed the center of Cupola Pond, removing core samples from a depth of 40 feet. According to their research, sediment has been collecting in the basin for 23,000 years, and the water tupelos have been around for much of that time. Geologist Stanley Harris speculates that the pond's longevity may result from a very slow but continuous dissolving of the underlying bedrock, with the gradual settlement of the depression keeping pace with the rate of sedimentation.

Cupola Pond still looks much as it must have to the first European settlers who encountered it. Nearly 500 water tupelos fill the five acres of water, some reaching a height of about 70 feet, and most having a diameter of between 10 and 26 inches. Many of the standing trees that are more than 100 years old have become hollowed out near their bases. Naturalist Max Hutchison, who has studied the pond extensively, notes that these cavities are home to a variety of animals, especially such invertebrates as butterflies, moths, dragonflies, and damselflies. The absence of fish in the isolated pond helps account for the abundance of aquatic insects.

As Hutchison has described, a skirt of green vegetation rings the base of most of the water tupelos a few inches above the surface of the water. This miniature flora consists of mosses and lichens, as well as such flowering plants as white-water horehound, rayless beggar's-tick, and pink St. John's-wort. Royal fern, fringed sedge, and other plants border the pond, and adjacent to it are low-lying woods (flatwoods) dominated by pin oaks. During summer, when the level of the pond is low, the flatwoods soil is very dry and full of fissures. When heavy rains cause the pond to overflow, however, the flatwoods become wet and soggy, because the hard soil drains poorly.

A short walk southeast of Cupola Pond, along a normally dry, bouldery streambed, another unusual habitat can be found. These are openings in the forest, called barrens, that contain a mixture of grasses and wildflowers: bluestems, poverty oat grass, wild indigo, goat's rue, flowering spurge, purple coneflower, rattlesnake master, and blazing stars. According to the Nature Conservancy, which is cataloging every habitat type in the country, barrens are an endangered feature of the American landscape.

Hercules Glade

Some people call them balds or knobs, whereas others refer to them as barrens. I prefer glade, which I define as a forest opening strewn with exposed rocks. The Missouri Ozarks have hundreds of glades, ranging in size from a few hundred square feet to several hundred acres. Although most glades are surrounded by dense, oak-dominated forests, there is little encroachment by the trees. For one thing, the soil in the glades is not nearly deep enough for oaks. For another, most glades face south and west, and the hot summer sun strikes them directly, creating an arid and generally inhospitable habitat. Not even summer thundershowers moisten the ground; the water runs rapidly off the shallow soil.

A few trees do grow where pockets of soil have accumulated in the glades. Red cedar is common, being able to survive the dry conditions because of its diminutive, needlelike leaves that lose little moisture through evaporation. Less usual are two small trees whose roots deeply penetrate the cracks in the rocky substrate to reach the limited supply of water: the smoke tree, with its blue-gray fruiting sprays, and the fringe tree, with its pendulous clusters of white flowers.

To thrive in the glade environment, plants must compensate for the lack of moisture between June and September. Annuals complete their life cycle in a short time, then rely on their seeds for their perpetuation the following year. Most do not take the trouble to survive summer. The dwarf leavenworthia mustard, for example, germinates in late fall, overwinters with a tiny cluster of leaves, flowers in late March, forms its seeds by early May, and then dies back before the heat of summer. The slender-leaved heliotrope, on the other hand, usually germinates in June, flowers in July and August, and forms its seeds in September. Although this may seem like a poor strategy, the plant thus takes advantage of a niche left by the annuals more typical of the glades. As a compensation, this heliotrope's very narrow leaves control water loss. With the rising temperatures of summer, the edges of the narrow leaves roll under and inward, exposing even less area to the sun's rays.

Perennials have to survive the entire growing season to build up enough food reserves, mostly starches, to last the winter and produce new leaves the next spring. (Annuals generally require fewer reserves to form their seeds.) To cope with midsummer conditions, flower-of-an-hour and false aloe have thick fleshy leaves for storing large quantities of water. Prickly pear cactus retains excess water in fleshy, padlike stems; there is no water loss from its leaves, which are modified into spines. The leaves of wild petunia are covered with a dense entanglement of hairs, a barrier against rapid evaporation of water. The food reserves of the Missouri black-eyed Susan accumulate in thickened stems just beneath the surface of the soil, where they are protected from the sun, wheras the nodding wild onion and purple oxalis produce underground bulbs.

In the Ozarks, glades are distinguished by rock type, and rock type affects the vegetation. Centered in southern Missouri and northern Arkansas, these mountains constitute one of the oldest exposed geological regions in the world. A billion years ago, a thick lava flow covered the area. After it cooled, a great mass of molten rock thrust upward, melting and absorbing some of the lava in its way. The molten rock cooled very slowly, forming a coarse-grained granite. After eons of erosion, this granite dome eventually sank beneath the sea and was covered by several thousand feet of sedimentary lime-stone and sandstone. Two subsequent movements, one 400 million years ago and another more recently, again raised the region's rock mass. Since then, erosive forces have continued their work. In places, the limestone and sandstone have been worn away, leaving the ancient granite exposed; elsewhere, sandstone or limestone dominates.

Perhaps the most pristine of all the glades is Hercules Glade in the wilderness of the same name in southwestern Missouri. Hercules is a limestone glade; what little soil it has averages only about four inches deep.

During winter and early spring, the glade is uncharacteristically wet, often with shallow pools of standing water, because of abundant rain and snow. The glade experiences less evaporation because of the higher humidity. May brings a combination of giant purple-flowered beardtongue, whose tubular flowers measure more than two inches long and are nearly as broad; blue false indigo, with densely packed elongated clusters of flowers; and Missouri evening primrose, whose bright yellow, saucer-sized flowers conceal the rest of the plant.

In summer, lack of rain, rapid drainage, and maximum evaporation create desertlike conditions in the glade. Plants that persist into summer often wilt, and animals either burrow into the shallow soil or retreat into the surrounding forest. A lull occurs in the blooming of wildflowers, and dense stands of tall grasses give Hercules Glade the look of a prairie. Big and little bluestem, prairie dropseed, Indian grass, and side-oats grama, all characteristic plants of the prairies of Kansas and Nebraska, dominate the scene. Seeds of these grasses, which probably arrive by way of the frequent, strong westerly winds, have a good chance of germinating in the absence of competition from shade-forming trees. In late August and September, the glade again becomes a sea of color—orange Missouri black-eyed Susans, cylindrical violet spikes of blazing stars, and crowded purple heads of the daisylike palafoxia.

Although the plant life is the star attraction of Hercules Glade, the animal life is also intriguing because of its western affinities. The scissor-tailed flycatcher can be seen on occasion, an attractive bird that is not uncommon in Oklahoma but just barely gets into Missouri. The gawky roadrunner, better known from Arizona, apparently arrived in southwestern Missouri around 1956, spread to many glades, and then started to decline after the rigorous winter of 1976–1977. Another western animal that sometimes races at high speed over the rocky terrain is the collared lizard, a pugnacious, foot-long reptile readily identified by the two black bands around its neck (pl. 21). Huge, hairy-legged tarantulas occasionally crawl over the stony surface, while underneath some of the rocks lurks a species of scorpion.

Plate 1. A view along the Talimena National Scenic Byway, Ouachita National Forest (Arkansas).

Plate 2. Mountain Stream, Ouachita National Forest (Arkansas).

Plate 3 (top). Buffalo National Wild and Scenic River, Ozark National Forest (Arkansas).

Plate 4 (bottom). Blanchard Springs Caverns, Ozark National Forest (Arkansas).

Plate 5 (top). Creekside wildflowers along Boreas Pass, Arapaho National Forest (Colorado).

Plate 6 (center). Marsh marigold, Grand Mesa National Forest (Colorado).

Plate 7 (bottom left). Western monkshood, Grand Mesa National Forest (Colorado).

Plate 8 (bottom right). Mule's ears, Grand Mesa National Forest (Colorado).

Plate 9 (top). Elk Range in autumn, Gunnison National Forest (Colorado).

Plate 10 (above). Blue columbine, the Colorado state flower, common in Gunnison National Forest (Colorado).

Plate 11. Wild aster, Mount Zirkle, Routt National Forest (Colorado).

Plate 12. Western pasque-flower, San Isabel National Forest (Colorado).

Plate 13 (left). Monument plant, Twin Peaks, San Isabel National Forest (Colorado).

Plate 14 (above). False Solomon's-seal, Twin Peaks, San Isabel National Forest (Colorado).

Plate 15 (top).
Spring wildflowers,
Yankee Boy Basin,
Uncompahgre
National Forest
(Colorado).

Plate 16 (right).
Mother bear and
cubs, Superior
National Forest
(Minnesota).

Plate 17. Pack of gray wolves on alert, Superior National Forest (Minnesota).

Plate 18. Purple fringeless orchid, Mark Twain National Forest (Missouri).

Plate 19. Prairie sunflower, Grasshopper Hollow, Mark Twain National Forest (Missouri).

Plate 20. Cupola Pond, Mark Twain National Forest (Missouri).

Plate 21 (above). Collared lizard, Hercules Glade, Mark Twain National Forest (Missouri).

Plate 22 (right). Mountain whitefish, Beaverhead National Forest (Montana).

Plate 23. Bitterroot plant, Bitterroot National Forest (Montana).

Plate 24. Blodgett Canyon, Bitterroot National Forest (Montana).

Plate 25 (top). Bellflower, Custer National Forest (Montana and Wyoming).

Plate 26 (above). Mountain lion, Flathead National Forest (Montana).

Plate 27 (right). Glacial lake, Lee Metcalf Wilderness, Gallatin National Forest (Montana).

Plate 28. Absaroka-Beartooth Wilderness, Gallatin National Forest (Montana).

Plate 29. Wild turkey, a common sight at Samuel R. Taylor National Forest (Nebraska).

Plate 30. Oceanspray, Santa Fe National Forest (New Mexico).

Plate 31 (right). Rocky mountain maple, Cañada Bonito, Santa Fe National Forest (New Mexico).

Plate 32 (below). Spearfish Canyon in autumn, Black Hills National Forest (South Dakota).

Plate 33. Red columbine, Black Hills National Forest (South Dakota).

Plate 35. Firecracker honeysuckle, Davy Crockett National Forest (Texas).

Plate 34. Obedience plant, Angelina National Forest (Texas).

Plate 36. Blue sage, Davy Crockett National Forest (Texas).

Plate 37. Ramshorn Mountain, Hoback Canyon, Bridger-Teton National Forest (Wyoming).

Plate 38. Mountain bluebell, Periodic Spring, Bridger-Teton National Forest (Wyoming).

Plate 39. Lichen along the Continental Divide, Medicine Bow National Forest (Wyoming).

Plate 40. Vedauwoo Rocks, Medicine Bow National Forest (Wyoming).

Plate 41. Brooks Lake, Absaroka Range, Shoshone National Forest (Wyoming).

Indian Creek

The most vivid memories of my boyhood trips to the Missouri Ozarks, across the Mississippi River from my southern Illinois home, are of the crystal-clear rivers and streams that flowed in rocky beds. Illinois terrain is often covered by thick layers of wind-blown soil, called loess, which slips into that region's streams, muddying the waters and forming a silty sediment. In contrast, erosion along an Ozark stream causes pebbles and boulders to tumble in, with only a minimum of soil. One such sparkling stream in a perfect hideaway is Indian Creek, which originates near the community of Willow Springs in south-central Missouri and meanders southwestward before entering the North Fork of the White River. The creek flows through private property and public forest land within the boundaries of Mark Twain National Forest.

Through countless centuries following the uplift of the Ozark Mountains 400 million years ago, Indian Creek has carved a twisting, narrow defile through layers of sandstone and dolomite, leaving 200-foot cliffs on either side. From the northern bluff tops, down across the creek, and up to the summits of the south bluffs, one can pass through nine plant communities in less than 600 feet.

The tops of the north cliffs are exposed to intense sunlight throughout the year. As a result, the soil is dry, and only a scrub forest occupies this desert-like habitat. The trees that are able to survive here—blackjack oak, post oak, and red cedar—are often gnarled and usually short statured, although they may be more than 100 years old. The two oaks have a heavy waxy coating, or cuticle, on their leaves that prevents excessive moisture loss, and their roots penetrate deeply into the crevices in the cliff. The leaves of the red cedar are tiny needles and scales that expose a minuscule surface to the sun. The only shrub here is the farkleberry, which has waxy, leathery leaves. Ground-level herbs are sparsely scattered, and there are many patches of bare rock.

Here and there on the north bluff tops are areas apparently too dry for any trees, even oaks and cedars. These glades, or barrens, contain prairie plants. Little bluestem and the six-foot-tall gama grass are common, as are pale coneflower and round-headed bush clover. The grasses depend on extensive root systems to survive, whereas the coneflower and bush clover have hairy leaves that inhibit excessive moisture loss.

From the north bluff tops, the hike down the south-facing slope toward Indian Creek takes you through a dry, upland forest. The angle of the slope moderates the impact of the summer sun's rays, but the plants still struggle to obtain enough moisture. Black hickories and chinquapin oaks are the dominant trees, and they grow taller and straighter than the oaks on the bluff top. Pink azaleas and fragrant sumacs share the shrub layer, and the understory plants, less scattered than those on the bluff top, include butterfly pea and an assortment of asters and goldenrods. The greater concentration of vegetation on the slope results in a swifter buildup of soil, as leaves fall and plants eventually die.

Toward the bottom of the slope, the intermittent shade cast by the opposing cliff allows a more moist woods to prevail. Red maples, Ohio buckeyes, basswoods, and bitternut hickories are able to grow here, sometimes more than 50 feet tall. Their crowns touch completely, closing the canopy and shading the understory. Spicebushes, bladdernuts, and pawpaws in the shrub layer add more shade. The abundant wildflowers beneath generally have thin leaves that permit each chlorophyll-bearing cell to receive enough of the minimal sunlight.

Most of the south-facing slopes flatten out into a terrace a few feet wide as they approach the creek. Here, sycamores, box elders, and hackberries grow, trees that tolerate annual flooding in permeable soil that does not remain wet for long periods. The terrace floor is colored yellow and white in early spring by the blossoms of common butterweed and white bulbous cress. During summer, dense colonies of touch-me-nots grow shoulder high.

The streambed itself is filled with gravel, cobbles, and boulders. In some places, the stream flows through narrow channels; in others, over wide, level beds of rock. Woody plants that often grow in the creek bottom, where they are continuously washed by the flowing stream, are the Ozark witch hazel, Ward's willow, smooth alder, and ninebark. Biologist Max Hutchison has observed damage on the trunks of many of these plants, where boulders slammed against them during torrential rains.

In many places, the cliff face on the south side of Indian Creek abuts the water's edge. The limestone rock immediately above the stream is concealed by overlapping layers of mosses and liverworts, and occasional delicate-leaved ferns, such as the southern maidenhair, gracefully arch over the water. In this zone grow some of the rarest flowering plants of the southern Missouri Ozarks—sullivantia, grass-of-Parnassus, white camas (of the lily family), barren strawberry, and lady's-slipper orchid.

Elsewhere the slope on the south side of the creek is less precipitous. Because this slope is never exposed to the unrelenting rays of the afternoon sun, moisture is abundant in the moderately deep soil. White oaks, slippery elms, and persimmons are the common trees, over a midlevel of flowering dogwoods and black haws. Scattered herbs include mayapple, Indian physic, and bottlebrush grass.

Finally, at the summits of the south bluffs are areas that are open to the sunlight and therefore dry, with poor soils. Shortleaf pines, black and red oaks, and black gums are common, as is sassafras. Somewhat of a surprise, because of the seemingly hostile environment, are three small kinds of fragile-looking orchids that grow in the rocky, acidic soil beneath the pines and oaks: the rattlesnake plantain orchid, small adder's mouth orchid, and twayblade orchid.

Kaintuck Hollow

When early nineteenth-century pioneers from Kentucky and adjacent southeastern states made their way into the Ozark Mountains of southern Missouri, the densely forested valleys sometimes reminded them of their prior homesteads in the Appalachians. One group of these settlers, mesmerized by the clear-flowing, spring-fed waters of a stream that ran through hills teeming with wild turkeys and other game, called their new home Kaintuck Hollow, because it looked so much like the land they had left. Most traces of these early inhabitants are gone, except for a few weathered tombstones, but the natural features and the name persist, now as part of the Mark Twain National Forest.

Beneath hills covered by shagbark hickories and white, black, northern red, and scarlet oaks, a tributary of Mill Creek runs northwestward through Kaintuck Hollow. The springs that feed it are often nearly choked with watercress, whereas the stream itself is lined with ninebark, witch hazel, and Carolina willow. A marsh at the northern end of the valley provides suitable habitat for small alder, buttonbush, cattails, the purple-flowered joe-pye weed, and other wetland plants. An attraction in the middle of the hollow is a land bridge, the result of a natural tunnel eroded through the dolomite bedrock. The tunnel is 175 feet long, 30 feet wide, and 10 feet high at its larger end, narrowing to 20 feet wide and six feet high.

West of the stream, trails maintained by the Forest Service wind through the upland forest and past rocky slopes that support a remarkable array of lichens. Lichens consist of an alga and a fungus growing together, each deriving benefits from the other. The alga, which contains the green pigment chlorophyll, is able to carry out photosynthesis, producing nutrients that the fungus needs but is unable to manufacture. To absorb the nutrients, the fungus penetrates the algal cells but does not seem to harm them. For its part, the fungus provides firm attachment to a rock, wood, or soil surface and protects the alga from desiccation, mechanical injury, and perhaps heat and cold.

The complex makeup of lichens is evident only under a microscope (the Swedish doctor Eric Acharius, in 1803, was apparently the first person to describe their structure). The fungal part consists of colorless cells strung together end to end in threadlike strands. The algal portion, consisting of single or clumped cells, is green or blue green and is either surrounded by the strands or sandwiched between layers of them. Lichen authority Mason Hale noted that two different kinds of algae may be present in some lichens. A lichen species is named for its fungus. In nature, the fungus never seems to be without its algal associate, although the alga may be found alone. Laboratory studies show that the fungus also can live separately, taking on a different form.

Lichens usually propagate vegetatively: the reproductive bodies are often nothing more than clumps of a few algal cells enclosed in a few strands of fungus, which ultimately erupt through, or break away from, the surface of the lichen, essentially forming a clone of the original plant. The fungal component may also reproduce sexually, however. The British soldier lichen, for example, takes its name from the red reproductive bodies that tip the ends of some of the lichen branches. These tips release spores that develop into new fungi, or sometimes the tips themselves break away to form new plants.

Four types of lichens include perhaps 20,000 species. Crustose lichens consist of a thin crust on rocks or on the bark of trees. They are often etched into the surface on which they grow. Some live in Antarctica, farther south than any other plants. Foliose lichens also usually lie flat on rocks or tree twigs or bark, but they are attached by means of special structures known as rhizoids. Fruticose lichens, whose bodies are upright and branched, resemble miniature shrubs. These grow not only on rocks and trees but also directly on soil, as does the fourth type, leprose lichens, which look like powders.

Many lichens, particularly crustose ones, grow from the center outward, in a circular pattern. The rate of growth is very slow, usually between one-two-hundredth-inch and one-fourth-inch per year. Some individual plants may be several thousand years old (one in Swedish Lapland was estimated to be 9,000 years old). Accordingly, lichens may be used to provide a minimum age for the surface on which they are growing, a rather complex science known as lichenometry.

Lichens grow in dry areas, moist rain forests, wetlands, and even in fresh or salt water. As a group they tolerate a remarkable range of temperatures, from as low as −130 degrees F to as high as 168 degrees F. Often the first organisms to establish themselves on freshly exposed surfaces, including bare rock, lichens play an important ecological role. Those living on rock secrete acidic substances that gradually break down the rock into soil particles, enabling other species of plants to gain a foothold.

Slaughter Sink

The Missouri Ozarks consist largely of porous limestone and dolomite over-lain by stronger rock, such as sandstone. In many places beneath the surface, water has dissolved the limestone or dolomite, creating caves and passages, such as Meramec Caverns and Onondaga Cave. When a hollow collapses, the overlying sandstone falls into it, leaving a basin known as a sink. Hundreds of sinks of various sizes are scattered throughout the Ozarks. A spectacular one is Slaughter Sink, which lies in Mark Twain National Forest, along the northern edge of the Ozark Mountains.

Slaughter Sink is nearly one-fourth-mile wide and 175 feet deep; its steep sides are dolomite. Overlooking the western rim is a sandstone promontory flanked by a dolomite pinnacle. Known as Chimney Rock, the 30-foot-tall pinnacle is perched on a 15-foot pedestal.

The floor of Slaughter Sink is nearly flat, except for a secondary, small sink at its northern end. When the basin floods after heavy rains, the water drains through this small sink, then meanders for nearly a mile in a subterranean stream until it emerges as bubbling Boiling Springs near the Gasconade River. The spring also delivers water from other sinks in the vicinity, pour-ing 42 million gallons into the river each day.

The hilly terrain that surrounds Slaughter Sink supports an upland forest of black, red, post, and blackjack oaks and red cedar. Flowering dogwood and redbud are common smaller species, providing colorful blossoms in early May. Sandstone boulders are strewn across the forest floor.

On a dry, west-facing slope above the sink is a two-acre dolomite glade, an open area sprinkled with limestone and chert pebbles. The only woody species that survive here are some red cedars, hop hornbeam, and a few shrubby, white-flowered ninebarks, members of the rose family. The glade is dominated instead by prairie grasses (big and little bluestem, side-oats grama, and switch grass) and prairie wildflowers (blazing star, wild petunia, Missouri coneflower, rosinweed, false boneset, horsetail milkweed, and pur-ple prairie clover).

A narrow strip of woods with red maple, black oak, blackjack oak, Car-olina buckthorn, and red cedar grows around the lip of Slaughter Sink, a rel-atively dry zone because of exposure to the afternoon sun. In early March, before the leaves come out, white flowers with five narrow petals adorn the branches of small serviceberry trees. Summer wildflowers include Indian physic, bee balm, yellow lousewort, and St.-Andrew's-cross. Broomrape, a four-inch-tall plant that produces a single, inch-long tubular white flower, is unusual among flowering species because it lacks leaves and chlorophyll. It obtains nutrients by attaching itself with rootlike threads to the roots of var-ious other plants.

The sink's steep dolomite walls sustain a variety of plants that lodge in the crevices of the rock. Most are species restricted to dolomite or limestone. They include a number of ferns—two kinds of purple cliffbrakes, dwarf lip fern, black spleenwort, and bulblet bladder fern. Near the base of the sink's walls, a terrace has built up from material washed down the slope. This terrace supports a forest of red and chestnut oaks, basswood, sugar maple, box elder, and black walnut. Spicebush and bladdernut are the dominant shrubs, and maidenhair fern, jack-in-the-pulpit, and yellow lady's-slipper orchid are among the wildflowers.

Most of the 40-acre floor of the sink is covered by vines and wildflowers, with scattered silver maples, black willows, and persimmons. The vines include summer grape, greenbrier, raccoon grape, moonseed, trumpet creeper, and poison ivy, and the principal wildflowers are spotted spurge, prickly sida, and false nettle. The most striking wildflower is green-dragon, which has an eight-inch-tall spike protruding from its compact flower.

Slaughter Sink is home to the extremely rare Salem cave crayfish, known only from a few caverns in the Ozarks. It spends most of its life in subterranean areas, and as a result of its adaptation, it is white and blind. Crayfish may be found in nearly every freshwater habitat in North America, from cold northern lakes to warm southern swamps, from water-saturated fields to fast-flowing streams. But owing to its secretive lifestyle, the Salem cave crayfish found in Slaughter Sink has seldom been observed.

NATIONAL FORESTS IN MONTANA

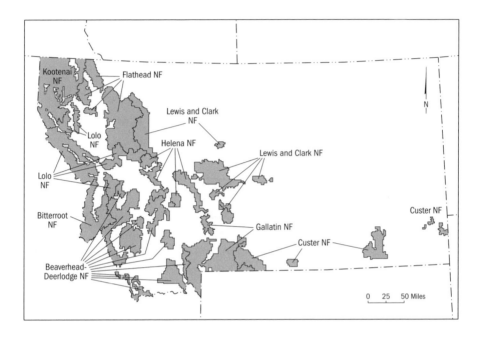

Until 1996, Montana had nine national forests, but in that year Beaverhead and Deerlodge were combined into one administrative unit. These forests are in Region 1 of the U.S. Forest Service.

Beaverhead-Deerlodge National Forest

SIZE AND LOCATION: Approximately 3.3 million acres in western and south-western Montana. Major access routes are Interstates 15 and 90, U.S. Highways 93 and 278, and State Routes 1, 29, 38, 41, 43, 55, 69, 87, and 287. District Ranger Stations: Butte, Deer Lodge, Dillon, Ennis, Philipsburg, Whitehall, Wisdom, and Wise River. Forest Supervisor's Office: 420 Barrett Street, Dillon, MT 59725, www.fs.fed.us/r1/b-d.

SPECIAL FACILITIES: Winter sports areas; boat ramps; swimming beaches.

SPECIAL ATTRACTION: Pioneer Mountains National Scenic Byway.

WILDERNESS AREAS: Anaconda-Pintler (157,993 acres, partly in the Bitterroot National Forest); Lee Metcalf (254,288 acres, partly in the Gallatin National Forest and partly on Bureau of Land Management land).

Until 1996, the Beaverhead and Deerlodge National Forests were separate entities, but today, they have been combined into one huge forest that spreads over 3.3 million acres in western and southwestern Montana.

The circle of mountains surrounding the town of Dillon, Montana, and encompassing much of the old Beaverhead National Forest has a major influence on the climate and vegetation of the area. The crest of the Madison Range forms the eastern boundary of the Beaverhead portion of the forest, and the Tobacco Root Range is at the north. Along the northwest, west, and southwest boundaries of the Beaverhead portion are the Bitterroot Mountains that contain the Continental Divide. Southeast of Dillon is the Gravelly Range. Between the city of Butte and the Bitterroot Mountains to the west are the impressive Pioneer Mountains.

With elevations ranging from 5,200 feet north of the Madison River near McAllister to 11,316-foot Hilgard Peak, the original Beaverhead National Forest displays a wide variety of vegetational communities. Sagebrush plains at the lower elevations quickly give rise to forests of lodgepole pines, Douglas firs, and Engelmann spruces as the elevation increases. Along the Continental Divide, forests of alpine larches, whitebark pines, and mountain heathers prevail.

Several species of large mammals live in the forest, including a sizable population of grizzly bears in the Madison Range. Other large mammals include gray wolves (fig. 16), white-tailed deer, mule deer, elk, moose, mountain goats, bighorn sheep, and black bears. Golden and bald eagles often soar overhead.

The national forest is a fisherman's paradise. Game fish eagerly sought are

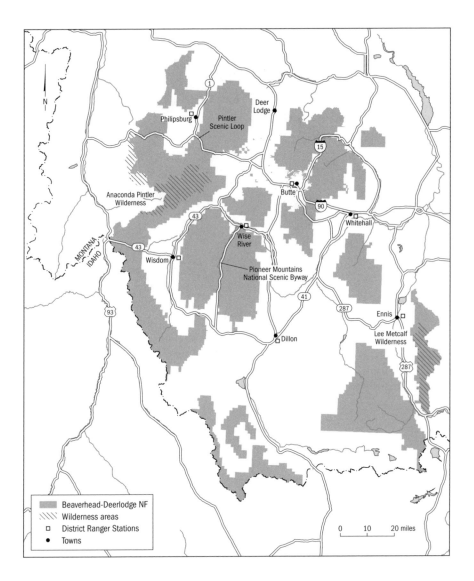

mountain whitefish (pl. 22), golden trout, cutthroat trout, rainbow trout, brown trout, brook trout, lake trout, arctic grayling, and burbot.

When you hike some of the trails in the Beaverhead part of the forest, you may be following the same route that Lewis and Clark traveled 200 years ago. Lemhi Pass, where Lewis and Clark crossed the Continental Divide, is on the border between the Beaverhead-Deerlodge and Salmon National Forests. The Sacagawea Memorial is nearby. To reach Lemhi Pass, take the Clark Canyon Reservoir exit off of Interstate 15 about 20 miles south of Dillon and proceed for about 25 miles west on County Road 324 and then another 10 miles on Forest Road 3909. A tiny campground sits near the pass. Lemhi Pass

Figure 16. At a distance, a coyote (top) might be confused with a wolf (bottom). But upon closer inspection, the differences are many.

is on the crest of the Beaverhead Mountains in the Bitterroot Range. On the east side of the Continental Divide, which forms the border between Montana and Idaho, the national forest part of the Bitterroot Range is only a few miles wide. Few points of interest exist in this narrow region except for the Continental Divide National Recreation Trail and a number of high mountain lakes, the deepest being Morrison Lake. After the Bitterroots make a broad curve to the east, with the Continental Divide still on the crest, the Beaverhead part of the national forest lies to the north of the Continental Divide and the Targhee National Forest lies to the south.

North of Lemhi Pass, however, a huge section of the national forest offers many opportunities to explore. The only east-to-west road that crosses this part of the national forest is State Route 43, which climbs from the community of Wisdom in the Big Hole Valley to Chief Joseph Pass on the Continental Divide and then into Idaho where it connects with U.S. Highway 93. Most of this stretch of highway is part of the designated Nez Perce National Historic Trail and the Lewis and Clark National Historic Trail. The pass has a breathtaking observation point, and you may join the Continental Divide National Recreation Trail here if you desire. Less than one mile north is the historic Lost Trail Pass. The final few miles of State Route 43 to Chief Joseph Pass are alongside Joseph Creek. Six miles east of Chief Joseph Pass, State

Route 43 crosses May Creek. The May Creek National Recreation Trail follows the creek westward to the Continental Divide. Miner Lake and Twin Lakes are popular fishing areas between Chief Joseph Pass and Lemhi Pass. Twin Lakes are one of the deepest in the forest, with depths up to 72 feet.

The Anaconda Range forms the northwest border of the Beaverhead-Deerlodge National Forest, with a part of the range in the Anaconda-Pintler Wilderness. This area, which has been extensively glaciated, has glacial cirques, broad U-shaped valleys, glacial moraines, and rugged peaks. Numerous streams and lakes are filled with snowmelt from peaks above timberline. The highest peaks in the wilderness are in the Beaverhead portion, with West Goat Peak, at 10,793 feet, 394 feet higher than East Goat Peak. Just east of the Continental Divide is Sawed Cabin on Pintler Creek near Sawed Cabin Lake. Martin "Seven Day" Johnson, a mountain man with legendary strength, built the cabin. This area can be reached via a hiking trail from a trailhead just north of the Pintler Lake Campground. Near the trailhead is Pintler Falls. The trail follows Pintler Creek to the Continental Divide, first passing through tranquil Pintler Meadows. Although the growing season along the Continental Divide is very short because of the brief summer season, attractive wildflowers include pasqueflower, alpine buttercup, shooting stars, and spring beauties. Mussigbrod Lake, the national forest's largest with 102 acres of sparkling water, is just outside the wilderness area.

The Pioneer Mountains comprise a large mass of land in the national forest between Interstate 15 to the east and Big Hole Valley to the west. The Pioneer Mountains National Scenic Byway passes through these mountains and is a grand way to see them. The scenic byway begins in the north at the community of Wise River and travels the entire length of the Pioneer Mountain Range. As you begin your ascent on the scenic byway from Wise River, you are in a dry vegetation zone of sage, but as the road climbs, coniferous forests prevail on these mountain slopes. For 26 miles, the scenic byway is paved as it makes its way to Crystal Park. Where the road crosses Stine Creek, you may join the 35-mile Pioneer Mountain Loop National Recreation Trail that proceeds into the West Pioneer Mountains. Tall mountain peaks, several of them more than 9,500 feet high, may be seen to the east of the scenic byway. The Fourth of July and Boulder Creek Campgrounds and the Pettingill parking area and trailhead have shorter and less strenuous hiking trails. At Gold Creek, the site of a past gold strike, there is a hiking trail into the East Pioneer Mountains. As the road continues to ascend, there is a splendid view from Lacy View and Grand Vista. Between Lacy View and Grand Vista is a side road along Wyman Creek that ends at lovely Odell Meadows. A four-mile hiking trail from the meadow follows Odell Creek to Odell Lake and a significant wetland. Just before ending at Odell Lake, another trail goes to Skull Creek Meadows Research Natural Area. Skull Creek Meadows are in a basin at the headwaters of Skull Creek and contain two magnificent fens that have a great abundance of peat and other mosses. Three rare sedges and the delicate hair grass have been recorded from the fens that are separated by a forest of lodgepole pines.

A side road off the scenic byway comes to Mono Creek Campground and, four miles beyond, are the ghost town of Coolidge and the abandoned Elkhorn Silver Mine. Crystal Park is named for the abundance of quartz crystals in the area, including purple crystals, violet crystals, and amethyst scepters.

Beyond Crystal Park, the road is unpaved as it descends to Elkhorn Hot Springs and the Maverick Mountain Ski Area. South of the ski area the scenic byway leaves the national forest and enters the sage-filled Grasshopper Valley to the village of Polaris and then to County Road 278 where the scenic byway ends.

The historic Charcoal Kilns are on the east side of the Pioneer Mountains. To reach them, take County Road 187 westward off of Interstate 15. This scenic road follows Trapper Creek to the settlement of Glendale and then on to the kilns and Canyon Creek Campground just beyond.

Several parts of the Beaverhead part of the forest lie east of Interstate 15 and the town of Dillon and southeast of Butte. A part of the Tobacco Root Range is in this region and contains several mountain peaks greater than 10,000 feet elevation, numerous high mountain lakes, and several abandoned mine sites. No paved roads cross this part of the Tobacco Root Range either from north to south or from east to west, but there are a few four-wheel-drive roads within this area. From Sheridan on State Route 287, there is a county road into the national forest to Mill Creek, Balanced Rock, and Branham Lakes campgrounds, and from Harrison on U.S. Highway 287, there is a lengthy road to Potosi Campground in the northeastern corner of this segment of the national forest.

Of biological interest is a rare Rocky Mountain peatland bog habitat along Leonard Creek a few miles west of McAllister. The Leonard Creek fen is at the headwaters of a tributary to Leonard Creek and is significant in that it is an organic floating mat that has developed over a former open-water basin. A floating mat, also known as a quaking bog, consists of unconsolidated peat that is extremely spongy underfoot and very dangerous for humans to be on. Most of the mat is flat, consisting of sphagnum, or peat moss, but occasional small mounds or hummocks occur here and there. The flat areas, known as moss lawns, also contain few-flowered spikerush and gray sedge. Standing water at the edge of the mat contains bog bean (a member of the gentian family), spatterdock (in the water lily family), bladder sedge, and a bladderwort. The upland forest adjacent to the mat is dominated by lodgepole pine, Engelmann spruce, whitebark pine, and birchleaf spiraea.

South of the Tobacco Root Range is the Gravelly Range, most of it in the Beaverhead portion of the forest. The Forest Service has prepared an auto tour known as the Gravelly Range Tour that penetrates some of the range. A brochure may be picked up at the district ranger station that explains the 24 numbered stops. Just east of stop number four, the Cottonwood Campground, is the Cottonwood Research Natural Area. This area has been set aside to preserve ridges and dry slopes covered by Idaho fescue and bluebunch wheatgrass and associated moist areas of juniper, sagebrush, and a Douglas fir community. The 128 acres have been enclosed by a fence for 55 years and are excellent undisturbed examples of these plant communities. Beneath the Douglas firs, the forest floor has a parklike appearance with the occurrence of pinegrass and basin wild rye.

The Gravelly Range Tour goes between 10,542-foot Black Butte one mile to the west and 9,880-foot Cave Mountain, 3.5 miles to the east. Cave Mountain is the site of another pristine proposed Research Natural Area where there is an excellent example of an alpine grassland dominated by Idaho fescue and interrupted by whitebark pine, subalpine fir, Engelmann spruce, and Douglas fir. Shrubby components of the area include shrubby cinquefoil, snowy willow, Wolf's willow, alpine gooseberry, shrubby goldenweed, and white mountain avens.

At the east side of the Gravelly Range are several lakes that attract fishermen and other water enthusiasts. From north to south are Wolf Lake, Cliff Lake, Hidden Lake, and Elk Lake. Extending for two miles west of the western edge of Cliff Lake is Cliff Lake Research Natural Area, one of the largest Research Natural Areas in any national forest. Consisting of 2,291 acres, Cliff Lake Research Natural Area supports forests of lodgepole pines and Douglas firs estimated to be more than 200 years old. Other forest types include Engelmann spruce and subalpine fir. Nearly half of the Research Natural Area is nonforested sagebrush habitat and foothills prairies, the latter dominated by Idaho fescue. Because of the ruggedness of the area, it has not suffered much from cattle grazing.

The Madison Range forms the easternmost unit of the Beaverhead part of the national forest, and most of it is in the Lee Metcalf Wilderness that is shared with the Gallatin National Forest and the Bureau of Land Management. This wilderness consists of four units. The Bureau of Land Management manages the Bear Trap Canyon Unit along the Madison River; it was the Bureau of Land Management's first designated wilderness. The Gallatin National Forest manages the Monument Mountain Unit, which lies on the northwestern boundary of Yellowstone National Park, and most of the Spanish Peaks Unit. The 141,000 acres of the Taylor-Hilgard Unit are managed entirely by the Beaverhead-Deerlodge National Forest. This unit contains Hilgard Peak, the highest point in the forest at 11,316 feet, and well-named landmarks such as Sphinx Mountain and Sawtooth Ridge.

Mountains surround the city of Butte, Montana, and many of these mountains are in the old Deerlodge part of the national forest. The northern end of the Tobacco Root Range lies to the southeast, and the Highland Mountains are due south. To the west is the Flint Creek Range, and the Anaconda Mountains are to the southwest. At the extreme western edge of the national forest are the Sapphire Mountains.

The Tobacco Root Range has several peaks that are in excess of 10,000 feet, with Hollister Mountain the highest at 10,604 feet. Lost Cabin Lake is nestled at the foot of Noble Peak on the north side within the national forest. Hiking trails enter this mountainous area from the end of a long back road southward out of the community of Cardwell at exit 256 on Interstate 90. This road parallels South Boulder Creek into the national forest.

At the northeast corner of the Deerlodge part of the forest is a small section of the forest with the ghost town of Elkhorn right in the center. A 12-mile-long gravel road north of State Route 69 brings you to the remains of this once-thriving town with a few brown-sided buildings still standing. A four-wheel-drive road east of Elkhorn crosses Radersburg Pass and the southern end of the Elkhorn Mountains.

The Boulder River cuts across the Beaverhead-Deerlodge National Forest a few miles north of Butte. The Forest Service maintains several campgrounds along the river. Interstate 15 cuts diagonally across this part of the national forest. From exit 138 on the interstate, a forest road leads to three campgrounds with nature trails. Hiking any of these trails better familiarizes you with the Beaverhead-Deerlodge National Forest. Also north of Butte in Elk Park is the Sheepshead Recreation Area, which offers accessible facilities for picnicking, fishing, and large groups. Reservations can be made through the Butte Ranger District.

Delmoe Lake is a large lake about six miles east of Butte and reached by a crooked dirt road from exit 283 off of Interstate 90. Pipestone Rock south of Delmoe Lake has rocks used by early inhabitants for making various implements. State Route 41 climbs over Pipestone Pass and around the southern side of Toll Mountain where there is a pleasant campground. The Highland Mountains begin about eight miles south of Butte. The mining ghost town of Highland City is worth a visit, but it can be reached only by a long circuitous route from Interstate 15, exit 93, at Melrose.

The remainder of the Deerlodge part of the national forest lies west of

Interstate 90. The Flint Creek Mountains are between Clark's Fork Valley to the east and Philipsburg Valley to the west. These mountains have numerous peaks, countless lakes, tumbling mountain streams, occasional waterfalls, and many mine ruins. State Route 1 outlines the southern and western boundaries of the Flint Creek Mountains and is part of the Pintler Scenic Loop. Large lakes such as Georgetown Lake, Echo Lake, and Fred Burr Lake have the usual Forest Service amenities. The dirt road to Fred Burr Lake follows North Fork Flint Creek over Fred Burr Pass. The Discovery Basin Ski Area south of Philipsburg is a popular winter destination.

The John Long Mountains protrude into the northern end of Philipsburg Valley. Although most of these mountains are in the national forest, there are no developed Forest Service facilities.

The Anaconda Mountains cut across the southern end of Philipsburg Valley. Much of this range is in the Anaconda-Pintler Wilderness that is shared with the Bitterroot National Forest.

The Sapphire Mountains are at the western edge of the Beaverhead-Deerlodge National Forest. State Route 38 crosses these mountains and is one of the prettiest roads in the forest. Despite being a state route, it is not paved, nor is it kept open during winter. The road climbs steadily out of the Philipsburg Valley to Skalkaho Pass at the boundary between the Beaverhead-Deerlodge and Bitterroot National Forests. You may wish to experience high mountain camping at Crystal Creek or Mud Lake campgrounds just before reaching the pass. About nine miles south of Skalkaho Pass, and reached only by trail, is the Sapphire Divide Research Natural Area. This extremely scenic region contains Fox Peak and Congdon Peak as well as one of the best stands of alpine larches to be found anywhere. Although there are occasional pure stands of alpine larch, other trees such as whitebark pine, Engelmann spruce, and subalpine fir are often interspersed. An interesting shrubby flora exists in the Research Natural Area, with a heavy emphasis in the heath family. This family is represented by Labrador tea, rusty menziesia, mountain heath, white azalea, mountain blueberry, and wintergreen.

One other road worth taking is County Road 348 west out of Philipsburg. This road also heads to the mountains. At Squaw Rock along Rock Creek, there is a Forest Service campground before the road enters the Lolo National Forest.

Bitterroot National Forest

SIZE AND LOCATION: Approximately 1.6 million acres in western Montana and eastern Idaho, south of Missoula and on both sides of Hamilton. Major access routes are Interstate 90, U.S. Highway 93, and Montana State Routes 38, 472 (East Fork Road), and 473. District Ranger Stations: Darby, Stevensville, and Sula. Forest Supervisor's Office: 1801 N. First Street, Hamilton, MT 59890, www.fs.fed.us/r1/bitterroot.

SPECIAL FACILITIES: Winter sports areas; boat ramps.

SPECIAL ATTRACTIONS: Nez Perce Auto Trail; Sula Deer, Elk, and Bighorn Driving Tour; Salmon National Wild and Scenic River; Selway National Wild and Scenic River.

WILDERNESS AREAS: Selway-Bitterroot (1,340,460 acres, partly in the Clearwater, Lolo, and Nez Perce National Forests); Anaconda-Pintler (157,993 acres, partly in the Beaverhead and Deerlodge National Forests); Frank Church River of No Return (2,300,000 acres, partly in the Boise, Challis, Nez Perce, Payette, and Salmon National Forests).

The Bitterroot National Forest and the Bitterroot Mountains are named for the bitterroot plant, a small wildflower with gorgeous, oversized flowers (pl. 23). With numerous glowing pink petals, the flowers of the bitterroot, which measure up to two inches across, open from late April to July. Bitterroot plants grow on open slopes, and their leaves appear as soon as the snow melts, only to wither before the flowering season is over. The root itself has been a major source of food since native inhabitants lived in the area, and Meriwether Lewis collected it in the Bitterroot Valley in 1806. The scientific name for bitterroot, *Lewisia rediviva*, commemorates this intrepid explorer.

The Bitterroot National Forest begins above the foothills of the Bitterroot River Valley and consists of two mountain ranges, the Bitterroot Mountains on the west side of the valley and the Sapphire Mountains on the east. Parts of the Selway National Wild and Scenic River and the Salmon National Wild and Scenic River are in the national forest. Elevations along the Salmon River are about 2,200 feet, about 3,200 feet in the Bitterroot Valley foothills, and at 10,157 feet on Trappers Peak. Nearly half of the acreage in the Bitterroot National Forest is in the three wilderness areas of the national forest. As you explore the forest, be on the lookout for mule deer, white-tailed deer, moose, elk, black bear, mountain lion, mountain goat, and bighorn sheep.

Roads are scarce within the national forest, partly because of the ruggedness of the topography and partly because so much of the area is in wilderness. The roads that do penetrate the forest permit you to see much of the national forest, with occasional short hikes available, or to access the longer and more rugged trails that lead into the wilderness areas. This region has long been the home area of earlier Salish Indians, of Chief Joseph and the Nez Perce, and of the Lewis and Clark Expeditions of 1805 and 1806.

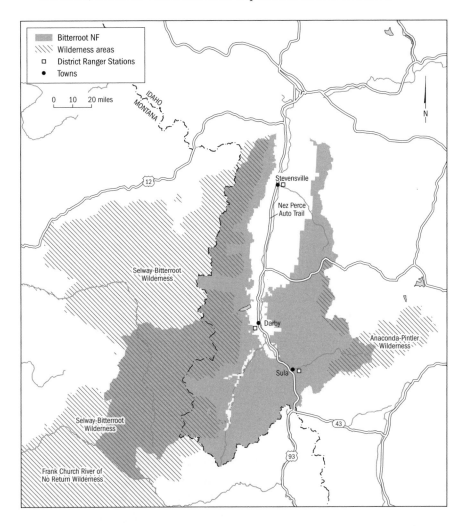

The Nez Perce Auto Trail begins off of State Route 473, about 20 miles southwest of Darby at the West Fork Ranger Station, and cuts into the heart of the Bitterroot Mountains, eventually going through a narrow corridor that separates the Selway-Bitterroot Wilderness from the Frank Church River of No Return Wilderness. This remarkable road eventually enters the Payette National Forest in Idaho or, if you desire, you may retrace your route at any time back to Darby. From the West Fork Ranger Station, the road follows Nez Perce Fork, passing the Fales Flat Campground that is across the river from Peyton Rock, a local landmark. At the campground notice several ponderosa pines where Indians of the past stripped away vertical shreds of outer bark to obtain the sweet cambium layer for food. Look for these scars about three feet above the base of the older trees. The scars may be as much as eight feet long and two feet wide. Prior to reaching Fales Flat Campground, watch for the remains of old bridges where the Nez Perce Trail crosses Little West Fork and Watchtower Creek.

For 15 miles the roadless area south of Fales Flat Campground is being considered as the Blue Joint Wilderness. This 65,370-acre wild area includes the narrow, rocky Razorback Ridge, large open meadows, and interesting rock formations such as Castle Rock and a natural arch just east of the confluence of Blue Joint and Jack the Ripper Creeks. Castle Rock is a volcanic plug with an elevation of 7,922 feet. Forests in this area range from Douglas fir and ponderosa pine at lower elevations to lodgepole pine at middle elevations to whitebark pine on the rocky ridgetops.

Beyond the Fales Flat Campground, the road makes a tortuous climb to 6,598-foot Nez Perce Pass. From the pass westward, the entire trail, no longer paved, is squeezed into a narrow corridor between the Selway-Bitterroot Wilderness to the north and the Frank Church River of No Return Wilderness to the south. The road descends from the pass and crosses several creeks before coming to its first side road, a rough 11-mile dirt road south to Hells Half Acre Mountain. The road twists over Hells Half Acre Saddle and Vance Creek Saddle before ending near Hells Half Acre Spring. For those wishing to do long-range hiking, a rugged trail to the south passes Devil's Point and climbs over Storm Ridge and eventually to the Selway River.

The Nez Perce Auto Trail past the side road follows Deep Creek, coming alongside a formation known as Old Warriors Face before reaching the Deep Creek Campground. Near the campground are the remains of a Civilian Conservation Corps camp that includes a stone water fountain, an incinerator, and several rock foundations. The bridge across Deep Creek is historic because its beautiful arched natural stone construction was accomplished by Lithuanian stone masons who assisted the Civilian Conservation Corps personnel. A half-mile-long road south of the Nez Perce Auto Trail leads to the Magruder Guard Station, a log building with beautiful interior work created by Ole Tangen, a Forest Service employee during the 1930s.

From the Magruder Ranger Building, the Nez Perce Auto Trail follows the wild and scenic Selway River, which has come from the south. After about four miles, the auto trail arrives at the Magruder Crossing Campground where the East Fork Creek joins the Selway River. A 12-mile-long side road north from Magruder Crossing to Paradise Campground is a must. The deep Selway River that the Paradise Road follows is lined with large western red cedars. To the west of the road is the Magruder Massacre Site. Lloyd Magruder and a few of his gold-mining friends were en route from Virginia City in 1863 carrying a considerable amount of gold dust. They were joined on their journey by four travelers who, four days later, attacked, robbed, and murdered Magruder and his friends at the site. Another friend of Magruder, Hill Beachy, followed the murderers to San Francisco and brought them back to Lewiston, Idaho, where they were tried and found guilty. They were the first people to be hanged legally in Idaho Territory. Several months before his death, Lloyd Magruder had agreed to represent Idaho Territory in the U.S. Congress. Three campgrounds sit along and at the end of the Paradise Road, and a pack bridge crosses the Selway River for a hiking trail to Spot Mountain. The road ends at the Paradise Campground. Those wishing to continue along the Selway River must do it by canoe. The river continues through the Bitterroot National Forest and into the Nez Perce National Forest.

To follow the Nez Perce Auto Trail, you cross the Selway River on a steel pony-truss bridge that was built in 1935 by the Civilian Conservation Corps. The road then proceeds southward along East Fork Creek, still in a very narrow corridor between the wilderness areas. The twisty road climbs over Haystack Saddle, Kim Creek Saddle, and Magruder Saddle before reaching Observation Point and Campground. The observation point is at an elevation of 7,620 feet and permits a good view of much of the southern end of the Bitterroot Mountains, including El Capitan.

The road descends to the old Salmon Mountain Ranger Station site that was established in 1911. The site is now the Salmon Mountain Base Camp, which is a trailhead for hunters during the hunting season. A one-mile trail southward to Salmon Mountain enters the proposed Salmon Mountain Research Natural Area where some of the best specimens of alpine larch in the country may be seen forming large parklike areas. You may also see ribbon forests of whitebark pine on the slopes of Salmon Mountain. Ribbon forests are elongated strips of trees that grow perpendicular to prevailing wind directions. These ribbons of forest alternate with narrow bands of wet subalpine meadows that are referred to as snow glades. Green fescue is the dominant grass in the meadows. Many of the whitebark pines have succumbed to the mountain pine beetle and a blister rust. The gray forms of these dead whitebark pines are referred to as ghost trees. Near the top of 8,944-foot Salmon Mountain are dwarf, gnarly trees that have withstood severe wind for several hundred years yet are only about two feet tall. A lookout tower is on the mountain summit. Look for mountain goats on the upper rocky slopes of the mountain.

The Nez Perce Auto Trail now climbs over Horse Heaven, so named because of the presence of extensive grasslands suitable for grazing. Horse Heaven Cabin is here. Although this one-room log cabin was built near Darby by the Civilian Conservation Corps in 1939, it was disassembled and then rebuilt at Horse Heaven, some 60 miles west of Darby. The Forest Service at the West Fork Ranger Station handles renting the cabin, which accommodates four persons.

The auto trail climbs to Sabe Vista where there are unsurpassable panoramic views. Also from the vista you may see the evidence of the destructive forest fire along Ladder Creek in 1988. From Sabe Vista the road drops 1,500 feet in two miles to Sabe Saddle and the border with the Nez Perce National Forest. One may continue on at this point, or return to Darby.

A few miles northwest of Darby is large Lake Como where there are campgrounds, boat ramps, swimming beaches, horse camps, and the seven-mile-long Lake Como National Recreation Loop Trail that encircles this attractive lake. One mile east of Lake Como along Rock Creek is the Rock Creek Wetland, an example of a Rocky Mountain peat bog unique in that it occurs at a relatively low elevation. The wetland, near the mouth of Rock Creek Canyon, has at its western edge an excellent example of a shrub carr community, a bog-type consisting mostly of wetland shrubs and sedges. Bog birch, a small shrub, is mixed in with lodgepole pine and Engelmann spruce above a substantial growth of bladder sedge, inland sedge, blue jointgrass, and nodding beggar's-ticks. The rare crested fern and depauperate sedge have been found here, as well. Sphagnum, or peat moss, is extensive.

Some of the finest scenery and natural features in the Bitterroot National Forest lie in a five-mile-wide gap between two sections of the Selway-Bitterroot Wilderness west of U.S. Highway 93 about halfway between Darby and Hamilton. Take County Road 76 west of U.S. Highway 93 for two miles until it reaches the Bitterroot National Forest. At this point, the forest road you are now on splits. The right fork takes you on one of the most scenic but crooked eight-mile roads to the Lost Horse Observation Point. Looking westward from the point, you see a beautiful U-shaped, glacier-carved valley in front of you with occasional glimpses of Lost Horse Creek at its base. The observation point is also at the eastern edge of the significant proposed Lower Lost Horse Canyon Research Natural Area. The 1,400 acres of this area preserve several forest types. Along the streams are forests of grand fir, Engelmann spruce, western red cedar, western yew, and black cottonwood. On the mid-slopes are fine open stands of ponderosa pine and Douglas fir, whereas on upper slopes are forests of whitebark pine and subalpine fir. You find such spectacular wildflowers as beargrass (not a true grass), twisted stalk, fawn lily, false bugbane, cow parsley, foamflower, and ovate-leaved trillium. If you take the left fork back at the forest road intersection, you drive through the lower part of the Research Natural Area and then along Lost Horse Creek for 10 miles to the Lost Horse Guard Station. The road forks, with the right fork going to Twin Lakes. To the north of Twin Lakes is a hiking trail over Lost Horse Pass and into the Nez Perce National Forest. The left fork in a mile comes to the Bear Creek Pass Campground. The area west of Twin Lakes and the Lost Horse Campground is the Upper Lost Horse Canyon Research Natural Area. This is a subalpine area of some old-growth forests of subalpine fir, Engelmann spruce, alpine larch, and lodgepole pine.

Four miles west of Hamilton is the Blodgett Canyon Campground and the trailhead for an exciting hike through Blodgett Canyon (pl. 24). After 10 miles, the trail climbs over Blodgett Pass and, after another 1.5 miles, ends at lovely Blodgett Lake. One mile south of the Blodgett Canyon Campground is a trailhead for a four-mile hike along Canyon Creek to Canyon Lake and Wyart Lake, two fine fishing lakes. Just before reaching Canyon Lake, enjoy the tumbling water of Canyon Falls. These trails pass through forests of lodgepole pine, ponderosa pine, Engelmann spruce, and Douglas fir, with an occasional Rocky Mountain maple and two species of juniper. Attractive shrubs include Lewis's mock orange, ninebark, white spiraea, snowberry, buffaloberry, and blue huckleberry. Wildflowers are abundant in season. Particularly attractive are heartleaf arnica, arrowleaf balsamroot, wild hyacinth, elegant mariposa lily, larkspur, leopard lily, yellowbells, Jacob's ladder, bluebells, and a couple of beardtongues.

North of Hamilton are many fine hiking trails that follow creeks east to west across this part of the Selway-Bitterroot Wilderness, including Big Creek Trail to the Big Creek Lakes, a short trail to St. Mary's Peak, Kootenai Creek Trail to three high mountain lakes, and Bass Creek Trail to Bass Lake.

Four miles north of Stevensville is the Charles Waters Campground where there is a physical fitness trail. An extremely crooked mountain dirt road from the campground goes to Bass Creek Overlook for a splendid view down into the creek.

Returning to the West Fork Ranger Station south of Darby, you may wish to go to the nearby Sam Billings Campground where you can hike a 4.5-mile trail to picturesque Boulder Creek Falls in the Selway-Bitterroot Wilderness. From the West Fork Ranger Station, continue southward on State Route 473 instead of taking the Nez Perce Auto Tour. This highway passes several Forest Service campgrounds and Painted Rocks Lake (not in the national forest) to the historic town of Alta and the Alta Ranger Station, the first ranger station in the United States. One mile north of the old station site is the Alta Pine Trail.

Several miles south of Darby, U.S. Highway 93 comes to Spring Gulch

Campground. A Forest Service side road takes you to Warm Springs and Crazy Creek campgrounds. From the latter campground, there is an exceptionally curvy dirt road up to Bear Creek Saddle. From here you may hike two miles to the restored lookout on Medicine Point that may be rented from the Forest Service as a cabin. As U.S. Highway 93 continues southward, it passes Indian Trees Campground where several of the mature ponderosa pines have had vertical strips of bark peeled off long ago for food by the local Indian population. The highway climbs to Lost Trail Pass on the border with both the Salmon and the Beaverhead National Forests. A peatland bog that fills a small glacial depression near the pass is worthy of a visit. Of particular interest in this bog is the uncommon great sundew and the attractive primrose monkey-flower. Western huckleberry is common here.

A section of the Wild and Scenic Salmon River forms part of the southwestern boundary of the Bitterroot National Forest. This section of the river offers white-water challenges at Rainier Rapids, Lantz Rapids, Devil's Teeth Rapids, and Little Devil's Teeth Rapids. The river may be reached from the Salmon National Forest in Idaho by driving to the Corn Creek Campground and hiking where you can put in your canoe and heading northward over Gunbarrel Rapids, or by hiking four miles from the Corn Creek campground to Horse Creek Campground at the edge of the Bitterroot National Forest.

The remainder of the Bitterroot National Forest lies east of U.S. Highway 93. Across the highway from the Indian Trees Campground is the Big Hole Battlefield National Recreation Trail that goes straight up to Gibbons Pass nearly four miles away. This is roughly a part of the same trail that Chief Joseph and the Nez Perce tribe traveled in 1877 in their attempt to reach Canada. Along the trail keep a lookout for scarred Indian trees and wagon ruts.

Two miles north of the Sula Ranger Station on U.S. Highway 93, State Route 472, the East Fork Road, plies into the eastern side of the Bitterroot National Forest as far as it can go. This route is part of the Forest Service's Sula Deer, Elk, and Bighorn Driving Tour, so named because it is a great route for the possibility of viewing wildlife. In winter you are likely to see elk grazing in the grassy meadows, but in summer, they move higher up into the mountains. Watch for bighorn sheep in the canyons.

State Route 472 is paved nearly to the East Fork Guard Station as it follows the Bitterroot River. Several dirt side roads to the south take you to trailheads for ventures into the Anaconda-Pintler Wilderness. One extremely crooked road from near the East Fork Guard Station climbs to McCart Lookout where the tower was restored in 1996 and 1997 for use as a rental cabin.

Three miles south of Hamilton, State Route 38 heads toward the Sapphire Mountains and the Skalkaho Game Refuge. The first side road to the north after entering the Bitterroot National Forest is one of the most crooked dirt roads found anywhere. After many miles and crossing Buck Horn Saddle, the road comes near Gird Point where the old lookout tower has been restored as a rental cabin. Just east of the side road is Centennial Nature Grove where old-growth ponderosa pines tower above Skalkaho Creek. A one-tenth-mile handicap-accessible trail permits visitors to see these beauties. The highway follows along the southeastern border of the Skalkaho Game Preserve, coming to spectacular Skalkaho Falls on Falls Creek. Just east of the falls is a two-mile hiking trail north through the game preserve to the junction with the Easthouse National Recreation Trail, a 23-mile route that circles the northeastern end of the game refuge and ends at Skalkaho Pass on State Route 38. If you choose not to hike these long trails, the highway from Skalkaho Falls climbs to Skalkaho Pass and into the Deerlodge National Forest.

Four miles north of Hamilton, the Willow Creek Road to the east eventually arrives at the Palisade Mountain National Recreation Trail. This scenic six-mile ridge trail has spectacular views of the Bitterroot Mountains before ending at Burnt Fork Lake. On either side of Skalkaho Pass, 350,000 acres of the Sapphire Mountains are being considered for designation as the Sapphire Wilderness.

Southeast of Stevensville, a series of state, county, and Forest Service roads bring the persistent traveler to Sawmill Creek where there is a fine natural area in the mountain foothills. This 155-acre site is a fine example of a bunchgrass prairie at the foot of a mountain slope. The prairie is dominated by bluebunch wheatgrass, Idaho fescue, and rough fescue, with a few scattered ponderosa pines, giving the area a savanna-like appearance. Sagebrush and juniper communities are in the adjacent lower mountain slopes, with Douglas fir forests higher up. Engelmann spruce and aspen are along the upper reaches of Sawmill Creek.

Custer National Forest

SIZE AND LOCATION: Approximately 1.2 million acres across southern Montana east of Cooke City and in several units in northwestern South Dakota. Major access routes are U.S. Highways 85, 212 (Beartooth All American Road), and 310, Montana State Routes 78, 323, 419, and 425, and South Dakota State Routes 7, 20, and 79. District Ranger Stations: Ashland and Red Lodge, Montana, and Camp Crooke, South Dakota. Forest Supervisor's Office: 1310 Main Street, Billings, MT 59105, www.fs.fed.us/r1/custer.

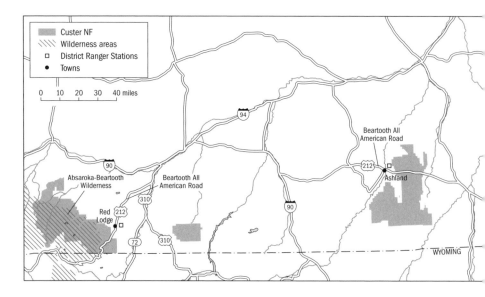

SPECIAL FACILITIES: Winter sports areas.

SPECIAL ATTRACTIONS: Beartooth All American Road; Granite Peak; Grasshopper Glacier; Slim Buttes; Pryor Mountains.

WILDERNESS AREA: Absaroka-Beartooth (943,626 acres, partly in the Gallatin and Shoshone National Forests).

Perhaps no other national forest has as much contrast as the Custer National Forest, which goes from dry prairie areas with conspicuous buttes in northwestern South Dakota to the Beartooth Mountains of southern Montana, one of the most rugged mountain ranges in the country.

A chain of three lofty mountain peaks, all reaching for the sky at elevations above 12,000 feet, is the dominant feature of the Beartooth Mountains at the western edge of the Custer National Forest. Surrounding these peaks are smaller but no less rugged mountains and high plateaus where an interesting array of plant and animal life may be found. Within the Beartooth are active glaciers, hundreds of lakes and waterfalls, narrow canyons, mountain meadows, and curious rock formations. One of the prettiest wildflowers in the Custer National Forest is the bell flower (pl. 25). The best way to get a general picture of the Beartooths is to take the spectacular Beartooth All American Road (U.S. Highway 212), at times a white-knuckle route, from Red Lodge to Cooke City. From Red Lodge to the Wyoming border, the highway is in the Custer National Forest, but where it drops into northern Wyoming, it is in the Shoshone National Forest. When it swings upward to Cooke City, Montana, the highway is in the Gallatin National Forest. The 60-

mile route from Red Lodge to Cooke City brings into view one spectacular vista after another. It circles around the southeastern and southern end of the Absaroka-Beartooth Wilderness. Although the highway was not completed until 1936, it follows the old Sheridan Trail that was staked out in 1882. The Beartooth All American Road enters the Custer National Forest near the Sheridan Campground. For several miles the highway follows Rock Creek to several more campgrounds before it begins a series of wild switchbacks up the very steep mountain slope. After several switchbacks, you should pull out at the Rock Creek Vista for scintillating views as well as for a rest from the tortuous drive. You may see bighorn sheep from the vista. In a few more miles, the Beartooth Highway crosses into Wyoming and into the Shoshone National Forest. (See also Shoshone National Forest.) After 35 miles of unbelievable scenery, the highway turns to the northwest and reenters Montana. Although the area around Cooke City, Montana, is in the Gallatin National Forest, you must go this way to reach two remarkable points of interest at the southern end of the Absaroka-Beartooth Wilderness in the Custer National Forest. From Cooke City, take a rough four-wheel-drive road northward for several miles, passing such picturesque lakes as Round Lake, Long Lake, and Star Lake. Just after Star Lake, the back road crosses into the Custer National Forest but deadends in less than one mile at the border of the wilderness. From here hike northward for four miles, passing Goose Lake, until you come to Iceberg Peak and the fascinating Grasshopper Glacier on the north side of the peak. This glacier, at 11,000 feet, is about one mile long and a half-mile wide and contains thousands of grasshoppers that were frozen in the glacial ice. The grasshoppers, identified as *Melanoplus spretus*, had been extinct for 200 years when they were discovered in the glacier in 1914 by J. P. Kimball, a mining geologist. A few miles east of Grasshopper Glacier are two other glaciers with embedded frozen grasshoppers.

Other accesses to the Absaroka-Beartooth Wilderness are all from the east and the north. A paved road westward out of Red Lodge follows the West Fork of Rock Creek to the Cross Country Ski Area and then to Wild Bill Lake, Basin Campground, the Basin Creek National Recreation Trail, and Cascade Campground. Several trails run into the wilderness from this drainage. North of the Rock Creek Forest Service Work Station is a continuous cliff of white limestone known as The Palisades. These ocean-deposited sedimentary rocks are the result of ancient inland seas followed by a period of mountain upbuilding more than 75 million years ago.

Another road off of State Route 78 follows East Rosebud Creek in a narrow corridor between two lobes of the Absaroka-Beartooth Wilderness. A picnic facility is available at small Sand Dunes on the west side of the creek. The road ends at East Rosebud Lake where there is a campground and several trails into the wilderness. If you continue along the hiking trail southwest of East Rosebud Lake, you are rewarded with several sparkling high mountain lakes as well as Charlie Falls and thunderous Impasse Falls.

Off of State Route 425 is another road through a narrow corridor into the wilderness. This road goes to lovely Emerald Lake, where there is a campground, and West Rosebud Lake. From the end of the road you may hike three miles to Mystic Lake and beyond to Island Lake, all wonderful high mountain lakes.

State Route 419 penetrates still another corridor into the wilderness, following Stillwater Creek to Woodbine Campground at the edge of the Absaroka-Beartooth Wilderness. Several old mines are along this road as well as an active platinum and palladium mine at the Stillwater.

Thirty-five miles east of Red Lodge are the Pryor Mountains, the lower half of which is in the Custer National Forest. The narrow, twisting Crooked Creek Canyon squeezes through high rocky canyon walls in a spectacular setting. A few miles southeast of the canyon is a special area, Lost Water Canyon Research Natural Area, protected by the Forest Service because it contains the easternmost stands of Douglas firs in the country. The vertical cliffs of this canyon consist of ancient Madison limestone. The plateau at the north end of the natural area is so exposed and windswept that the subalpine firs that occur there are stunted. On the drier south slopes of the canyon are stands of limber pine. Many fossils are deposited in the limestone cliffs. The last mile to the Research Natural Area must be hiked.

In the heart of the Pryor Mountains are several ice caves, one at the southeastern corner of Red Pryor Mountain and one in Red Pryor Mountain. The southeastern corner of the Prior Mountains in the Custer National Forest adjoins the Pryor Mountain Wild Horse Range. Several roads serve the eastern side of the Pryor Mountains, but you need a four-wheel-drive vehicle to follow them all the way through the national forest.

In eastern Montana, east and south of Ashland, are small mountains that rise up out of the grassy plains. Much of this region is in the Custer National Forest. Northeast of Ashland is the Cook Mountains Hiking and Riding Area, and south of Ashland are King Mountain and Tongue River Breaks Hiking and Riding Areas. No major roads serve these areas. U.S. Highway 212 bisects this part of the national forest from east to west, and Forest Highway 51 bisects it from north to south.

Near the southeastern corner of the Tongue River Hiking and Riding Area is the Poker Jim Research Natural Area. This 363-acre site consists of nearly level beds of soft sandstone, shales, and lignite. Several of the lignite beds have burned, resulting in large beds of scoria and clinkers. On this surface is an undisturbed mixture of ponderosa pine, sagebrush, and grasses. Botanists have identified four major habitats in the Research Natural Area. The driest sites are dominated by a cover of bluebunch wheatgrass with skunkbush, Spanish bayonet (a type of yucca), and creeping juniper present. In sites with a little more moisture, the principal plants are ponderosa pine, Idaho fescue, snowberry, serviceberry, and creeping Oregon grape. Mesic areas are on the lower north slopes where the most common species are ponderosa pine, Douglas hawthorn, chokecherry, and serviceberry. Along the streams is a riparian community of eastern cottonwood, Douglas hawthorn, and chokecherry. Pronghorns are common here, as are mountain cottontails, whitetailed jackrabbits, swift foxes, coyotes, and prairie dogs.

Just south of Poker Jim Research Natural Area is Poker Jim Lookout. From the Fort Howes Forest Service Facility on Forest Highway 51, Taylor Road crosses along the southern end of this district of the national forest, providing access to a lookout on Diamond Butte. Several buttes that rise up out of the flatland in the national forest have been used as landmarks for travelers for many years.

Three smaller units of the Custer National Forest are in extreme eastern Montana. Nine square miles of the highly scenic Chalk Buttes are there, and a road from the buttes climbs to Trenk Pass in the Chalk Buttes. From the pass is a four-wheel-drive road westward across the heart of the Chalk Buttes, passing an old cemetery before exiting the national forest. The Chalk Buttes are noted for their pale limestone formations.

A few miles northeast of the Chalk Buttes are the Ekalaka Hills where exposed limestone formations again occur. A campground sits at Ekalaka Park at the south end of the national forest land, and a picnic facility is available at MacNabb Pond on the eastern side. State Route 323 from Ekalaka crosses through the national forest.

The easternmost part of the Custer National Forest in Montana is in the Long Pines region. Several paved roads surround this 80-square-mile unit of the Custer National Forest, and Forest Road 3117 goes directly through the center of the Long Pines from north to south. A lookout sits next to the forest road at Tri Point, and a campground is at Lantis Springs. At the south end is impressive Capitol Rock, which has a fanciful resemblance to the U.S.

Capitol. The site of an old Civilian Conservation Corps camp is just off of Iron Road on the Montana–South Dakota state line.

In extreme northwestern South Dakota are five isolated units of the Custer National Forest. Two noncontiguous units in the Short Pine Hills occupy only a total of nine square miles. Moreau Peak (3,889 feet) and Lone Mountain (3,828 feet) are the main formations, but there are no developed Forest Service facilities.

To the north of the Short Pine Hills are units of the national forest that encompass the South Cave Hills and the North Cave Hills. From the Bullock Road (County Road 733), there is a forest road that enter the North Cave Hills through Fuller Canyon. The Picnic Spring Campground is near Fuller Pass. The rocky narrow defile at Riley Pass was part of the 3-B stagecoach route serving Belle Fourche, Buffalo, and Baker. Wheel ruts worn in the sandstone are still visible.

A group of narrow rock formations known as the Slim Buttes make up the fifth South Dakota unit of the Custer National Forest. This is the largest and most scenic of all the units. State Route 20 crosses the Slim Buttes at Riva Gap where there is a campground. Nearby are The Castles, picturesque rock formations that have been designated as National Natural Landmarks. Just east of the national forest boundary is the site of the Battle of Slim Buttes.

Near the southern end of the Slim Buttes area is a narrow road that climbs over J B Pass before coming to State Route 79. South of this junction, State Route 79 climbs over scenic Summit Pass before leaving the Custer National Forest. Several four-wheel-drive roads are in the Slim Buttes area.

Flathead National Forest

SIZE AND LOCATION: 2.3 million acres in northwestern Montana, south and west of Glacier National Park. Major access routes are U.S. Highways 2 and 93 and State Routes 35, 83, and 206. District Ranger Stations: Bigfork, Hungry Horse, and Whitefish. Forest Supervisor's Office: 1935 3rd Avenue, Kalispell, MT 59901, www.fs.fed.us/r1/flathead.

SPECIAL FACILITIES: Winter sports areas.

SPECIAL ATTRACTIONS: Jewel Basin Hiking Area; National Wild and Scenic Flathead River; Hungry Horse Reservoir.

WILDERNESS AREAS: Mission Mountains (73,877 acres); Great Bear (286,700 acres); Bob Marshall (1,009,356 acres). The Bob Marshall, Great Bear, and

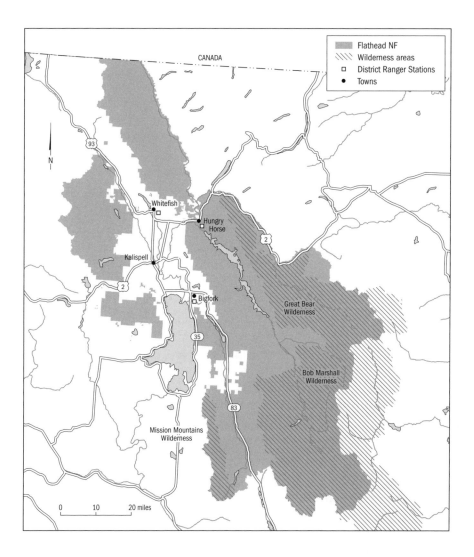

Scapegoat Wilderness areas adjoin one another and are jointly managed by the Flathead, Lolo, Helena, and Lewis and Clark National Forests.

The Flathead National Forest epitomizes the northern Rocky Mountains of the United States. Here are rugged peaks, alpine lakes, scintillating waterfalls, sparkling mountain streams, wildflower-laden meadows, vast river valleys, towering old-growth coniferous forests, and one of the last places in the lower 48 states where all the wildlife present when Lewis and Clark explored the west are still present in their natural environment.

All of these are primary features of the wilderness areas that make up 48 percent of the Flathead National Forest. Great Bear Wilderness, Bob Marshall

Wilderness, and Scapegoat Wilderness are connected north to south to provide for a vast wildland that is home to grizzly bears, black bears, mountain goats, mountain lions (pl. 26), bighorn sheep, gray wolves, elk, moose, and deer. These wilderness areas include part of the Continental Divide, capped by the spectacular Chinese Wall, a 22-mile-long limestone escarpment that rises 1,000 feet above the forest floor below. Just south of the Chinese Wall, in the heart of the wilderness, the mountains are so reminiscent of Switzerland that they have been referred to as the Flathead Alps.

The western edge of the Bob Marshall Wilderness includes the majestic snowcapped peaks of the Swan Mountain Range. At lower elevations, the forests are dominated by ponderosa pine, western larch, Douglas fir, and lodgepole pine, whereas higher up on the mountains are alpine larch, alpine fir, and whitebark pine. Wetter sites support Engelmann spruce. Here and there are aspen and paper birch. Bullet Nose Mountain, reached by hiking the trail along Una Creek, is the site for large deposits of fossil trilobites. Big Salmon Lake, five miles long, is one primary destination for fishermen. A number of waterfalls in the wilderness, including Big Salmon Falls, Dean Falls, Needle Falls, and Barrier Falls, are worth seeking in the wilderness. North of Big Salmon Lake is Mud Lake and the significant Mud Lake Fen along the margins of the lake. The dominant plant in the fen is hairy-fruited sedge, although the more conspicuous plants are hoary willow, bog bean, and dwarf birch, and some relatively rare plants, from sundews, to orchids, to sedges. The forest that surrounds Mud Lake is dominated by Engelmann spruce and subalpine fir. The Wild and Scenic Smith Fork of the Flathead River, with a mix of gentle relaxing and hair-raising rapids, cuts through the Bob Marshall Wilderness.

Between the northern end of Bob Marshall Wilderness and Glacier National Park is the Great Bear Wilderness. Fifty miles of the Middle Fork of the Wild and Scenic Flathead River flow through this wilderness. U.S. Highway 2, which passes between Glacier National Park and Great Bear Wilderness, has several access points for hikes into the wilderness.

Mission Mountain Wilderness, a narrow region a few miles west of the Bob Marshall Wilderness, extends 30 miles north to south but is only seven miles at its widest point. Many lakes dot the wilderness, including Gray Wolf Lake, High Park Lake, and Crystal Lake at the southern end. Because of the extreme ruggedness that includes vertical cliff faces and tricky talus slopes, few trails are available in the wilderness.

One of the most popular hiking areas in any national forest is the Jewel Basin Hiking Area at the north end of the Swan Mountain Range east of Kalispell and Hungry Horse Reservoir. The hiking area is not wilderness but is managed specifically for hiking, with both motor vehicles and pack and saddle stock prohibited. The area includes 15,349 acres of high country dotted with 25 lakes that glisten like gemstones in the sunlight. Hiking trails connect most of the lakes. The lakes range in size from tiny 1.8-acre West Jewel Lake to 50.5-acre Upper Black Lake. The lakes in the basin are known and managed for native cutthroat trout. Mount Aeneas, on the western side of Jewel Basin, has an elevation of 7,528 feet. The easiest access to Jewel Basin is from the Kalispell area, drive west on State Route 83 to Echo Lake Road, then the Echo Lake Road four miles to a T. The right fork at the T leads to the Jewel Basin Road and seven miles to the trailhead. After the T, the road quickly turns to gravel. The last five miles are steep, but the road is generally well maintained.

In the heart of the Flathead National Forest is huge Hungry Horse Reservoir that was constructed by the Bureau of Reclamation. The reservoir is 34 miles long and 3.5 miles wide at its widest point. Maximum depth of the reservoir is 500 feet. Fishing for cutthroat trout, Dolly Varden trout, grayling, and whitefish is popular. Many campgrounds are adjacent to the reservoir. Southeast of Hungry Horse Reservoir up the Spotted Bear River drainage is the Bent Flat Fen. The large fen and associated peatlands is one of the best examples in the area. This is classified as a rich patterned fen, which means that it contains a network of pools and ridges oriented perpendicular to one another. Several spikerushes, hardstem bulrush, and Buxbaum's sedge are common components of the fen, with scattered gnarly specimens of shrubby cinquefoil, dwarf birch, and even dwarf forms of Engelmann spruce. This fen contains a high percentage of rare Montana plants, including yellow lady's-slipper orchid, sparrow's-egg lady's-slipper orchid, great sundew, tufted bulrush, and green-keeled cotton grass.

Several lookouts and cabins in the Flathead National Forest have been renovated into rental cabins. Contact the district ranger stations for reservations. Other points of interest include Martin Falls outside the northern end of Miller Creek Demonstration Forest, Tally Lake, and the Salish Mountains on the western edge of the national forest, all located west of the town of Whitefish. The Miller Creek Demonstration Forest is a 5,500-acre tract where some of the earliest research on forest management started. Self-guided tours of the demonstation forest explain the effects of clear-cutting, broadcast burning, and wildfire. Popular ski areas are Big Mountain Ski Area north of Whitefish and Blacktail Mountain Ski Area south of the town of Lakeside. The self-guided Danny On Memorial Trail is a popular hike that begins at the Big Mountain Ski Base and goes for six miles to the Summit House at the ski peak, where there is a Forest Service interpretive center. You may take the gondola to the top and walk the three miles back down to the ski base after the snow clears, usually between early July and Labor Day. If you are so inclined, there is an additional loop of 3.8 miles through Flower Point where there are gorgeous alpine wildflowers.

In addition to the recreation opportunities described, this national forest contains some excellent examples of how forests can be managed for quality water, wildlife habitat, and wood for building materials. Northwest Montana contains the largest concentration of western larch, or tamarack, trees in the west. The needles of these majestic trees turn a bright yellow in fall and rival the colors of New England. The occurrence of western larch is closely tied to past fires. Fires planned by the forest managers or naturally occurring fires started by lightning have created and continue to create the ideal conditions for a new forest of valuable western larch and many other plants to become established in the ashes of the fires.

Gallatin National Forest

SIZE AND LOCATION: Approximately 1.8 million acres in south-central and southwestern Montana, north and northwest of Yellowstone National Park. Major access routes are Interstate 90, U.S. Highways 20, 89, 191, 212, and 287, and State Routes 86, 298, 411, 419, and 420. District Ranger Stations: Big Timber, Bozeman, Gardiner, Livingston, and West Yellowstone. Forest Supervisor's Office: 10 E. Babcock Street, Bozeman, MT 59715, www.fs.fed .us/r1/gallatin.

SPECIAL FACILITIES: Winter sports areas.

SPECIAL ATTRACTIONS: Beartooth Highway; Madison River Canyon Earthquake Area; Hebgen Lake; Gallatin Petrified Forest; Gallatin Canyon; Hyalite Canyon.

WILDERNESS AREAS: Absaroka-Beartooth (943,626 acres, partly in the Custer and Shoshone National Forests); Lee Metcalf (254,288 acres, partly in the Beaverhead National Forest).

Persons visiting Yellowstone National Park who do not venture into the adjacent Gallatin National Forest are missing a marvelous area of mountain peaks, glacial lakes, rugged valleys, dense coniferous forests, and tumbling waterfalls. The Gallatin National Forest borders Yellowstone National Park on the north and northwest and then extends northward past Bozeman and Livingston for several miles. The Yellowstone, Gallatin, and Boulder Rivers penetrate the national forest. Several mountain ranges are within the Gallatin National Forest.

Stretching northward from Bozeman for 27 miles is the Bridger Range, named for scout and frontiersman Jim Bridger who brought early settlers into the Gallatin Valley over the Bozeman Trail. These mountains have a bedrock that is predominantly limestone, with several areas of bare rock as well as areas with deep glacial soil. Several peaks are above 9,000 feet in elevation, including the highest, Hardscrabble Peak and Sacagawea Peak, both at 9,665 feet. Both of these can be hiked from the Fairy Lake Campground. The campground is situated next to beautiful Fairy Lake. The Bridger Bowl Ski Area is on the eastern side of the Bridger Range and may be reached over a short road off of State Route 86. State Route 411 provides access to the western side of the Bridger Mountains. In the northern part of the mountain range is scenic Flathead Pass that can be reached by a road off of State Route 86. A pretty wildflower meadow is at the pass. The Bridger Mountains National Recreation Trail begins at the M Picnic Area a short distance northeast of Bozeman and penetrates the lower two-thirds of the Bridger Mountain Range for approximately 24 miles.

Several miles northeast of Livingston are the Crazy Mountains. This is an isolated range that rises above the adjacent plains and that has several peaks above 10,000 feet, with Crazy Peak topping out at 11,214 feet. Big Timber Canyon cuts a deep and scenic gorge across the lower part of the Crazy Mountain Range. County Road 25 enters the gorge from the east, terminating at Half Moon Campground. A hiking trail continues through the western end of Big Timber Canyon from the campground. Upper Big Timber Falls is a short distance west of the campground, and there are several fine, small lakes in the area. Several roads off of U.S. Highway 89 lead into the western side of the Crazy Mountains. The northern part of the Crazy Mountains is in the Lewis and Clark National Forest.

The remainder of the Gallatin National Forest lies south of Interstate 90. A few miles southwest of Bozeman is Hyalite Creek. The creek, which enters the national forest five miles south of Bozeman, has carved scenic Hyalite Canyon. Langohr Campground is at the southern end of the canyon. By following Hyalite Creek for four miles southeast from the campground, one comes to Hyalite Reservoir. The area around Hyalite Creek is mostly high alpine, with scenic steep canyons, sparkling mountain streams, several lakes, and beautiful waterfalls. Palisade Falls drops over a sheer cliff wall. A half-mile interpretive trail to the falls has additional signs in braille for the blind.

Five-and-a-half miles southwest of Hyalite Reservoir is the Wheeler Ridge Research Natural Area, a 640-acre tract at the northern end of the Gallatin Range. This beautiful area preserves an extensive old-growth stand of whitebark pine and subalpine fir at elevations between 8,000 and 8,800 feet. A significant feature of the area is the whitebark pine stand. The pines here are

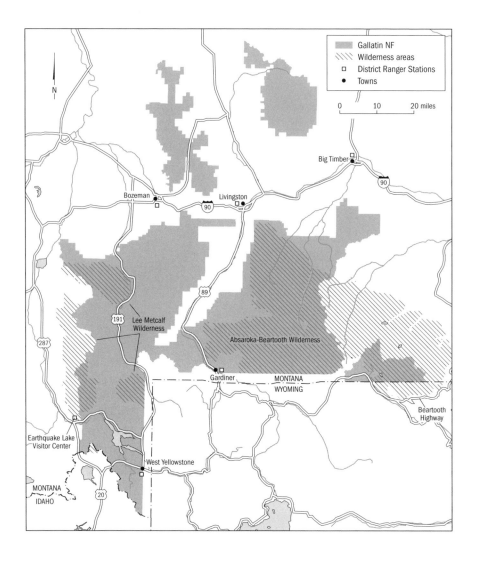

straighter, larger, and taller than any place else in the northern Rocky Mountains. Below 8,000 feet in the natural area are forests of lodgepole pines, with occasional Douglas firs as much as 100 feet tall interspersed. Most of the mountain streams have occasional wet sedge meadows alongside them. The region has large talus piles in the vicinity of steep-walled cirque basins. The area may be reached via a hiking trail from the end of Forest Road 980 at Big Bear Lake.

Three units of the Lee Metcalf Wilderness are partly or entirely in the Gallatin National Forest along the forest's western side. The Spanish Peaks Unit, named for the snowclad Spanish Peaks, is 22 miles southwest of Bozeman and 20 miles northwest of Yellowstone National Park. U.S. Highway 191 from

Bozeman to West Yellowstone circles around the eastern side of the Spanish Peaks. Gallatin Peak (11,015 feet), Wilson Peak (10,700 feet), and Beacon Point (10,248 feet) are capped with snow year-round. Numerous glacial lakes, waterfalls, mountain meadows, and bare rock surfaces dot the area (pl. 27). Fishing for trout in the mountain streams is popular and usually rewarding. Rich Douglas fir forests are found on the lower slopes of the mountains, and gnarly, windswept alpine firs grow on the most exposed ridgetops. Engelmann spruce is common along most of the waterways. In fall, color is added to the forests by the scarlet of mountain maples, the red of mountain ash, and the yellows of aspen and alder. Alpine tundra vegetation usually is above the 9,000-foot elevation, just above a zone of subalpine firs and whitebark pines. All of the lakes in this unit of the wilderness are glacial with the exception of Lava Lake, which was formed following a massive landslide. Cascade Creek Trail and Spanish Creek Trail are the most popular of several trails in the wilderness. Cascade Creek Trail starts five miles above the Squaw Creek Ranger Station on the western side of the Gallatin River. The trail follows Cascade Creek to Lava Lake and Table Mountain. Spanish Creek Trail follows the South Fork of Spanish Creek and Little Hellroaring Creek up to Falls Creek, Big and Little Brother lakes, and Mirror and Spanish lakes.

East of U.S. Highway 191, between the Spanish Peaks and Hyalite Creek, is an area of interesting rock formations and picturesque lakes. A forest road follows Squaw Creek past a formation known as Spire Rock where there is a campground. Garnet Mountain is a mile south of the campground.

U.S. Highway 191 between Big Sky Meadow Village at the south end of the Spanish Peaks and the northwest corner of Yellowstone National Park follows the Gallatin River and skirts the western edge of Elkhorn Ridge. Just outside of Big Sky Meadow Village is Soldier Chapel, built by Nelson Story to honor his son and others killed in wars. The chapel is now used for weddings, church services, and a place to sit and enjoy solitude. Along U.S. Highway 191 are trailheads at the Porcupine Guard Station for a pleasant hike along Frist Creek, at Rainbow Point for a hike to the north end of Elkhorn Ridge, and at Cinnamon Guard Station for a hike westward into the Cinnamon Mountains.

South of the Spanish Peaks is the Taylor-Hilgard Unit of the Lee Metcalf Wilderness. The western half of this wilderness is in the Beaverhead National Forest. The highest peaks in the Taylor-Hilgard Unit straddle the border between the Beaverhead-Deerlodge and Gallatin National Forests. Hilgard Peak is the highest, at 11,316 feet, with Dutchman Peak, Echo Peak, Imp Peak, Nutters Cathedral Peak, Koch Peak, and Shedhorn Mountain, listed from south to north, somewhat lower. The Gallatin side has several fine lakes in the southern part of the wilderness.

U.S. Highway 287 follows the Madison River and curves just below the Lee Metcalf Wilderness at Raynolds Pass. Just east of the pass is the spectacular Madison River Canyon Earthquake Area, a tract of 37,800 acres set aside as a special geological area. Stop at the Earthquake Lake Visitor Center to learn about the remarkable earthquake that occurred on August 17, 1959, and changed the face of the Earth forever in the area. Walk to the Memorial Boulder and Overlook on top of The Slide. The slide across the Madison River consists of enough rocks and earth to fill California's Rose Bowl 10 times and 20 percent of the Panama Canal. The slide completely blocked the river, forming Earthquake Lake. Upstream from the slide and Earthquake Lake, violent shocks lowered the northern shore of Hebgen Lake, sending huge waves up and down the lake and over the dam. Miraculously, the dam held. Large sections of old U.S. Highway 287 fell into the lake. Five miles of the old highway are under the lake. Just beyond the Beaver Creek Campground is Refuge Point, a high knoll that was a haven for many of the campers who survived the earthquake. On the night of the earthquake, about 150 people were camped in the Madison River Canyon. Because the slide blocked the west end of the canyon and Hebgen Dam blocked the east end, many of the campers took refuge on the knoll. A Forest Service Smokejumpers Rescue Team began the evacuation of the campers, some of them injured. In all, 28 persons lost their lives in the canyon. A trail from Refuge Point leads southward to Ghost Village where buildings, now deserted, still stand. At the Cabin Creek Campground, you may walk through the campground where several families were stranded during the earthquake. Just east of the campground is Hebgen Dam. During the earthquake, the land surrounding the dam dropped about nine feet. The first shock of the earthquake occurred at 11:37 P.M. and was felt in eight states. Scientists estimated the power was equal to 2,500 atomic bombs and registered 7.1 on the Richter scale.

U.S. Highway 287 continues along the north shore of Hebgen Lake, passing through Red Canyon and joining U.S. Highway 191 just outside of Yellowstone National Park. You may wish to take the five-mile-long side road up Red Canyon to Red Canyon Trail. This half-mile trail goes to the Red Canyon Scarp.

U.S. Highway 20 crosses the Gallatin National Forest between West Yellowstone and Targhee Pass, climbing over Henrys Lake Mountain at the pass. South of West Yellowstone is the Rendezvous Ski Trail Area.

Seven miles north of the east end of Hebgen Lake is the Monument Mountain Unit of the Lee Metcalf Wilderness. Monument Mountain and Snowslide Mountain, both above 10,000 feet, are near the center of the wilderness, although Cabin Creek Peak on the western edge of the wilderness is the highest at 10,572 feet.

One of the most remarkable areas in any national forest is the Gallatin Petrified Forest where the petrified trunks of trees have been left in an upright position. More interesting is that several layers of forests have been buried on top of one another by a series of volcanic flows. Whereas the living forests in the area today consist of lodgepole pine, Engelmann spruce, Douglas fir, whitebark pine, junipers, and aspens, the petrified trees belong to a sequoia, a redwood, walnuts, hickories, birches, breadfruits, and acacias. Accesses to the Gallatin Petrified Forest, which covers nearly 26 square miles, is from the Tom Miner Campground that is nestled along Trail Creek or from a trailhead at the end of a back road several miles southwest of the Big Creek Ranger Station. Tom Miner Campground and Big Creek Ranger Station are located on roads west of U.S. Highway 89. This highway follows the Yellowstone River from Livingston to Gardiner and the northern entrance to Yellowstone National Park. Along the way is Yankee Jim Picnic Area where there is also a boat ramp and fishing area for the Yellowstone River. Farther along the highway are similar facilities at La Duke Spring. The trail at this latter area follows Bassett Creek for nearly four miles. At Gardiner, U.S. Highway 89 enters Yellowstone National Park, but there are gravel and dirt roads north and east of town that wind through scenic country in the Gallatin National Forest until they end at the border of the Absaroka-Beartooth Wilderness where there are trails into the wilderness (pl. 28). Most of the eastern and southeastern parts of the Gallatin National Forest are in the vast wilderness, except for the area around Cooke City. The easternmost part of the wilderness is in the Custer National Forest.

The Absaroka-Beartooth Wilderness of nearly one million acres is a continuous land mass except for a narrow corridor along the Boulder River that penetrates deep into the heart of the wilderness. The Boulder River also separates the two major mountain ranges that make up the wilderness, the Absaroka Mountains and the Beartooth Mountains. The Absarokas are volcanic in origin and date back 63 million years. On the other hand, the Beartooths are granitic and are more than 600 million years old. The Absarokas have steep, rocky ridges, rolling mountains and foothills, and landflows, with streams that meander through glacial valleys. The Beartooths have high jagged mountain peaks, deep canyons, and high plateaus of tundra. Much of the area of the Beartooths, because of the high elevations, are rocky and barren. Montana's highest peak, Granite Peak, at 12,599 feet, is in the Beartooths, along with other peaks in excess of 12,000 feet. Granite Peak is actually a series of peaks that join to form a semicircle and is on the border between the Gallatin and Custer National Forests. Peaks in the Absarokas are fewer and lower, with 11,206-foot Mount Cowan the highest. More than 340 lakes and innumerable streams grace the wilderness.

Road access to the wilderness is limited. The Boulder River Road is in a narrow corridor between the two mountain ranges. It begins as State Route 298 in Big Timber. At the boundary of the Gallatin National Forest is the spectacular Natural Bridge and Falls. A gushing waterfall plummets over a sheer rock cliff immediately to the left of a marvelous natural bridge that has been carved out of rock through countless years of erosive action. South of the falls and bridge, the road follows Boulder River, with Contact Mountain, one of the Beartooths, on the east and West Boulder Plateau to the west. Campgrounds are at Falls Creek, Big Beaver, Aspen, Chippy Park, Hells Canyon, and Hicks Park. The road has good surface as far as the Box Canyon trailhead. Beyond this, the road is much rougher and extends for several miles to Independence Peak, which is totally surrounded by the wilderness.

The only other accesses to the Absaroka-Beartooth Wilderness in the Gallatin National Forest are around Cooke City. This community is reached via U.S. Highway 212, the Beartooth Highway, which extends for 60 miles from Red Lodge, Montana, to Cooke City, dipping for a while southward into Wyoming. Only the last nine miles of this scenic highway are in the Gallatin National Forest, with Cooke City about midway in the nine-mile stretch. Three miles east of Cooke City, the Beartooth Highway climbs over Colter Pass, with the Chief Joseph Campground nearby. Scars from the 1988 wildfire that swept over the pass mar the area. At Colter Pass, there is a short trail to Kersey Lake, which is surrounded by a series of high mountain wetlands. From near the Cooke City Cemetery, a side road you may wish to take follows Miller Creek over Bull of the Woods Pass to Crown Butte. To the east of Crown Butte is Chimney Rock. At Crown Butte a trail runs over Daisy Pass and eventually into the Absaroka-Beartooth Wilderness. Another side road from the cemetery leads into the Custer National Forest.

Tom Miner Basin

In southern Montana, off U.S. Highway 89, Forest Service Road 63 curls for about 12 miles through Tom Miner Basin and ends at the Tom Miner campground. From there, a hiking trail framed by rugged mountains follows a creek westward on a steady, 2.5-mile climb to Buffalo Horn Pass. At the pass another trail leads north toward 10,269-foot Ramshorn Peak, passing mountain meadows dotted with Indian paintbrushes, red beardtongues, and yellow mule's-ears. Along the mountain streams and ridges grow lodgepole and whitebark pines, Engelmann spruce, Douglas fir, and quaking aspen. A more startling sight, however, are the standing petrified tree trunks and stumps, some more than eight feet tall and four feet across, that are interspersed in the living vegetation. Other fossilized logs are strewn across the terrain.

This is the northern end of Gallatin Petrified Forest, which extends for 50 miles from Montana's Gallatin National Forest southward into Yellowstone National Park, Wyoming. To the southeast there is a second petrified forest of standing trees in the bluffs along the Lamar River, a few miles from its confluence with the Yellowstone. Trapper Jim Bridger, tongue in cheek, described those trees during the 1830s, reporting "peetrified birds a sittin' on peetrified trees a singin' peetrified songs in the peetrified air."

The two petrified forests were apparently formed following the same volcanic eruptions, sometime during the Eocene, the geological epoch from about 60 million to 40 million years ago. Underground water, often hot, picked up silicon dioxide (quartz) from the layers of volcanic material and circulated through trees buried by the eruptions. The water filled the cavities of the empty wood cells and deposited minerals on the tough cell walls. As a result, even microscopic details of the wood were preserved in stone.

William Henry Holmes, a member of the U.S. Geological and Geographical Survey of the Territories in the 1870s, was the first to write about the fossil formations of the Yellowstone area in some detail. He described the petrified trees as resembling pillars of some ancient temple. After observing that standing trees were exposed at different levels on some cliffs, Holmes suggested that 10 or more forests had been buried in successive layers of volcanic ash.

A puzzle soon emerged, however. In 1899, Frank Hall Knowlton, a paleobotanist with the U.S. Geological Survey, identified nearly 150 species of plants in the petrified forests. Whereas the living forests of the Yellowstone area are now dominated by coniferous plants, relatively few of these, mostly pines and redwoods, were in the fossil forests. Instead, Knowlton discovered a remarkable diversity of species. Many of the trees, such as sycamores, magnolias, chestnuts, maples, dogwoods, walnuts, hickories, elms, and willows, were apparently relatives of similar hardwood species that live in temperate regions today. Others seem to have been closely related to plants now typical of the tropics and subtropics. These included avocados, figs, greenheart ocoteas, acacias, and breadfruits. (More recently, botanist L. H. Fisk examined the pollen grains buried in the layers containing the fossil trees and was able to add other temperate species to the ancient flora: larch, spruce, yew, holly, alder, basswood, and ash.)

Botanists who have studied the area have commented on this unusual mixture of temperate and tropical species, which apparently grew side by side. Princeton geologist and paleobotanist Erling Dorf investigated the petrified forests nearly seven decades ago and theorized that during the Eocene, the topography of the Yellowstone region consisted of a series of broad, flat river valleys separated by gently rolling hills that rose some 1,500 feet above the valley floors. The overall climate in the area, he suggested, was essentially the same as that found on the Gulf Coast of the southeastern United States today, ranging from subtropical in the valleys to warm and temperate in the surrounding hills. As a result, he concluded, both tropical and temperate species could have overlapped, especially in the valleys.

Then, Dorf continued, volcanoes began to erupt east and northeast of the region, showering rocks and ash over the forests in the valleys and burying the tree trunks up to a height of 15 feet. Basing his ideas on a contemporary study of forest regeneration, Dorf estimated that about 200 years after the volcanoes erupted a new forest began to grow. The new growth took root in the volcanic material that had buried the previous forest. Measuring the growth rings of some of the standing fossil stumps, Dorf calculated that the new forest was about 500 years old when another series of volcanic eruptions buried it. This process, he concluded, happened again and again, with at least 27 distinct forests of petrified trees formed in layer-cake fashion over 20,000 years. Dorf estimated that these layers, now being exposed through erosion, are more than 1,600 feet deep.

Like most researchers, Dorf took the presence of standing trunks and stumps as evidence that the fossil forests had been buried in place by volcanic debris. Recently, however, Georgia State University geologist William J. Fritz, studying the effects of the Mount St. Helens eruptions in Washington in 1980, found that tree stumps were often transported and deposited vertically by mud flows. He wondered if this had also happened in the Yellowstone area during the Eocene, which would explain why temperate and tropical species were mixed together.

Analyzing the material in which the fossil forests had been buried, Fritz found that only about 10 percent of it was made up of volcanic rock. The remainder, containing most of the fossilized wood, consists primarily of rounded conglomerate rock, which only could have been deposited by sedimentary processes in a volcanic environment. Fritz also noted that, contrary to Dorf's hypothesis of gently rolling hills, some geologists today believe that during the Eocene the mountains in the Yellowstone region rose 8,000 to 10,000 feet above the valley floors.

Fritz hypothesizes, therefore, that mud flows originated high on the flanks of volcanoes, where coniferous and other cool-temperate trees grew. Fed by rain associated with the volcanic eruptions, and probably by melting snow cover, the flows uprooted trees in the higher elevations and transported them down into the hotter valleys. Fritz suggests that some of the uprooted stumps were deposited vertically amid the tropical species, which were being covered in place by the volcanic rock and ash.

The trees transported by the mud flows would probably have had their roots and branches broken off and their bark stripped and replaced by a mud casing. They also may have been deposited upside down. Stumps buried in place, instead, could have retained well-developed root systems and would more likely be upright. A good test of Fritz's hypothesis is to see if the trees buried in place are the tropical species, whereas those that appear to have been transported are the temperate species.

In studying the petrified forests, Fritz also came to challenge Dorf's hypothesis that many successive forests were buried one above the other in layer-cake fashion. Rock layers containing standing petrified trees usually cannot be followed for any distance. The layers are interrupted by zones of sedimentary rock in places where, Fritz believes, air-fall ash, wood, leaves, and plant litter were washed into streams and lakes.

Helena National Forest

SIZE AND LOCATION: Approximately 975,000 acres in west-central Montana on all sides of Helena. Major access routes are Interstate 15, U.S. Highways 2 and 287, and State Routes 200, 279, 280, and 284. District Ranger Stations: Helena, Lincoln, and Townsend. Forest Supervisor's Office: 2880 Skyway Drive, Helena, MT 59601, www.fs.fed.us/r1/helena.

SPECIAL FACILITIES: Winter sports areas; boat ramps; swimming beaches.

SPECIAL ATTRACTION: Gates of the Mountain.

WILDERNESS AREAS: Scapegoat (239,936 acres, partly in the Lewis and Clark National Forest); Gates of the Mountain (28,562 acres).

The Helena National Forest typifies the Wild West, with mountain peaks, bubbling streams, coniferous forests, ghost towns, and old mine sites. The Lewis and Clark Expedition passed through this region, making camp on the banks of the Missouri River on July 19, 1805. In his journal for that date, Meriwether Lewis wrote "This evening we entered the most remarkable cliffs that we have yet seen. These cliffs rise from the water's edge on either side perpendicularly to the height of 1,200 feet . . . from the singular appearance of this place, I called it the Gates of the Rocky Mountains." Today, the Meriwether Campground is located where Lewis and Clark camped. An excursion boat through the Gates of the Mountain leaves during summer from Hilger Landing, about 17 miles north of Helena. Unusual rock formations in the Madison limestone that make up the cliffs are visible during this scenic boat tour. In one place are vertical falls, called the Devil's Slide. At another place above Fielder Gulch you see where a mountaintop slipped some 250 years ago. Rows of holes in the cliffs reveal the nesting colonies of bank swallows. The boat stops at the Meriwether and Coulter campgrounds for anyone wishing to get out and explore. As you hike the trail from the campgrounds on the east side of the river, you may see spiral trees whose trunks have been twisted mysteriously. Between the two campgrounds are numerous fossils of crinoids, brachiopods, gastropods, and corals, the remains of organisms that lived in a shallow warm sea that covered this area more than 350 million years ago. North of the Meriwether Campground are paintings made by prehistoric Indians on the rock walls. If you wander up Mann Gulch, you still see the effects of a great forest fire in early August 1949. During the course of fighting the fire, 12 smokejumpers and a Forest Service guard were trapped by the fire and fatally burned. A memorial with the names of these men may be seen at the Meriwether Campground. Gates of the Mountain Wilderness lies to the east of the campgrounds, with the Meriwether Canyon Trail offering a scenic adventure into the heart of the wilderness. The highest elevation in the wilderness, Moon Mountain at 7,980 feet, is at the northeastern corner of the area.

State Route 280 leaves from the northeastern part of Helena and crosses over Hauser Lake before entering the Helena National Forest. Once in the forest, the highway comes to York, the site of an old gold placer camp in 1866. Originally called New York, the town grew to about 1,000 residents by the 1880s, and an estimated five million dollars in gold was extracted from the area. As you drive the road, notice low rocks with large holes and small caves along Trout Creek. The rock is cobblestone and contains round cobbles cemented together with calcium carbonate, a claylike cement. The holes and small caves develop where weathering has eroded the clay. Beyond the Vigilante Campground the road enters scenic and rugged Trout Creek Canyon and into the Big Belt Mountains. For three miles the road is surrounded by steep-walled Madison limestone, some of the cliffs soaring 1,000 feet above the road. At the Hanging Valley Vista you overlook Magpie Creek and get a striking view of Trout Creek Canyon. A good but rugged hiking trail from the northern end of the Vigilante Campground runs into Hanging Valley. If you look to the west, you see a flat-topped rocky overhang known as The Shelf. Water flowing over the shelf intermittently freezes and thaws, forming the overhang. At another point west of the road you see limestone rocks with many small, narrow channels. These are The Flutes, which are grooves on the rock surface caused by water passing over the face of the rock and dissolving calcium carbonate, which forms the small channels.

If you return to the townsite of York, you may take another interesting

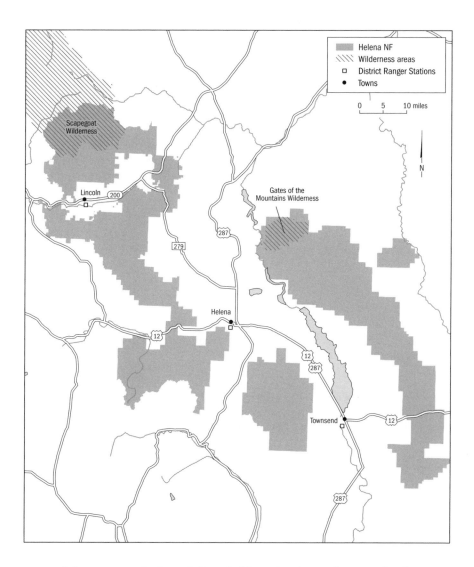

road that goes north through Dry Gulch to the tiny settlement of Nelson,
dating back to the 1880s. From Nelson, take the road to the northeast. You
pass the site of the Checkerland Ranger Station before reaching Refrigerator
Canyon. The ranger station, no longer present, was in operation from 1913
to 1972. Refrigerator Canyon is north of the road with a scenic trail through
it that continues into the Gates of the Mountain Wilderness. The narrow
canyon is only 15 feet wide with rock walls reaching more than 100 feet upward. Because of the cool breezes that often funnel through this gorge, it is
well named as Refrigerator Canyon. A nice assemblage of spring wildflowers
decorates the trail.

The Big Belt Mountains are oriented from the northwest to the southeast from the Gates of the Mountain Wilderness nearly to U.S. Highway 12. Several roads and trails are in the mountains to such places as Avalanche Butte, Hellgate Gulch, Bilk Mountain, and Boulder Baldy, but the only developed Forest Service recreation sites are near the southeastern end of the mountains at Gypsy Lake, Skidway, and Deep Creek. Four-wheel-drive roads climb over such intriguing places as Duck Creek Pass and Klondike Pass.

Much of the Elkhorn Mountains southeast of Helena and east of Interstate 15 are in the Helena National Forest, with thousands of acres designated as the Elkhorn Wildlife Management Unit. Although several roads enter this area, there are no Forest Service recreation sites.

The Continental Divide crosses the Boulder Mountains southwest of Helena. The many roads in this regions led at one time to the several mines, the remains of which you may still observe in the mountain slopes. Take U.S. Highway 12 westward out of Helena for about seven miles and then Tenmile Creek Road southward. At the Tenmile Campground is the informative Tenmile Environmental Educational Trail. Four miles farther south is the ghost town of Rimini, surrounded by abandoned mines. Park Lake and campground, just a short distance southeast of Rimini, is best reached by a road from the Clary exit on Interstate 15.

U.S. Highway 12 west of Helena crosses the Continental Divide at McDonald Pass where there are panoramic views of the surrounding area. For an interesting alternate route over the Continental Divide, take the Sweeny Creek Road northward from U.S. 12 and across from the Tenmile Creek Road. The road climbs along Sweeny Creek and over Priest Pass on the Continental Divide, descending down the western side of the divide to the junction with a paved road that is alongside Dry Creek. If you choose to take a short hike along Sweeny Creek, you encounter lodgepole pine, Douglas fir, ponderosa pine, aspen, and chokecherry above a rich shrub layer of wild rose, kinnikinnick, snowberry, snowbrush, serviceberry, buffalo berry, and birchleaf spiraea. Grassy openings in the woods consist mostly of blue bunch wheat grass.

After crossing Priest Pass and coming to the T in the road, the left turn at the T brings you to U.S. Highway 12. The right fork takes you up over historic Mullan Pass. The pass is named for Capt. John Mullan who started a wagon road in 1848 to cross the Continental Divide between Fort Benton on the Missouri River to Walla Walla, Washington. The road beyond Mullan Pass proceeds to Roundtop Mountain and then out of the forest to Marysville, another historic mining town that now has a winter sports area.

Northwest of Marysville is 7,600-foot Granite Butte. Northwest of this prominent landmark along the eastern end of a forest road is the proposed Granite Butte Research Natural Area. This natural area preserves a typical mountain grassland with some interesting forested areas adjacent. The grassland is dominated by Idaho fescue, rough fescue, and bluebunch wheatgrass. Downslope are majestic old-growth forests of subalpine fir, lodgepole pine, and Douglas fir. Higher up in the area where wind-deposited snowdrifts are prevalent during winter are ribbon forests of whitebark pine. (See Bitterroot National Forest for a description of ribbon forests.)

State Route 200 cuts diagonally across the northwestern part of the Helena National Forest and the Lewis and Clark Mountains, crossing the Continental Divide at Rogers Pass. Where State Route 200 leaves the western end of the Helena National Forest is Blackfoot Canyon, a scenic gorge along the Blackfoot River. The forested area north of State Route 200 and the town of Lincoln is in grizzly bear country. The wildest portion of the national forest is the Scapegoat Wilderness at the northwestern corner of the Helena National Forest. Red Mountain, at the southern edge of the wilderness, is the highest elevation in the wilderness at 9,411 feet. The higher elevations contain harsh alpine meadows above heavily forested mountain slopes. Connected to the eastern end of the Scapegoat Wilderness is another wild area known as the Silver King–Falls Creek Roadless Area. This region contains a high windblown ridge above Alice Creek where the Lewis and Clark Expedition found themselves on July 7, 1806. In his journal, Captain Lewis noted that he was "delighted at discovering that this was the dividing ridge between the waters of the Columbia and Missouri rivers."

Red Mountain is at one corner of the proposed Red Mountain Research Natural Area that preserves a large windswept alpine tundra ecosystem. Large red and light-colored limestone argillite boulders are scattered across the tundra. Alpine larch, whitebark pine, lodgepole pine, and limber pine grow here, but in very dwarf, gnarly forms. Red Mountain is one of the easternmost locations for alpine larch in the United States. The mountain has been deeply scoured by alpine glaciers so that cirque headwalls and basins and U-shaped valleys are common.

Six miles east of Red Mountain in the Silver King–Falls Creek Roadless Area, is Indian Meadows, an extensive opening with an abundance of grasses and wildflowers. Copper Creek and Snowbank Campgrounds are adjacent to the meadows and accessible by the Copper Creek Road off of State Route 200. Part of Indian Meadows is a designated Research Natural Area that occupies a glaciated bench bordered by low hills. Of particular interest are depressions on the bench that contain either ponds, marshes, or fens. Geyer's willow is common around much of the wetlands. Uncommon species are two kinds of sundews and pale sedge.

Kootenai National Forest

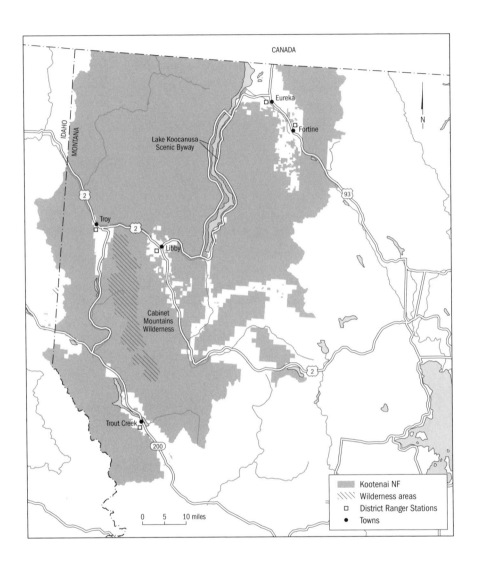

SIZE AND LOCATION: Approximately 2.2 million acres in northwestern Montana and a portion of adjacent Idaho, extending to the Canadian border. Major access routes are U.S. Highways 2 and 93 and State Routes 37, 56, 200, 202, 508, and 567. District Ranger Stations: Eureka, Fortine, Libby, Troy, and Trout Creek. Forest Supervisor's Office: 1101 U.S. Highway 2 West, Libby, MT 59923, www.fs.fed.us/r1/kootenai.

SPECIAL FACILITIES: Winter sports areas; boat ramps.

SPECIAL ATTRACTIONS: Northwest Peak Scenic Area; Ten Lakes Scenic Area; Ross Creek Cedars Scenic Area; Lake Koocanusa Scenic Byway.

WILDERNESS AREA: Cabinet Mountains (94,272 acres).

Mountains surround the northwestern Montana town of Libby, and most of these mountains are in the Kootenai National Forest. The Purcell Mountains occupy the northwestern fourth of the national forest, with the Salish Mountains and the Whitefish Mountains in the east. The wild and woolly Cabinet Mountains are south of Libby, with part of this range in the Cabinet Mountains Wilderness. The Kootenai and Clarks Fork Rivers cross the national forest, each river being dammed in places to create large reservoirs. The Kootenai has been dammed to form Lake Koocanusa, whereas Clarks Ford has two dams, one forming Cabinet Gorge Reservoir and the other creating Noxon Reservoir.

Possibly the two most beautiful parts of the Kootenai National Forest have been designated as scenic areas. The Northwest Peak Scenic Area is in the extreme northwest corner of the Kootenai National Forest in the Selkirk Mountain Range, extending to within a half-mile of the Canadian border. The western part of this scenic area is in the Kaniksu National Forest. Northwest Peak, at 7,705 feet, is the dominant feature of the area and may be reached by a hiking trail off of Forest Road 538. Several small lakes are in the scenic area, most of them at the base of rocky peaks amid dense alpine forests.

The Ten Lakes Scenic Area is east of the Tobacco Plains and U.S. Highway 93 in the Whitefish Mountains at the northeastern corner of the Kootenai National Forest. This 15,700-acre scenic area is dominated by a high ridge of the Whitefish Mountains, which have been shaped by alpine glaciers. Deep glacial cirques and rim-rocked basins protect the 10 lakes in the area. At the highest elevations in the scenic area are whitebark pine and alpine larch, with Engelmann spruce, subalpine fir, Douglas fir, and lodgepole pine at lower elevations. The scenic area is one of the best places in Montana to observe summer alpine wildflowers. Bighorn sheep are common, and grizzly bears inhabit the area. The setting for Ernest Thompson Seton's wildlife stories is in the Ten Lakes Scenic Area. Blacktail Trail from the end of Burma Road leads between Poorman and Green Mountains, passing a few mine shafts and old mining machinery. The Bluebird Basin Trail goes through lush forests and alpine meadows. You may access it from the Therriault Lakes just outside the southeastern corner of the scenic area.

The Kootenai River flows across the northern edge of Libby. About 12 miles to the west along U.S. Highway 2 are the spectacular Kootenai Falls that thunder into the river. North of the falls is the main mass of the Purcell Mountains, extending for nearly 40 miles to the Canadian border. These mountains consist of numerous small peaks, countless streams and creeks, dozens of lakes, and a scattering of waterfalls. Most of the region is under the influence of mild, wet, coastal weather during fall, winter, and spring.

During summer, occasional thunderstorms interrupt the sunshine. Because of the great amount of moisture the area receives, much of the forest is covered by lush vegetation and some outstanding old-growth forests. State Route 508 and Forest Road 68 form a spectacular loop through the Purcell Mountains and provide many accesses into the depths of the Kootenai National Forest.

Four miles east of Libby is a forest road from St. Anthony that goes northward into the Purcell Mountains and brings you to the Skyline National Recreation Trail. You may hike this scenic trail for many miles as it connects several mountain peaks with elevations ranging from 5,495-foot Saddle Mountain to 6,772-foot O'Brien Mountain. The trail ends along State Route 508 a few miles north of Yaak Falls.

After passing Kootenai Falls and the village of Troy, U.S. Highway 2 takes a northwesterly course. Five miles before the highway reaches Idaho, State Route 508 branches off and enters the Kootenai National Forest, initially coming to scenic Yaak Falls, an impressive cascade along the Yaak River. The falls has a pleasant campground. North of Yaak Falls is a side road that follows the east side of the Yaak River to Kilbrennan Lake, a good fishing lake with a small campground adjacent as well as a boat launch area. Off the Kilbrennan Lake Road is another paved road that goes to smaller Loon Lake with another small campground. Loon Lake is surrounded in all directions by 5,000-foot-plus mountain peaks.

State Route 508 continues to follow the Yaak River through great scenery and past campgrounds at Red Top, Whitetail, and Pete Creek. Along Pete Creek is a good place to look for moose. A forest road northward from the Pete Creek Campground follows Pete Creek to pastoral Pete Creek Meadows. By following the West Fork of the Yaak River from here, you come to the Northwest Peak Scenic Area.

Three miles east of Pete Creek Campground is the tiny community of Yaak. Forest Road 92 from Yaak to the north eventually comes to the Lower and Upper Falls of the West Fork of the Yaak River. Observation areas at both falls provide for spectacular views. If you are adventurous, take back roads northward from the Upper Falls to Dusty Peak, within a half-mile of British Columbia, where there is a fine stand of Douglas firs with a carpet of pinegrass on the forest floor.

State Route 508 from Yaak heads southward along the South Fork of the Yaak River and becomes Forest Road 68 for your return to Libby, but there are several things to see and do along the way. By taking Forest Road 746 back to the north, you come to Vinal Lake and Hoskins Lake before reaching Caribou Campground. Vinal Lake is not only a picturesque fishing lake but also where you may hike the Vinal McHenry Boulder National Recreation Trail or drive the Turner Falls Road, and either route brings you to a lovely waterfall on Vinal Creek. Nearby are the pleasant Fish Lakes. Hoskins Lake—actually two lakes surrounded by marshes, alders, and willows—is the site for a proposed Research Natural Area, a wintering range for mule deer and white-tailed deer, and the location of an osprey nest. The main feature of the natural area is a forest of mature Douglas firs and western larches.

Forest Road 68 passes Turner Mountain Ski Area on its way to Libby. The road also passes a back road to the Big Creek Baldy Mountain lookout, which may be rented as a cabin.

East of Libby, State Route 37 follows the Kootenai River to Libby Dam where the river has been impounded to form Lake Koocanusa, which extends for 90 miles northward into British Columbia. The lake cuts a narrow gorge between the Purcell Mountains to the west and the Salish Mountains to the east. The Lake Koocanusa Scenic Byway forms a 67-mile loop around the lake, with State Route 37 following the eastern edge of the lake and Forest Road 228 the western edge. By staying on State Route 37 on the east side, you pass the Koocanusa Marina. From the marina is a side road along Cripple Horse Creek for many miles to the Twin Meadows Guard Station. The guard station is situated above the seven-mile-long Flathead Tunnel that was drilled through Elk Mountain for the Burlington Northern Railroad. This is the second longest railroad tunnel in the western hemisphere. From the guard station is a back road to Bowen Lake on the border with the Flathead National Forest. Near the lake is Bowen Creek Fen, a significant subalpine peatland. The fen is covered with sphagnum that forms hummocks as much as 10 inches high. An active beaver pond is at the south end of the wetland. Bog birch and hoary willow are common shrubs, with several sedges and marsh cinquefoil also present. Botanists have found the uncommon kidney-leaved violet along the creek nearby.

Where State Route 37 crosses Tenmile Creek, there is a three-mile trail to picturesque Tenmile Falls. A little farther along is the geologically interesting Tenmile Talus Slope. Before Forest Road 228 branches off of State Route 37 and crosses Lake Koocanusa for its return along the west side of the lake to Libby Dam, a short side road leads to Sutton Creek Falls.

If you elect to stay on State Route 37 instead of crossing the lake, you come to the community of Rexford and the Rexford Branch Campground. East of the campground is a series of strange rock formations known as the Hoodoos. This area is at the edge of the Tobacco Plains where the climate is dry enough to support cacti and other arid-loving species.

If you cross the lake and take Forest Road 228 along the west side of the lake, at one point a land mass protrudes into the lake between the road and lake. This is a marvelous area known as the Big Creek Research Natural Area. This 320-acre tract has an old-growth forest of western larch, ponderosa pine, and Douglas fir with occasional pockets of giant western red cedars. Some of the largest larches and pines, which have diameters of 2.5 feet, are estimated to be about 400 years old. A number of smaller trees and shrubs are in the area, including Rocky Mountain maple, snowberry, bearberry, red osier, Utah honeysuckle, wild currant, serviceberry, birchleaf spiraea, and hairy ceanothus. Flowers that bloom in season are bluebead lily, bedstraw, pipsissewa, rattlesnake plantain orchid, round-leaved violet, wintergreen, and sweet cicely.

Southwest of Libby are the Cabinet Mountains, with 94,272 acres of the mountains in the Cabinet Mountains Wilderness. In the wilderness you encounter high rocky peaks, groves of giant western red cedars, deep blue lakes, clear cool streams, and several waterfalls. Many animals live in the wilderness, including lynx, cougars, snowshoe hares, pine squirrels, bobcats, weasels, porcupines, mink, grizzly bears, mountain goats, pikas, moose, and wolverines. Cutthroat, rainbow, and brown trout have been introduced into many of the lakes. State Route 56 parallels the western side of the wilderness. Just south of Bull Lake, a side road to the west goes to the Ross Creek Cedar Grove Scenic Area. Along the banks of Ross Creek are giant western red cedars, some of them more than 175 feet tall with a diameter of about eight feet. This lush woods also has large specimens of Douglas firs and an understory filled with ferns and wildflowers. A one-mile interpretive trail takes the visitor through the area. The trail also passes interesting formations such as the Rock Slide, Cedar Chimney, the Wrestlers, Fairy Den, and The Twins.

South of Ross Creek, State Route 56 follows Bull River to the Clarks Fork River and a road junction with State Route 200. You pass the historical Bull River Ranger Station. Where the two highways meet, the Clark Fork River has been impounded to form the Cabinet Gorge Reservoir and, a few miles southeast, the Noxon Reservoir.

State Route 482 south from Libby has side roads that provide access to the eastern side of the Cabinet Mountain Wilderness. From the Libby Creek Campground, a good road follows Libby Creek to Howard Lake and Campground. Just before reaching Howard Lake is an area where you may try your luck at panning for gold. Tepee Lake east of State Route 482 is a small, deep pond surrounded by a floating mat of vegetation. Among the plants found in the lake are the great sundew, three-way sedge, and several species of sedge.

At the far southeastern corner of the Kootenai National Forest, the Forest Service maintains a campground on the shore of McGregor Lake.

Lewis and Clark National Forest

SIZE AND LOCATION: Approximately two million aces in west-central Montana, west and southeast of Great Falls. Major access routes are U.S. Highways 12, 87, 89 (King Hill Scenic Byway), and 191 and State Routes 3, 259, and 360. District Ranger Stations: Choteau, Harlowton, Neihart, Stanford, White Sulphur Springs. Forest Supervisor's Office: 1101 15th Street North, Missoula, MT 59403, www.fs.fed/r1/lewisclark.

SPECIAL FACILITIES: Winter sports areas; boat ramps.

SPECIAL ATTRACTIONS: Chinese Wall; King Hill Scenic Byway.

WILDERNESS AREAS: Bob Marshall (1,009,356 acres, partly in the Flathead National Forest); Scapegoat (239,936 acres, partly in the Flathead National Forest).

Lewis and Clark National Forest has two distinct divisions, and each has very different characteristics. West of Great Falls is the Rocky Mountain Division where the mountains are extremely rugged and where access is limited. To the south and east of Great Falls is the Jefferson Division, which consists of six mountain ranges, all more gentle in character and more readily accessible to visitors.

The Rocky Mountain Division is dominated by two vast wilderness areas, the Bob Marshall Wilderness and the Scapegoat Wilderness. These wildernesses make up nearly 60 percent of the Lewis and Clark National Forest.

If you have not seen the Chinese Wall in the Bob Marshall Wilderness, you have not seen one of the most spectacular geological formations in the world. Unfortunately, most people never see it because it requires a hike of at least 30 miles to get to it. The Chinese Wall forms the 15-mile border between the Lewis and Clark and Flathead National Forests. Its vertical cliffs rise a spectacular 1,000 feet above the land below. The limestone cliffs, more than five million years old, are interrupted at one place, known as Spotted Bear Pass. If you decide to hike or horseback ride to the wall, perhaps the best route is from the end of Headquarters Creek Road. Just before the end of the road, however, take time to enjoy Mills Falls. The trail first follows Headquarters Creek to the Gates Park Guard Station, about the midway point to the wall. From the guard station, follow the trail along Rock Creek. This trail eventually crosses the Chinese Wall at Spotted Bear Pass and into the Flathead National Forest.

More than one million acres are in the Bob Marshall Wilderness, and there is more to be enjoyed than the Chinese Wall. Kevan Mountain is notable for its abundance of trilobite fossils, prehistoric marine arthropods that lived 200 million to 300 million years ago. Prairie Reef is a rugged 8,868-foot peak that offers one of the best views of the Chinese Wall.

Adjoining the Bob Marshall Wilderness to the southeast is the Scapegoat Wilderness. The limestone wall on the side of Scapegoat Mountain is an extension of the Chinese Wall but is a smaller version of it. Several trails lead into the Scapegoat Wilderness from the east side. A trail along the South Fork begins at the South Fork Campground; a trail along Straight Creek and over Elbow Pass to Cigarette Rock begins at the Benchmark Campground. Cigarette Rock is named for its distinctive form.

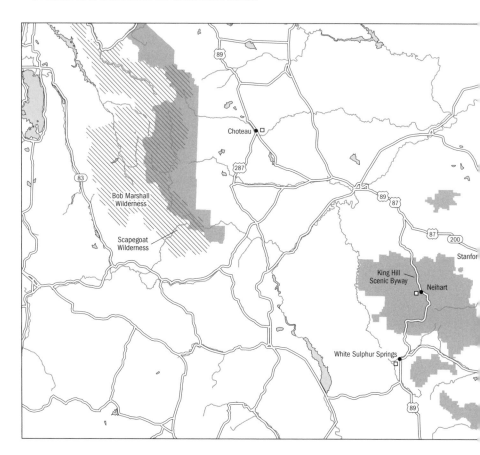

Outside the Scapegoat Wilderness are two waterfalls worth seeing. Double Falls is a pair of spectacular falls two miles west of Ford Creek Resort. It may be reached by the Petty Ford Creek Trail that begins outside the national forest and is four miles from the trailhead. Cataract Falls is in the Horse Mountain area. A short trail from the Elk Creek Road leads to the falls, which are on Cataract Creek, a tributary to Elk Creek.

Sun River Gorge runs along the east side of the wilderness. It is a scenic canyon where Indian pictographs are found on some of the walls of the overhanging cliffs.

Gibson Reservoir, found where the Sun River was impounded, extends from the eastern edge of the Bob Marshall Wilderness for six miles. The Gibson Overlook at the east end permits a grand overview of the reservoir and surrounding areas. County Road 144 westward from Chouteau enters the Lewis and Clark National Forest after about 25 miles. The Cave Mountain Campground is south of Cave Mountain where there are ice caves in the mountain year-round. A few miles southwest of the campground is attractive Mill Falls, reached by a road that follows the South Fork of the Teton River. The road continues north of Cave Mountain to a parking lot for snowmobiling and the Teton Pass Ski Area.

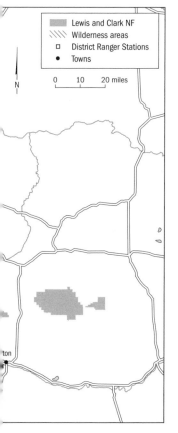

Lewis and Clark NF
Wilderness areas
District Ranger Stations
Towns

0   10   20 miles

N

East and southeast of Great Falls are six noncontiguous units of the Jefferson Division of the Lewis and Clark National Forest, each centered around a mountain range. Due east of Great Falls are the Highland Mountains. The national forest manages about 75 square miles of this range, with a dozen peaks between 6,000 and 7,670-foot Highwood Baldy. The only Forest Service facility in this unit is the Thain Creek Campground.

The largest unit of the Jefferson Division includes the Little Belt Mountains southeast of Great Falls. To get an overview of this area, drive the King Hill Scenic Byway (U.S. Highway 89), which bisects the region. The scenic byway enters the Lewis and Clark National Forest from the north near the nearly ghost town of Monarch. The highway follows Belt Creek, at first following a narrow route between Sun Mountain and Monarch Mountain. Just east of Sun Mountain is Paine Gulch, a narrow, steep-sided defile that rises nearly 1,000 feet on either side. The 2,500-acre gulch is a proposed Research Natural Area where there are forests of Douglas firs, limber pines, and lodgepole pines that alternate with alpine and subalpine meadows. Numerous seeps, springs, and sinkholes occur in the area. An information center is available at the Balk Creek Forest Service Station just north of the Aspen Campground. The highway continues southward past the old Neihart Cemetery before reaching the mining ghost town of Neihart. Dozens of old mines are northeast of town, with back roads leading to all of them, including the famous Glory Hole Mine. Between Neihart and the

Many Pines Campground is a short trail to impressive Memorial Falls on Memorial Creek. A dirt road running east along Jefferson Creek brings you near Tepee Butte (8,248 feet), with the final two miles to the peak available only to hikers. As the scenic byway climbs toward Kings Hill Pass, it passes the Silvercrest Winter Recreation Area and comes to the Showdown Winter Sports Area just west of the pass. The Porphyry Peak Lookout is nearby.

South of Kings Hill Pass, the scenic byway comes alongside Sheep Creek and follows this creek due west. Along Jumping Creek, which enters Sheep Creek, is a pleasant campground.

Large parts of the Little Belt Mountains are both west and east of U.S. Highway 89. Points of interest west of U.S. 89 are the Deep Creek Loop National Recreation Trail, Lick Creek Cave southwest of Crown Butte, the Tenderfoot Experimental Forest, campgrounds along Logging Creek and Moose Creek, and several boat camps along the crooked Smith River, which forms the western edge of the Little Belt Mountains unit. The Tenderfoot Creek Experimental Forest consists of 8,870 acres at the headwaters of Tenderfoot Creek in the Little Belt Mountains. The area is mostly covered by dense stands of lodgepole pines, with occasional wet meadows and grassy slopes. The research at the station is designed to develop methods for sustaining lodgepole pine communities east of the Continental Divide. Lick Creek Cave is in porous Madison limestone and features the Dome Room.

East of U.S. Highway 89 are Tollgate Mountain and the site of the old mining town of Yogo, Yogo Gulch Sapphire Mine, and the site of Barker-Hughesville silver-mining camp of the 1870s. The Yogo Gulch Sapphire Mine produced more than 10 million dollars worth of arguably the world's most beautiful sapphires. The South Fork of the Judith River flows through beautiful limestone canyons and along the western end of Russell Point. Just south of Russell Point is the restored Jake Hoover's sod and log cabin where western artist Charles Russell lived with the old trapper from 1880 to 1882. The subjects of many of Russell's paintings are from this area.

At the southeastern end of the Little Belt Mountains Unit are the double peaks of Daisy Notch and several gorges that have been carved by creeks, such as Daisy Narrows, Nevada Narrows, and Morrisy Coulee Narrows. Several natural springs are in this area, as well.

Just south of the Little Belt Mountains are the Castle Mountains, which form the centerpiece of another small unit of the Lewis and Clark National Forest. The unit features several mountain peaks, particularly on the western side of the range, with Elk Peak (8,566 feet) and Wapiti Peak (8,552 feet) the highest. Grasshopper Creek and Richardson Creek Campgrounds are at the northwestern edge of this unit. Just outside the northern boundary about three miles southeast of Castle Mountain is the mining ghost town of Castle where a few old buildings still stand. The town had a population in excess of 6,000 in 1900.

The southernmost mountain range in the Lewis and Clark National Forest is the Crazy Mountains, most of which is in the Gallatin National Forest.

Two units of the Lewis and Clark National Forest at the far eastern edge of the national forest encompass the Big Snowy Mountains and the Little Snowy Mountains. At the western end of the Big Snowy Mountains, near the top of a mountain, is a perpetual ice cave. You may hike to it from Crystal Lake. Along the way you should keep on the lookout for fossils embedded in the cliffs. East of Crystal Lake is a series of waterfalls known as the Crystal Cascades. You may see several of these sites by hiking the Crystal Lake Loop National Recreation Trail. Several other caves are in the Big Snowy Range, including Devil's Chute Cave. No Forest Service recreation areas are available in the Little Snowy Mountains.

Lolo National Forest

SIZE AND LOCATION: Approximately 2.1 million acres in southwestern Montana, surrounding the city of Missoula. Major access routes are Interstate 90, U.S. Highways 10 and 12, and State Routes 28, 135, and 200. District Ranger Stations: Huson, Missoula, Plains, Seeley Lake, and Superior. Forest Supervisor's Office: Building 24, Fort Missoula, Missoula, MT 59804, www.fs.fed .us/r1/lolo.

SPECIAL FACILITIES: Winter sports areas; boat ramps.

SPECIAL ATTRACTIONS: Lolo Trail; Rattlesnake Wilderness National Recreation Area; Sheep Mountain Bog.

WILDERNESS AREAS: Rattlesnake (32,976 acres); Welcome Creek (28,135 acres); Selway-Bitterroot (1,340,460 acres, partly in the Clearwater and Bitterroot National Forests).

Traveling U.S. Highway 12 from Missoula, Montana, to the Idaho border, important events in the history of the great northwest are recalled. The highway closely follows the historic Lolo Trail that Nez Perce Indians used to reach the Weippe Prairie hunting area in Idaho from Montana's Bitterroot Valley. Although the trail was 120 miles long, only the 29-mile stretch of U.S. Highway 12 from the village of Lolo, which is situated a few miles south of Missoula, to Lolo Pass on the Montana-Idaho border winds in and out of the Lolo National Forest.

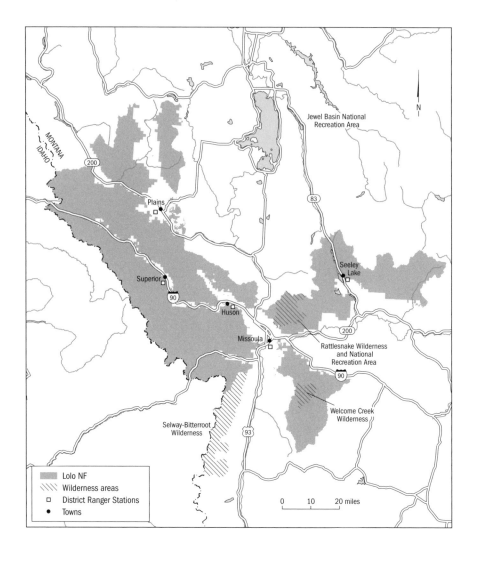

Jewel Basin National
Recreation Area

N

MONTANA
IDAHO

200

Plains

83

Seeley
Lake

Superior

90

Huson

Missoula

200

Rattlesnake Wilderness
and National
Recreation Area

90

Welcome Creek
Wilderness

Selway-Bitterroot
Wilderness

93

Lolo NF
Wilderness areas
District Ranger Stations
Towns

0   10   20 miles

In 1805, Lewis and Clark used this trail on their famous expedition to the Pacific and returned by the same route one year later. At Lolo Hot Springs, members of the Lewis and Clark Corps of Discovery bathed in the springs that are still there today. South of Lolo Hot Springs, the Lolo Trail climbs to Lolo Pass and into the Clearwater National Forest of Idaho. The Lolo Pass Winter Sports Area is at the pass. In 1877, Chief Joseph and his Nez Perce tribe escaped the pursuit of Gen. O. O. Howard's militia by using the same route. Just above the village of Lolo is the site of old Fort Fizzle. The fort was hastily constructed when they thought that Chief Joseph and his men were heading this way. The U.S. military sent soldiers from Missoula to construct Fort Fizzle in an effort to apprehend Chief Joseph and his men. But the encounter never happened because the Indians, in a surprise move, took a different route around the fort. Today a path runs to the Soldier's Corral at Fort Fizzle, which is a replica of the breastworks built in 1877.

The Lolo Trail is but a small part of the vast 2.1 million acres of the Lolo National Forest, which stretches many miles on either side of Missoula in southwestern Montana. The western segment of the Lolo National Forest features the incredibly scenic Bitterroot, Coeur d'Alene, and Cabinet Mountain Ranges, and is bisected by the rushing Clarks Fork River. Interstate 90 provides easy access to this part of the forest.

South of Fort Fizzle is the Selway-Bitterroot Wilderness, but only the small northwestern corner of the wilderness is in the Lolo National Forest. At the edge of the wilderness is the Carlton Ridge Research Natural Area, which preserves parklike groves of alpine larch and whitebark pine at elevations around 8,000 feet. At lower elevations are forests of subalpine fir and Engelmann spruce. The parklike groves are rare for the northern Rocky Mountains.

Interstate 90 stays alongside the Clarks Fork River from Missoula to the Idaho border, and between Interstate 90 and U.S. Highway 12 are several interesting areas in the Lolo National Forest. Immediately west of Missoula is the Blue Mountain Recreation Area with several hiking trails, including the Blue Mountain National Recreation Trail that goes to the Blue Mountain Lookout. Along Interstate 90 just northwest of Missoula is the Forest Fire Research Lab and Aerial Smokejumpers Facility where visitors are welcome.

Three miles south of Interstate 90 along County Road 343 is Montana's largest pine tree, a majestic ponderosa pine that has been living along Fish Creek for hundreds of years. Between Interstate 90 and State Route 12 on Williams Peak, Thompson Peak, and Up Up Mountain are lookout towers. Forest Highway 250 is an exciting route from Superior, on Interstate 90, over Hoodoo Pass and into the Clearwater National Forest. This good surfaced road follows Trout Creek for most of the way, with campgrounds at Trout Creek and Heart Lake, the latter on a side road off of Forest Highway 250. Forest Highway 221 from St. Regis follows Little Joe Creek and South Little Joe Creek to Moore Lake Campground and the nearby spectacular rock slide known as Little Joe Slide. Interstate 90 leaves Montana and the Lolo National Forest where it climbs over Lookout Pass.

The Lolo National Forest north of Interstate 90 and west of Missoula is even more extensive. In the vast region are the Coeur d'Alene Mountains, the Cabinet Mountains, and the Ninemile Divide. Two miles west of Huson on Interstate 90, County Road 457 goes to the Ninemile Ranger Station and the Ninemile Remount Depot National Historic Site. The Remount Depot was built in the 1930s by the Civilian Conservation Corps as a pack stock breeding and training facility. Still a working ranch, it is now the location for the Ninemile Wildlands Training Center where classes and expedition courses are held. North of the Ranger Station is an interpretive nature trail at Grand Menard Picnic Area. From the picnic area are roads to the Civilian Conservation Corps campsite and a fishing pond and campground at Kreis Pond.

Ninemile Divide is a mountainous area between the Ninemile Ranger Station and Interstate 90 to the west. In the heart of the Ninemile Divide is one of the finest observation points in the national forest at Stark Mountain Vista and Lookout, reached either by a hiking trail or a hazardous back road.

At St. Regis on Interstate 90, where Clarks Fork River abruptly curves to the east, State Route 135 follows the river until the highway connects with State Route 200. Along the way is Cascade Falls, one of the most scenic waterfalls in the Lolo National Forest. A mile-long trail leads to the falls from the Cascade Campground. West of Cascade Falls, where Squaw Creek enters the Clarks Fork River, is an unusual habitat on a rocky, south slope that supports dry vegetation. The trees on the dry slope are mostly ponderosa pines and Douglas firs, with bunchgrasses growing on the forest floor. This is the proposed Squaw Creek Research Natural Area.

Interstate 90 and State Route 200 both follow a northwesterly course through the Lolo National Forest. Located between these two highways are the Coeur d'Alene Mountains. Cabin City Campground has an interpretive nature trail, and Gold Rush Campground has a more rugged hiking trail. Eddy Mountain Lookout to the east offers a panoramic view, but it can only be reached by a treacherous four-wheel-drive road.

Near the community of Haugen on Interstate 90 is the Savenac Nursery. Since 1901, the Forest Service has operated a 100-acre nursery capable of producing 12 million ponderosa pine, western white pine, Douglas fir, Engelmann spruce, western red cedar, and lodgepole pine annually.

Ten miles west of the Savenac Nursery, at the townsite of Taft, Forest Road 286 follows Randolph Creek northward and then westward around Taft Peak, following a historic route over Mullan Pass and into the Coeur d'Alene National Forest. This is part of the old military road taken by Capt. John Mullan in 1862 on his 624-mile route between Fort Benton, Montana, and Walla Walla, Washington. Forest Highway 7 from the town of Thompson Falls enters the Bitterroot Mountains west of town and climbs over Thompson Pass and into the Coeur d'Alene National Forest.

As you drive State Route 200 westward toward Thompson Falls, a pullout along the highway on the north side of Clarks Fork River is a great place to look for bighorn sheep on Koo-Koo-Sint Ridge to the north. Koo-Koo-Sint is the Indian name for David Thompson, a fur trader who came to the area in 1809. Koo-Koo-Sint is translated as "the man who looks at stars." The ridge separates Clarks Fork River and Thompson River. County Road 56 is a scenic route along Thompson River that passes the old Copper King Mine on the way to the Clark Memorial. The Sundance Ridge Trail follows a rocky ridgetop for about four miles to Priscilla Peak Lookout and then continues for several miles into the Cabinet Mountains.

Adjacent to Priscilla Peak is the proposed Barktable Ridge Research Natural Area, which has an old-growth forest of mountain hemlocks that is an eastern extension of the range of this handsome species. Growing with the mountain hemlocks are western larch, ponderosa pine, western white pine, whitebark pine, lodgepole pine, Douglas fir, subalpine fir, and Engelmann spruce. The top of Barktable Ridge is flat, but it has extremely steep sides.

Just north of the town of Thompson Falls is a scenic trail that squeezes through Weber Gulch. Also north of town is a forest highway to an overlook above spectacular Graves Creek Falls. A little farther north of Thompson Falls is a forest road that is possibly the most crooked road in the national forest as it climbs to the Cougar Peak Lookout. A magnificent panoramic view awaits you if you can make it to the top. In the vicinity is Petty Creek Research Natural Area, 310 acres of Douglas fir, western larch, grand fir, and lodgepole pine at elevations between 4,000 and 5,000 feet.

The Lolo National Forest is also east of Missoula on both sides of Interstate 90. On Lewis and Clark's return trip from the Pacific, the expedition decided to split into two routes from Traveler's Rest, which is a short distance south of present-day Missoula. On July 1, 1806, Clark and a few men proceeded southward along the Bitterroot River over the same route they had used the year before, whereas Lewis went northeast along the Blackfoot River with nine soldiers, 17 horses, and Lewis' dog Seaman, a black Newfoundland. The two groups were scheduled to meet in August at the Missouri River. When Lewis's party got to what is now called the Monture Creek, they camped. Lewis named the creek Seaman's Creek after his dog, but the name was later changed to commemorate George Monture, a U.S. Army scout. The Forest Service maintains a campground and service facility along Monture Creek. You may reach the campground via a road to the north out of Ovando from State Route 200.

To explore the northeastern part of the Lolo National Forest, take State Route 83 northward from Clearwater Junction on State Route 200. This road brings you to a series of large lakes along the Clearwater River. In order, from south to north, you come to Salmon Lake, Seeley Lake, Lake Inez, Lake Alva, Rainy Lake, Summit Lake, and Clearwater Lake. The Forest Service maintains campgrounds and boat ramps at most of the lakes. North of Seeley Lake is the Clearwater Canoe Trail. If you want a more secluded lake, take a forest road to Lake Elsina at the edge of the national forest where there is a campground.

East of State Route 83 is the Swan Range and the Bob Marshall Wilderness in the Flathead National Forest, but between the highway and the wilderness is a part of the Lolo National Forest where you find Morrell Falls and the Morrell Falls National Recreation Trail, the Morrell Mountain Lookout, and the Pyramid Pass trailhead where there is a trail into the Bob Marshall Wilderness. A 10-mile round-trip hiking trail goes to the Morrell Mountain Lookout from the end of a forest road. On the west side of Pyramid Peak in the Swan Range in the Lolo National Forest is the Pyramid Creek Research Natural Area, which has several old-growth stands of western larch, as well as nice specimens of subalpine fir, Engelmann spruce, whitebark pine, and Douglas fir.

Immediately north of Missoula is the Rattlesnake Wilderness and the Rattlesnake Wilderness National Recreation Area. The western side of the wilderness has gentle mountain slopes, whereas the eastern side has steeper slopes. West of the Rattlesnake Wilderness is the Montana Snowbowl, and south of the Rattlesnake Wilderness National Recreation Area is the Marshall Ski Area. Southeast of the wilderness is Sheep Mountain and a significant wetland.

Within two miles east of Missoula are scenic Pattee Canyon and Crazy Canyon with trails and a campground. Seven miles south of Pattee Canyon along Plant Creek at the northern end of the Sapphire Mountains is a mature stand of 300-year-old western larches and Douglas firs. Engelmann spruces occur along the creek. This area has been set aside as a Research Natural Area.

Several miles south of Pattee and Crazy Canyons are the Sapphire Mountains and the Welcome Creek Wilderness. County Road 102 follows around the eastern edge of the wilderness. The wilderness is only nine miles long and seven miles wide, but it is composed of steep ridges and narrow valleys. Welcome Creek has plenty of trout.

Beyond the south end of the wilderness, the road becomes Forest Highway 102 and continues along the eastern side of the Sapphire Mountains as it follows scenic Rock Creek until it enters the Deerlodge National Forest. This road goes along the west side of Little Hogback Ridge and past Rock Creek Cabin and the Hogback Homestead. The Rock Creek Cabin was reconstructed in 1997 with original sawn-cedar shingles, a rebuilt front porch, and a new back porch. Stanley Lukens, a forest ranger, and his wife lived in the cabin for two to three months a year over a 10-year period. Hogback Homestead was reconstructed in a grassy meadow between the Sapphire and John Long Mountains. The nostalgic two-story cottage has three bedrooms and facilities for cooking. The homestead may be rented from the Forest Service for a daily use fee. Good hiking trails off of Rock Creek run into several gulches and on adjacent ridges.

Sheep Mountain Bog

Waterlogged wetlands with at least a 15-inch layer of peat accumulation are referred to as peatlands. Peat is the dead remains of plants that incompletely decompose because of the water-soaked conditions. The peat usually consists of sphagnum and other mosses, but it may also include the remains of sedges and other aquatic plants. As the layer of peat thickens, oxygen and nutrients necessary for plant growth drop. Plants that grow in layers of peat depend on external sources from the atmosphere or from in-flowing mineral-rich water for their nutrients.

Because of their requirements for cool, humid climates with excessive precipitation, peatlands are commonly found at low elevation boreal areas of Alaska, Canada, eastern Europe, and Siberia. Canadian biologist S. C. Zoltai estimates that 12 percent of Canada's land mass is peatland. The water-holding organic matter of a peatland may permit a peatland to persist for hundreds of years. The decay of organic matter is limited by the anaerobic, acidic, and nutrient-poor conditions of a peatland.

Peatlands are rare in the northern Rocky Mountains of the United States because of the extreme environmental conditions that they require. Because peatlands are scarce in these mountains, the plants that live in them are often rare. In their study of wetlands of Idaho, biologists Robert J. Bursik and Robert K. Moseley indicate that up to 15 percent of the plants in Idaho that are considered rare are found only in peatlands.

Most peatlands in the northern Rocky Mountains occur west of the Continental Divide where Pacific air masses account for a greater amount of precipitation than the amount that reaches the eastern side of the Continental Divide. In areas where midsummer drought is common, however, the growth of peat-forming mosses is limited.

Peatlands may be classified as either bogs or fens, and the differences between these two types have been obscured and often confused in the past. Currently accepted definitions are based on the source of incoming water and nutrients. As peat accumulates and is raised above the water table, bog conditions prevail. Bog plants receive water and nutrients only from precipitation. On the other hand, fens are usually developed on more or less flat surfaces and are concave or only slightly elevated. Fen plants receive their nutrients from water that has trickled through mineral-bearing soil and bedrock or from water that has run off of adjacent uplands and into a creek before entering the fen. Marshes are wetlands with standing water for all or part of the year and the soil is well aerated and rich in minerals. Sedge meadows occur in shallow basins and are drier than bogs and fens and have much fewer accumulations of organic matter.

Sheep Mountain Fen is a peatland in the Lolo National Forest of Montana that occurs as the centerpiece of a 105-acre area that has been designated as the Sheep Mountain Bog Research Natural Area. By applying the definitions described above, the bog in the Research Natural Area is actually a fen that occupies a cirque basin. Studies have revealed organic deposits in the fen that include pollen and spores as well as layers of volcanic ash. The fen itself is covered by a substantial accumulation of snow every winter.

Surrounding the fen are forested mountain slopes. The lower slopes are dominated by Engelmann spruce, Douglas fir, and lodgepole pine, with subalpine fir and whitebark pine on the upper slopes of the Research Natural Area. The Research Natural Area includes 102 acres of forest and three acres of fen.

Sheep Mountain is in grizzly bear country, and black bear, cougar, lynx, mountain goat, and mule deer are common residents as well. The northern bog lemming is known from the general vicinity.

Sheep Mountain is located 12 miles northeast of Missoula and less than three miles southeast of a corner of the Rattlesnake Wilderness. You may reach it via a one-fourth-mile hike from the end of a forest road off of State Route 200, 10 miles east of the Bonner exit of Interstate 90.

Sphagnum forms a nearly continuous mat across most of the fen, with occasional flowering plants protruding from the mat, including the poor sedge. Other plants in the mat are alpine laurel, marsh spike rush, Chamisso's cotton grass, and sticky tofieldia.

A marshy habitat surrounds the three-acre fen and supports a diverse flora of shrubs, wildflowers, grasses, and sedges. Shrubby species are subalpine spiraea, smooth wintergreen, western huckleberry, and Sitka alder. Canby's lovage and false hellebore are common wildflowers. The most abundant grass is bluejoint grass, with beaked sedge the prominent sedge.

The forests in the Research Natural Area that surround the fen have a dominance of Douglas fir, subalpine fir, lodgepole pine, whitebark pine, and Engelmann spruce. A well-developed shrub layer has developed, particularly where adequate moisture is available. Shrubs and small trees in the area are chokecherry, bitter cherry, ocean spray, shiny-leaved spiraea, western serviceberry, snowberry, blue elderberry, mallow ninebark, red mountain heather, and globe huckleberry. On the forest floor are smooth wood rush, one-sided wintergreen, arrowleaf balsamroot, pointed mariposa lily, false Solomon's-seal, Wilcox's penstemon, and round-leaved alumroot. The large and impressive beargrass is found here and there, as is the western woodsia fern.

NATIONAL FORESTS IN NEBRASKA

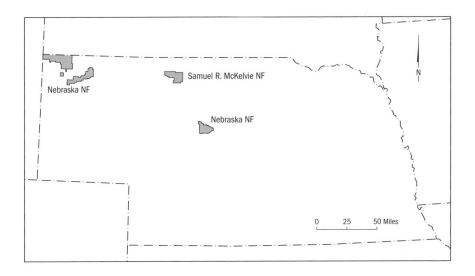

The national forests of Nebraska have undergone several administrative changes in their history. In 1902, two reconstructed forested areas were designated by President Theodore Roosevelt as the Dismal River Forest Reserve and the Niobrara Forest Reserve. These two forest reserves in the sandhills of central Nebraska were combined in 1908 to form the Nebraska National Forest. In 1960, a natural forested area in west-central Nebraska was added to the Nebraska National Forest as the Pine Ridge District. In 1971, however, Congress designated the Niobrara District as the nation's 155th national forest, calling it the Samuel R. McKelvie National Forest in honor of Nebraska's 28th governor. Today, the Nebraska National Forest and the Samuel R. McKelvie National Forest are administered from the same supervisor's office. The national forests of Nebraska are in Region 2 of the U.S. Forest Service.

Nebraska National Forest

SIZE AND LOCATION: 235,000 acres in two districts, one in central Nebraska and the other in northwestern Nebraska. Major access routes are U.S. Highways 20, 83, and 385 and State Highways 2 and 97. District Ranger Stations: Chadron and Halsey. Forest Supervisor's Office: 125 N. Main Street, Chadron, NE 69337, www.fs.fed.us/r2/nebraska.

SPECIAL FACILITIES: Swimming pool; tennis courts.

SPECIAL ATTRACTIONS: Bessey Nursery; Signal Hill Research Natural Area.

WILDERNESS AREAS: Soldier Creek (7,800 acres).

The Nebraska National Forest is most unusual in that the forest in one of the two districts is reconstructed. Prior to 1903, the central part of Nebraska was typical Great Plains, with trees found only along waterways. University of

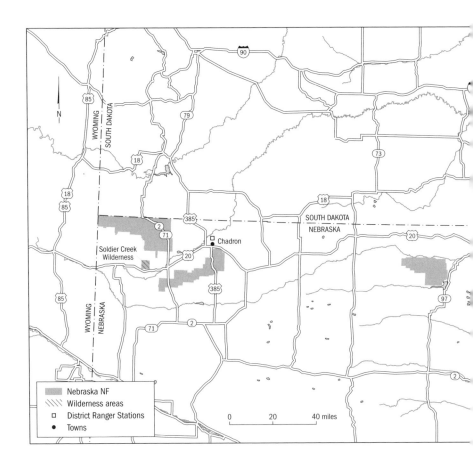

Nebraska botany professor Charles Bessey got the idea that trees should be able to survive and grow in forest stands in the sandy soil of central Nebraska. In 1903, millions of tree seedlings were set out in an area around Halsey. These seedlings did grow and mature, developing into a reconstructed forest that makes up half of the Nebraska National Forest known as the Bessey District.

The Bessey District, which covers a region of about 140 square miles, is in an area of sand where sand dunes and blowouts are common. State Route 2 runs along the northern edge of the forest. The highway is separated from the forest by the Middle Loop River.

At the northeast corner of the Bessey District is the world-famous Bessey Nursery. Although the nursery was established to provide trees for the Bessey Ranger District, it is active today in producing trees for other national forests in Region 2 and for rural landowners. In the past, the nursery has had as many as 20 million tree seedlings at one time, and five or six million seedlings are moved to permanent locations each year. Current production

is about 2.5 million seedlings annually on 46 acres of irrigated seedbeds and in a 4,000-square-foot greenhouse. Trees in production in 2004 include eastern red cedar, Rocky Mountain juniper, ponderosa pine, jack pine, Colorado blue spruce, green ash, wild black cherry, hackberry, bur oak, northern red oak, and swamp white oak. Shrubs such as chokeberry, sand cherry, American plum, and skunkbush sumac are also grown. Visitors are welcome to tour the facility.

Adjacent to the nursery is the Bessey Recreation Complex that has campsites, tennis courts, a softball field, and a swimming pool, the latter very rare in national forests. Circle Road and Natick Road are two loop drives that cover much of the Bessey District. A short spur road goes to the Scott Lookout Tower. Visitors who climb to the top of the tower can get an overall view of the sandhills habitat and the reconstructed forest. The Scott Lookout National Recreation Trail connects the tower with the Bessey Recreation Complex campground.

At the southeastern corner of the district is the Whitetail Campground on the banks of the Dismal River, and near the southwest corner of the district is the Signal Hill Research Natural Area. This 700-acre tract preserves a typical section of the Nebraska sandhills and is hilly with several sand dunes. Typical grass species that cover much of the ground are sand lovegrass, sandhill bluestem, prairie sand reed, little bluestem, blowout grass, Indian grass, switch grass, and sandhill muhly, interspersed with chokeberry, American plum, New Jersey tea, sand cherry, lead plant, Arkansas rose, and a species of yucca. Watch for wild turkeys (pl. 29) in the area.

Many back roads cross the Bessey District, but because most of them are in sandy terrain, do not try to drive them without a four-wheel-drive vehicle. If you drive or hike in the district from March to May, be on the lookout for greater prairies chickens and sharp-tailed grouse. During May and June, the endemic and showy blowout penstemon bursts into full flower.

In western Nebraska south and southwest of Chadron is the Pine Ridge Ranger District, which is completely different from the Bessey District. Much of the Pine Ridge area is a steep escarpment between areas that are more or less flat. Known as the Pine Ridge Escarpment, the region has several buttes with steep drop-offs. The predominant tree in the area is ponderosa pine, a species not found in too many places in Nebraska.

The Pine Ridge District consists of three noncontiguous areas, each comprising 8,000 acres or more, and eight tiny disjunct units that have no recreation facilities. The largest unit is immediately south of Chadron and surrounds Chadron State Park. U.S. Highway 385 bisects the area from north to south, and Kings Canyon Road is a scenic side road off of U.S. Highway 385 that winds through the eastern part of the unit. At the Cliffs Picnic Area are hiking trails into this pretty part of the national forest. Red Cloud Campground is near the southern end of this unit just west of U.S. Highway 385. Spotted Tail Trail is a nice loop hiking trail near the junction of U.S. Highway 385 and Kings Canyon Road.

Another unit of the Pine Ridge District is five miles west of U.S. Highway 385. The Pine Ridge National Recreation Area, an extremely rugged area of buttes and dense forests of ponderosa pine, covers 6,600 acres of this region. Motorized vehicles are not permitted in the recreation area. A part of the Pine Ridge Trail winds through this entire unit of the national forest, beginning near Coffee Mill Butte on the eastern side. The trail crosses several creeks, including East Oak and West Oak. A spur trail runs to the Roberts Campground at the northern end of the unit.

Several miles to the west, on the western side of Fort Robinson State Park, is Nebraska's only wilderness area, the Soldier Creek Wilderness. Trooper Trail, which winds over ridges of ponderosa pine and grassy uplands, begins at the western edge of Fort Robinson State Park. Along the western end of the 4.5-mile trail are two windmills that have pumped continuously since around 1900.

Samuel R. McKelvie National Forest

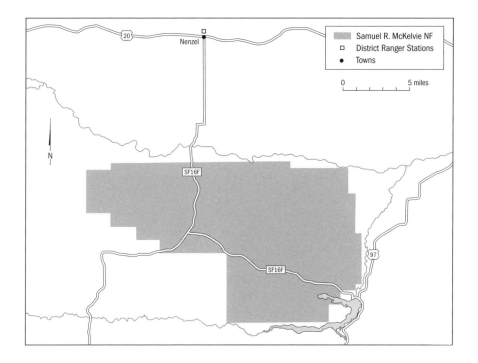

SIZE AND LOCATION: 115,703 acres in north-central Nebraska. Major access routes are State Route 16F and County Road 5. District Ranger Station: Nenzel. Forest Supervisor's Office: 125 N. Main Street, Chadron, NE 69337, www.fs.fed.us/r2/nebraska.

Samuel R. McKelvie National Forest, like the Bessey District of the Nebraska National Forest, contains a reconstructed forest on rolling sandy hills and sand dunes. The planting of trees in the forest in 1903 along the Niobrara River was the result of Charles Bessey, University of Nebraska botany professor, who pushed for the establishment of forested areas in the Great Plains. Today, about 2,300 acres of the national forest have been planted successfully with nearly 750,000 trees. The success rates for the plantings have been best for ponderosa pine, followed by eastern red cedar, jack pine, and the nonnative Scot's pine. In addition, numerous shrubs with edible berries have been planted for their wildlife value.

The only recreation area in the national forest is the Steer Creek site at the southwestern corner of the forest. The Bluejay Hiking Trail makes a loop around the campground. Numerous four-wheel-drive roads crisscross the national forest, providing a firsthand view of the Nebraska sandhills. White-tailed deer, mule deer, pronghorns, and wild turkeys live in the area. Just outside the southeastern corner of the national forest is the large Merritt Reservoir and the Merritt Reservoir Wildlife Management Area, where there are several campgrounds and boat ramps. A neck of the reservoir enters the national forest.

NATIONAL FORESTS IN
NEW MEXICO

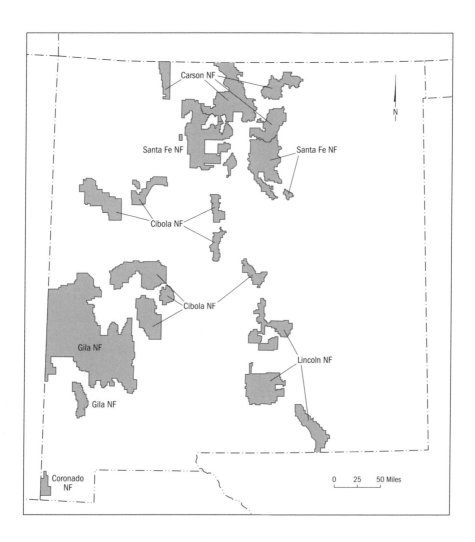

Five national forests are in the state of New Mexico. They are in the South-western Region, Region 3, of the U.S. Forest Service. The administrative office of Region 3 is at 333 Broadway, SE, Albuquerque, NM 87102. The national forests in Arizona are also in Region 3, and they were described in another volume in this series.

Carson National Forest

SIZE AND LOCATION: 1,507,000 acres in north-central New Mexico on either side of Taos, extending northward to the Colorado border. Major access routes are Interstate 25, U.S. Highways 64, 84, and 285, and State Routes 38 (Enchanted National Scenic Byway), 75, 76, 111, 150, 518, 519, 522, 527, 554, and 563. District Ranger Stations: Bloomfield, Canjilon, El Rito, Peñasco, Questa, and Tres Piedras. Forest Supervisor's Office: 208 Cruz Alta Road, Taos, NM 87571, www.fs.fed.us/r3/carson.

SPECIAL FACILITIES: Winter sports area; boat ramps; ATV trails.

SPECIAL ATTRACTIONS: Rio Grande National Wild and Scenic River; Enchanted National Scenic Byway; Clayton Pass; Tajique Canyon.

WILDERNESS AREAS: Latir Peak (20,000 acres); Wheeler Peak, (19,661 acres); Cruces Basin (18,000 acres); Pecos (223,333 acres, partly in the Santa Fe National Forest); Chama River Canyon (50,300 acres, partly in the Santa Fe National Forest).

Approximately 135 million years ago, a great geological uplift formed the Sangre de Cristo Mountains, and this range is the backbone of the Carson National Forest. The Sangre de Cristo Mountains begin where the western edge of the Great Plains ends, and the range is nearly 200 miles long, from a point between Santa Fe and Las Vegas, New Mexico, at the south end to Salida, Colorado, at the north end. The mountains contain Wheeler Peak, the highest summit in all of New Mexico and the southwestern United States at 13,161 feet. Wheeler Peak is in the Wheeler Peak Wilderness, and the top of it and adjacent mountain peaks above 10,000 feet have alpine tundra vegetation.

Taos Mountain range extends north to south in the Sangre de Cristo Mountains and includes Latir Peak Wilderness, Wheeler Peak Wilderness, and the area in between the two wildernesses. Latir Peak Wilderness is at the northern end of this district in a very remote and little-visited area. Most of the wilderness is densely forested except for alpine tundra areas on the highest mountain peaks: Virsylvia (12,829 feet), Cabresto (12,448 feet), Baldy (12,046 feet), and Pinnacle (11,900 feet). Latir Peak, at 12,708 feet, is on private property just north of the wilderness. New Mexico State Route 563 follows the southern boundary of the wilderness and has a side road to Cabresto Lake. From the campground at the lake, there are hiking trails into the wilderness.

Wheeler Peak Wilderness is dominated by 13,161-foot Wheeler Peak near the southern edge of the wilderness, although several other peaks 12,000 feet or higher also are in the wilderness. State Route 150 parallels the northwestern border of the wilderness, ending at the Twining Campground after passing three other campgrounds. Southeast of the campground is the popular Taos Ski Valley. Trails from State Route 150 climb up through several side canyons north of the highway in nonwilderness forest land. Five miles of the Rio Grande National Wild and Scenic River are along the edge of the Carson National Forest a few miles north of Taos. In this area the river flows through a deep gorge with a Forest Service campground and overlook at the end of a forest road.

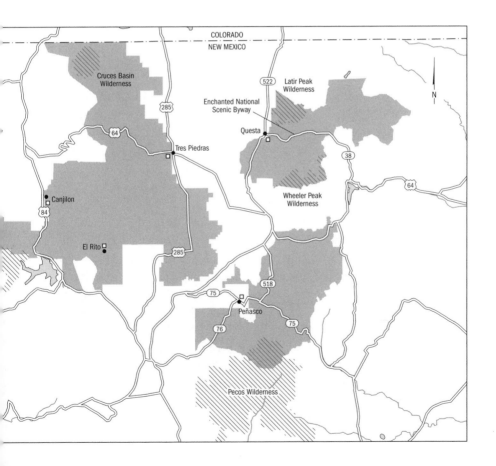

State Route 38 follows the Red River and crosses the Carson National Forest between the Latir Peak and Wheeler Peak Wildernesses. This is the Enchanted National Scenic Byway, and it extends from Questa on the west to Eagles Nest on the east. At the Columbine Campground, you may pick up the Columbine-Twining National Recreation Trail that connects the Columbine and Twining Campgrounds. As the scenic byway continues eastward, there are several campgrounds, the Red River Ski Area, and the Enchanted Forest Ski Area. From the Elephant Rock Campground is a 1.5-mile trail to an interesting formation called Elephant Rock. The highway crosses Clayton Pass where there is a small forest of bristlecone pines. Between the Red River and Enchanted Forest ski areas, State Route 578 branches south off of State Route 38 and follows the Red River past Fourth of July Canyon and Black Copper Canyon before ending at Bull of the Woods Mountain. From the end of the road, a hiking trail into the Wheeler Peak Wilderness passes Lost Lake and Horseshoe Lake on its way to Wheeler Peak. The lower 3.5 miles of the Red River have been designated a National Wild and Scenic River as it passes through a deep and scenic gorge.

East of the Latir Peak Wilderness is the 100,000-acre Valle Vidal Unit of the national forest. This huge area, now managed for wildlife, was a gift from the Pennzoil Company in 1982. The Cimarron Mountains, a part of the Sangre de Cristo Range, extend from north to south through the area. Forest Road 1950 crosses the entire width of the Valle Vidal Unit with Cimarron and McCrystal Campgrounds, Shuree Picnic Area, and Vista Grande along the way. The Shuree Picnic Area is near a series of ponds, and there is a nature trail at the McCrystal Campground. From Vista Grande there are spectacular views in all directions. A huge elk herd lives in this unit, as well as mule deer, black bear, mountain lion, and Merriam's turkey. In the northern part of the Valle Vidal Unit are the Little Costello Research Natural Area and the McCrystal Meadows Research Natural Area. Little Costello is a windswept area of dry, alpine tundra fringed by Engelmann spruce and bristlecone pine. McCrystal Meadows is actually a marsh dotted with small glacial lakes.

A huge section of the Carson National Forest lies south and southeast of Taos. The Santa Fe Range of the Sangre de Cristo Mountains is in this region. U.S. Highway 64 and State Routes 75, 76, and 518 are great routes to access this part of the Carson National Forest. U.S. Highway 64 eastward from Taos follows the Rio Fernando across the northern part of this district. At the El Nogal Picnic Area is the Devisadero Loop Trail that follows a ridgeline to 8,304-foot Devisadero Peak and back. Where the trail is on the south slope of the mountain, the vegetation is composed of piñon pine, juniper, and Gambel oak. On the cooler north slopes, the forest consists of Douglas fir and white fir. From Devisadero Peak you get spectacular views of Taos, the

Rio Grande Gorge, San Antonio Mountain, Taos Pueblo, and Wheeler Peak. Five campgrounds sit along U.S. Highway 64 before the highway leaves the national forest, a few miles north of Angel Fire, and a popular ski area sits outside the national forest.

State Route 76, known as the "high road to Taos," begins from the Indian community of Truchas and continues northward through the national forest to another Indian community, Trampas, to the junction with State Route 75. From this junction, State Route 75 is a scenic route that goes past three campgrounds and the Sipapu Ski Area. A side road up Junta Canyon accesses two more campgrounds, Duran Canyon and Upper La Junta, and continues to the Arellano Research Natural Area. Arellano Canyon is east of La Cueva Peak, and 641 acres of this canyon have exceptionally fine examples of old-growth forests and are in the Research Natural Area. Tall trees of Engelmann spruce, corkbark fir, white fir, Douglas fir, and ponderosa pine are present, as well as some great gnarly bristlecone pines. You can hike to the canyon from the north along an old jeep road. From the community of Ranchos de Taos, State Route 518 goes southward for nearly 20 miles to its junction with State Route 35. Along the way are the Pot Creek Cultural Site and a great observation point on U.S. Hill. From U.S. Hill there is a forest road eastward through Gallegos Canyon to Gallegos Peak. A trail to the top of the peak, which is at an elevation of 10,500 feet, rewards you with a view to the south of Jicarita Peak in the Pecos Wilderness. A beautiful mountain meadow filled with wildflowers during spring and summer is just below the summit. Pot Creek is an ancient Anasazi occupation area where the local Tiwa Indians lived between A.D. 1,100 and 1,320. One may visit the site from late June to early September, viewing a reconstructed Anasazi dwelling and remains of the Indians' irrigation system on a one-mile trail. The Pot Creek site today is home to a forest of piñon pine and juniper, but evidence indicates that during Anasazi times, it was mostly grassland. Four small adobe dwellings and an eight-room pueblo remain. The site is part of the Southern Methodist University Archaeological Field School.

Two miles south of U.S. Hill is Amole Canyon. It offers great hiking and cross-country skiing trails.

The southern end of this district of the national forest is occupied by the Pecos Wilderness, part of which is in the Santa Fe National Forest. The best features of the wilderness in the Carson National Forest are 12,835-foot Jicarita Peak, 12,266-foot Little Jicarita Peak, Hidden Lake, Horseshoe Lake, Trampas Lake, and the San Leonardo Lakes. Several mountain streams run from north to south through the wilderness. The highest forested elevations are dotted with Engelmann spruce, corkbark fir, white fir, Douglas fir, ponderosa pine, limber pine, and bristlecone pine. Aspen groves occur here and there. Rocky Mountain bighorn sheep are residents of the wilderness, as well as black bear, elk, and mule deer.

The western half of the Carson National Forest lies west of U.S. Highway 285, extending from the Colorado state line south for about 50 miles. The Brazos Mountain Range occupies most of this district of the national forest. Within the range are a few mountain peaks that soar above 10,000 feet. San Antonio Mountain (10,908 feet) is only three miles west of U.S. Highway 285 just nine miles south of the Colorado border. Banco Julian (10,413 feet) is several miles east of San Antonio Peak and can be reached by following Forest Road 87 west of San Antonio Mountain. West of Banco Julian is the small Cruces Basin Wilderness. Because of its isolation, this wilderness has some of the most undisturbed areas in the national forest. The wilderness has no mountain peaks, but Brazos Ridge on the western side is spectacular. Several fine fishing streams are in the wilderness. Forest Road 87 circles around the southern and western parts of the wilderness. The Lagunitas Campground is in the vicinity of several small lakes. Brazos Ridge Overlook is a great spot for taking photos.

U.S. Highway 64 from Tres Piedras crosses this district of the national forest. Hopewell Lake is along the way, and the campground there has the highest elevation of any in the national forest.

North of the town of Abiquiu and the Ghost Ranch Conference Center on U.S. Highway 84 is a pretty part of the forest that includes the beautiful and serene Canjilon Lakes. Literally dozens of small, high mountain lakes dot the region, and campgrounds are available at Lower Canjilon Lakes and Upper Canjilon Lakes. The myriad lakes lie just southeast of Canjilon Mountain, whose 10,913-foot summit overlooks the serene Canjilon Meadows to the north. Many small lakes lie in the meadows as well. In the vicinity is secluded Trout Lake and campground, which are surrounded by many small lakes.

Just north of the Ghost Ranch Conference Center is the spectacular Echo Amphitheater on the west side of U.S. Highway 84. This natural rock amphitheater is a geological marvel. Mesa de las Viejas is a huge formation west of Echo Amphitheater that can be reached only by a circuitous route that leaves U.S. Highway 84 about seven miles north of the Echo Amphitheater Picnic Area. By following the road westward, you may drive up on the north side of the mesa, or you may take a southern branch road that takes you to two great vistas, Rim Vista and West Rim Vista. The mesa lies at the upper end of the Chama River Canyon Wilderness, most of which is in the Santa Fe National Forest.

The Jicarilla Ranger District is the westernmost unit of the Carson National Forest, reaching the Colorado state line at its northern extremity. The district covers 180 square miles and is a long, narrow strip 11 miles wide at its widest point and 33 miles long. It consists of a few large mesas and several canyons, but there are no high mountain peaks. U.S. Highway 64 crosses through the center of the district. The Forest Service maintains Cedar Springs and Gas Buggy Campgrounds in the southern section of the region, and Buzzard Park Campground in the northern part.

Clayton Pass

East of Costilla, New Mexico, just south of the Colorado border, a narrow Forest Service road climbs over the Sangre de Cristo Mountains, going past Clayton Pass, a 300-acre tract managed by the Carson National Forest as a Research Natural Area. Less than a mile from a small parking lot at the Clayton horse corral, which borders the southern edge of the natural area, the terrain ascends rapidly from 10,000 feet to an 11,000-foot ridge top, which runs northeast. The lowest zone is occupied by a meadow of oat grass, Thurber fescue, and bluegrass, with an occasional bristlecone pine seedling. As the elevation increases, however, these bristlecones become more abundant, shading out the grasses. Finally, where the slope is extremely steep and rocky, a nearly pure stand of the trees dominates the landscape.

Bristlecone pines receive their common name from the often firm bristles that protrude from the cones. I have previously described the grotesquely twisted bristlecone pines that grow on bleak, windswept peaks in the White Mountains of eastern California. One of those gnarled specimens, barely 35 feet tall, is 4,600 years old. Others, equally ancient, grow on Wheeler Peak, in eastern Nevada. Although botanists have known for more than a century that bristlecone pines also grow in the Rocky Mountains, from central Colorado to northern New Mexico, as well as in Utah and in an isolated stand on the San Francisco Peaks north of Flagstaff, Arizona, none of these other places seem to yield trees of such remarkable longevity.

William Dunmire and his colleagues from the New Mexico field office of the Nature Conservancy, for example, determined that none of the trees at Clayton Pass are more than about 300 years old. Yet they found one noteworthy pine more than 60 feet tall, whereas another behemoth had a circumference of 11 feet measured 4.5 feet above the ground. Among the conditions supporting this vigorous growth are about 30 inches of annual rainfall, which comes mostly between May and October, and the relatively nourishing mixture of orange, chert-bearing sedimentary rock and loamy soil. Although temperatures are low—ranging from 54 degrees F during July to a low 16 degrees F in January—up to 70 inches of snowfall insulate the ground during winter months.

Beginning in 1966, D. K. Bailey of the National Oceanic and Administrative Environmental Research Laboratory, whose interest is the dating of ancient trees, tried to locate bristlecone pines in Colorado that might approach the age of those in California. After searching for five years in vain, Bailey eventually went to Nevada to see some of the ancient bristlecones firsthand. In addition to the twisted growth of the Nevada trees, he noticed one other obvious difference from the Rocky Mountain specimens. The needles of the Nevada bristlecone pines lacked the white, sticky resin that covered the needles of the Colorado trees. This difference had never been detected before, because the resin is not obvious on dried specimens and apparently few botanists had ever compared living specimens of the Colorado and Nevada bristlecone pines.

Suspecting that perhaps the bristlecone pines from the two regions were not the same species, Bailey began looking for other consistent differences between them. He took detailed notes on every bristlecone pine stand known in Colorado, New Mexico, Utah, Nevada, and California and made several revealing observations. The needles of the Rocky Mountain trees (Colorado and New Mexico) have a single dark line extending the length of the underside, whereas the needles from Utah, Nevada, and California trees have two dark lines. After they are about two years old, the needles of the Rocky Mountain trees tend to curve outward, giving the branch a soft appearance; the needles from the western trees remain straight, so that the branches seem coarser. The needles of the Rocky Mountain trees stay on the trees for 10 to 15 years; those of the more western trees persist for 25 to 30 years. When fresh, the needles of the Rocky Mountain trees have a pungent turpentine-like odor; needles from western trees lack this odor.

The cones of the Rocky Mountain trees are very sticky and resinous, and they are difficult to handle because their bristles are very prickly and point directly outward. The cones from the western trees are less sticky and the bristles turn downward and are less prickly. Bailey also noticed that cones from the Rocky Mountain trees can stand on end because their bases are flattened; western cones, which have very rounded bases, topple over.

Apart from compiling structural differences among bristlecone pines, Bailey checked their ranges and ecology. When he plotted the distribution of the hundreds of trees he had examined, he found that all of the trees with Rocky Mountain characteristics are confined to Colorado and New Mexico, along with the single stand in Arizona, whereas all the others are in Utah, Nevada, and California. No overlap occurs. In fact, a distance of 160 miles separates the Rocky Mountain trees from the more western populations. Interestingly, all of the western trees grow in areas where limestone or dolomite lies under the soil; the Rocky Mountain trees live in more acid soils. And the western trees receive less than 30 inches of annual precipitation, most of it in winter, whereas the Rocky Mountain trees receive up to 45 inches, most of it in summer.

Bailey found that the western trees are often more than 3,000 years old, whereas the Rocky Mountain trees never exceed 2,000 years. He concluded that there were two species of bristlecone pines that had distinguishing features and separate habitats. He thus named the longer-living western species *Pinus longaeva,* reserving the traditional name *P. aristata* for the Rocky Mountain populations.

Cibola National Forest

SIZE AND LOCATION: 1,687,720 acres in central and west-central New Mexico. Major access routes are Interstates 25 and 40, U.S. Highways 60 and 85, and State Routes 12, 14, 42, 55, 107, 114, 131, 165, 337, 400, 536 (Sandia Crest National Scenic Byway), and 547. District Ranger Stations: Grants, Magdalena, Mountainair, and Tijeras. Forest Supervisor's Office: 2113 Osuna Road NE, Suite A, Albuquerque, NM 87113, www.fs.fed.us/r3/cibola.

SPECIAL FACILITIES: Winter sports areas.

SPECIAL ATTRACTIONS: Sandia Peak Aerial Tramway; Sandia Crest National Scenic Byway; Continental Divide Loop Route; Paxton Cone; Sandia Cienega; Tajique Canyon.

WILDERNESS AREAS: Withington (19,075 acres); Sandia Mountains (37,232 acres); Manzano Mountain (36,970 acres); Apache Kid (44,530 acres).

The vast and spread-out Cibola National Forest consists of several mountain ranges on all sides of Albuquerque. Because of its nearness to this metropolis, the Sandia Mountains are the most popular and most visited in the national forest. The spectacular granite west face looms 5,700 feet above Albuquerque. The east slope is more gentle and heavily forested with occasional exposed limestone. At the western base of the Sandias is semidesert vegetation, whereas on the crest, alpine vegetation occurs. In the upper parts of the mountain are forests of Douglas fir and corkbark fir, the latter with an ashy gray bark that has the feel of cork. The mountains, which are sacred to Pueblo Indians, were home to the Rocky Mountain bighorn sheep until the 1990s.

An interesting way to reach the top of the Sandia Mountains is by taking the Sandia Peak Tramway, about a 20-minute ride that provides great panoramic scenery, particularly at night when the lights of Albuquerque fill the valley. Another way to reach the mountaintop is to drive the Sandia Crest National Scenic Byway. From the town of Tijeras on Interstate 40, take State Route 14 northward for 5.5 miles to State Route 536. This 13.5-mile scenic byway winds its way to the summit of San-dia Mountain where there is a restaurant. As soon as the scenic byway enters the national forest, there is a short side road (Forest Road 190) to the Cienega Canyon Picnic Area. A handicap-accessible nature trail leads into the cienaga, which is a high mountain marsh. A rugged 2.75-mile hiking trail goes from the picnic area to the Crest Trail, climbing 1,600 feet in elevation. At the Doc Long Picnic Area, the original wood structure built by the Civilian Conservation Corps during the 1930s remains. The scenic byway then twists around Tecolote Peak. At Tree Spring there is a nice trail into the Sandia Mountains Wilderness. The

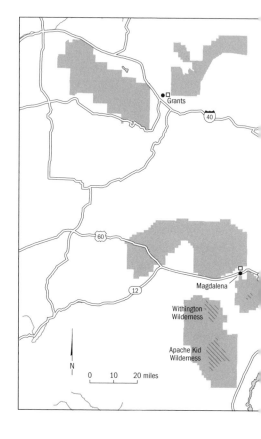

wilderness is mostly on the western edge of the Sandia Mountains. The Crest Trail follows the crest of the mountain for 22 miles. Beyond Tree Spring is the Sandia Peak Ski Area. A lift here, usually open also during summer, takes you to the Upper Tram Terminal. From the Balsam Glade Picnic Area along the scenic byway, State Route 165 enters Capulin Canyon and then follows Las Huertas Creek. Sandia Cave is along this road. In the cave have been found remains dating back 1,000 years. The road eventually exits the national forest and circles westward to Placita and eventually Bernalillo and Interstate 25. The scenic byway terminates at Sandia Crest where there is a restaurant and a nature trail.

Another route into the Sandia Mountains is from a forest road that enters the national forest off of Tramway Boulevard in northeastern Albuquerque. This forest road splits, the right branch going to La Cueva Picnic Area, the left branch going to the Juan Tabo Picnic Area. Both areas have

hiking trails, the La Luz Trail from Juan Tabo the most popular as it climbs an incredible 3,300 feet in eight miles, ending at the Upper Tram Terminal.

South of Tijeras and Interstate 40 are the smaller Manzanita Mountains, the northern extension of the Manzano Mountains. State Route 337 enters this section of the Cibola National Forest at the Tijeras Pueblo Ruins. The road then passes through Cedro Canyon, with several picnic areas along the way. A three-mile-long road climbs to the lookout on Cedro Peak.

The southern part of the Manzano Mountains lies south of the Isleta Indian Reservation. The mountains were inhabited during the Pueblo period until the seventeenth century. The western side of the mountains is precipitous, so most approaches are from the eastern side from U.S. Highway 60 and State Routes 55 and 131. The Manzano Mountains Wilderness is along the western side of the range with Manzano Peak the highest elevation in the wilderness at 10,098 feet. The mountains are unique in being one of the few places in New Mexico for bigtooth maple. The maples may be readily seen in Tajique Canyon and Torreon Canyon. Fourth of July Campground is at the end of the road up Tajique Canyon. All along the eastern edge of the wilderness are canyons and trailheads. A forest road from the community of Manzano goes to the top of Capilla Peak.

Southeast of the Manzano Mountains and west of the town of Corona is a small volcanic mountain range known as the Gallinas. The highest point in this range is Gallinas Peak, which has an elevation of 8,637 feet. Southeast of the peak is Red Cloud Canyon where there are old iron and fluorite-copper mine sites. Three miles southwest of Corona is a forest highway that climbs to the top of Gallinas Peak. Another road goes up Red Cloud Canyon past some of the abandoned mines. Northwest of Gallinas Peak in the Cibola National Forest is Pueblo Blanco Canyon in the Mesa de los Jumanos. You may still see evidence of the Pueblo Blanco here.

Although the Sandia Mountains loom mightily above Albuquerque, they do not have the highest elevation in the Cibola National Forest. That distinction belongs to Mount Taylor, its summit at 11,301 feet. This mountain, west of Albuquerque and near Grants, is in the San Mateo Mountains that are capped in places by up to 300 feet of lava. Mount Taylor and its nearby neighbor, La Mosca, are relatively easy to get to. Lobo Canyon Road, which is State Route 547, climbs into the San Mateo Mountains and comes within four miles of La Mosca Peak and the summit of Mount Taylor. A jeep trail and then a hiking trail take you to each summit. The peaks have fabulous stands of Engelmann spruce.

As you drive the Lobo Canyon Road, you see signs of geologically recent activity with lava fields and pumice. Although there is a lot of Forest Service land north of Mount Taylor, there are no developed Forest Service recreation sites.

West of Grants and located between Interstate 40 and State Route 53 are the Zuni Mountains. This range is roughly 80 miles from east to west and 40 miles from north to south. Lava fields sit at the extreme southern end of the mountains. The Continental Divide runs through the Zunis, and the Continental Divide Loop Auto Tour stays near the crest for a while. The road is in an area that was heavily logged in the past and old logging railroad routes can be seen here and there. This tour begins and ends in Grants. Leaving town on Forest Highway 49, the road enters the national forest and Zuni Canyon. An old logging shack can be seen to the south, and lava beds are now in evidence. The volcanic Paxton Cone is in the vicinity. Near the cone, Forest Highway 50 branches off and heads northwest, paralleling the Continental Divide. Many natural springs dot the area. If you do not want to stay on the auto loop for its entirety, you may take Forest Highway 480 back to Grants. This road passes south of a conspicuous granite dome called Mount Sedgwick. The Ojo Redondo Campground is 1.5 miles southwest of Mount Sedgwick. If you stay on the Continental Divide Auto Tour and Forest Highway 50, you wander in and out of Forest Service land for several miles. When the highway climbs onto McKenzie Ridge, it is on national forest property for several miles. McGaffey Lake is a popular fishing lake, and there is a campground nearby.

After leaving McGaffey Lake, the Continental Divide Tour is now State Route 400, and this road heads north to Fort Wingate and Interstate 40 where you may complete the loop by returning to Grants.

West of Socorro are four units of the Cibola National Forest, each centered around a mountain range. The largest of these districts contains the San Mateo Mountains, but this San Mateo Range is not the same one that Mount Taylor is in. Interstate 25 is at the eastern edge of this district, and State Routes 78 and 52 are at the western edge. At the north end is Withington Wilderness, one of the most remote areas in New Mexico. Steep-walled canyons make this wilderness difficult to traverse. Mount Withington is just outside the western edge of the wilderness and the summit can be reached via a long forest road that originates at State Route 52 and goes through Bear Trap Canyon. West Red Canyon Road is another long forest road that goes from State Route 52 to the Withington Wilderness.

In the southern half of the San Mateo Mountains is the larger Apache Kid Wilderness. History recalls that the marauding Apache Kid, who had been raiding local ranchers, was killed by the distraught ranchers. His grave, at the center of the wilderness, is marked by a blazed tree. Four different trails wind their way to the Apache Kid Grave Site. The wilderness is full of mountain peaks, dry ridges, narrow canyons, and countless springs. 10,139-foot San Mateo Peak is the highest in the wilderness. A forest road from U.S. Highway 85 follows Nogal Canyon to the southeastern edge of the Apache Kid Wilderness where the Springtime Campground is located.

Northeast of the San Mateo Mountains are the Magdalena Mountains where gold, silver, lead, and zinc have been mined in the past. Several roads penetrate this part of the national forest. State Route 114 leaves the southern edge of the community of Magdalena and enters the northern Magdalena Mountains. A forest road from State Route 107 near Magdalena Peak provides access to the mining ghost town of Kelly, which had a post office until 1945. Today, all that remains of Kelly are a church, some rock foundations, stone walls, and a mining scaffold. Several lead mines operated in the area. In the 1900s, rocks in a tailings pile were found to contain zinc carbonate, or smithsonite. This discovery opened up a prosperous zinc-mining activity that lasted until the smithsonite was depleted in 1931. Magdalena Peak has the profile of a woman when viewed from a certain angle.

The most popular route in this region is through Water Canyon at the northeastern corner of the district. After coming to the Water Canyon Campground, the route becomes a dirt road as it makes a circuitous climb up to 10,783-foot South Baldy Peak, the highest elevation in the Magdalenas.

Immediately north of Magdalena is a low range known as the Bear Mountains. This is a very rocky area that has had violent volcanic activity in the past. Because the highest elevations in the Bear Mountains are around 8,300 feet, the forests are nearly all composed of piñon pines and junipers. Forest Highway 354 follows the entire eastern edge of this part of the national forest, and State Route 52 crosses the national forest diagonally from Magdalena northwest through Dry Lake and Corkscrew Canyons. The Bear Mountains have no Forest Service recreation areas.

West of the Bear Mountains is the small Datil Mountain Range. This rocky mountain consists of tuff, conglomerates, and sandstone. Madre Mountain (9,556 feet) and Davenport Peak (9,354 feet), the latter with a lookout, are the highest parts of this range. U.S. Highway 60 cuts across this district of the Cibola National Forest and is a very scenic route as it passes through White Horse Canyon. West of Davenport Peak is a low jagged ridge known as the Sawtooth Mountains. A jeep road goes up the western side of this ridge. Ponderosa pines are at the highest elevations in the Datil Mountains, above an extensive forest of piñon pines and junipers.

Sandia Cienega

The rugged Sandia Mountains rise abruptly east of Albuquerque, New Mexico, extending from their 6,000-foot base to 10,678-foot Sandia Crest in just 2.7 miles and then tapering off somewhat more gradually. Because it faces the rays of the hot afternoon sun, the steep western side of the range is sparsely vegetated, with many bare patches of rock and sand. Many of the plants that grow there—gnarled piñon pines and Rocky Mountain junipers, cholla and prickly pear cacti, yuccas, agaves, bear grass, sagebrush, rabbit brush, snakeweed, and stunted asters—are typical of the Chihuahuan Desert, whose main expanse is farther south. The eastern slopes, in contrast, are heavily forested with conifers. Driving along the scenic highway that snakes up the eastern side to Sandia Crest, one passes through several zones of vegetation, from semidesert at the lowest elevations to Hudsonian conifer forest at the crest.

Several springs and groundwater seeps emerge here and there on the eastern slopes, giving rise to unexpected wetlands in the midst of dense forests. Where standing water remains throughout the year, trees do not take hold, and the marshes that form may range in size from just a few square feet to several acres. In the western United States, such mountainside marshes are referred to as cienagas (from the Spanish *ciénaga*). The Sandia Cienega, about three acres in size, is located on Forest Road 190, just a short detour off the route to Sandia Crest. It falls within Cibola National Forest, which has constructed a wheelchair-accessible trail along the cienaga's east end. (Clumpy vegetation and standing water more than a foot deep can make walking hazardous, so visitors are encouraged to keep to the trail.)

At an elevation of about 7,600 feet, Sandia Cienega slopes very gently from the west to the east. Two spring-fed rivulets form its northern and southern borders, although seepage from the bedrock creates areas of standing water throughout the year, particularly in the eastern and southern sections of the cienaga.

Sandia Cienega is completely surrounded by forest. The rivulet on the northern edge is lined with box elder, bluestem willow, and New Mexico locust. The viny western virgin's-bower, or Rocky Mountain clematis, scrambles over some of these trees. The flowing water of the rivulet is home to the American brooklime, a small blue-flowered aquatic that belongs to the snapdragon family.

The gentle south-facing slope beyond the rivulet is forested with piñon pine, Rocky Mountain juniper, alligator juniper (with its square-plated bark that resembles the scaly back of an alligator), ponderosa pine, and Gambel's oak. Beneath them grow several wildflowers, including blue beardtongue, scarlet beardtongue, white cress, rose geranium, and Mexican squawroot. The squawroot is a six-inch-tall, stubby plant that, lacking chlorophyll, survives by parasitizing the roots of pines and oaks. During summer, yellowish flowers form in the axils of its dull brown scale leaves.

The north-facing slopes on the south side of the cienaga are much more moist and are densely forested with white fir and Douglas fir. Understory plants include white baneberry, red columbine, larkspur, and mountain bluebells.

At the east end of the cienaga, the wetland merges into hardwood forest that overlooks a picnic area. Similar woods appear on the west end, beyond the adjacent road. The hardwood trees are box elder, quaking aspen, and chokeberry, with hop tree the main component of the shrub layer. One of the common wildflowers is the starry Solomon's-seal. Because of the disturbance created by the presence of the trail, road, and picnic area, many weeds of European origin have invaded the eastern and western edges of the cienaga. Two kinds of brome grasses, sow thistle, and prickly lettuce are common.

Where the cienaga remains wet throughout the year but not inundated, the native species include a leafy aster with blue flower heads that open during fall, mountain avens with flowers bearing five yellow petals followed by spherical fruits that are soft and prickly, white rock cress, tall forget-me-not, fall-flowering mountain goldenrod, and golden glow. Golden glow is a five-foot-tall perennial that is related to black-eyed Susan except that it does not have the characteristic dark center of a black-eyed Susan and has deeply lobed leaves.

Toward the eastern end of the cienaga, but extending westward along the north and south sides, standing groundwater is ideal habitat for woolly sedge, small-fruited bulrush, flat-stemmed rush, smooth horsetail, blue joint grass, canary grass, marsh manna grass, water bent grass, and marsh blue-grass. These are all native species that require wet soil and often standing water for at least a part of the year.

Nearly all the trees and wildflowers that live in the piñon–juniper forest and in the Douglas fir forest on the slopes near the cienaga are restricted to the western United States. In contrast, most of the species that live in the cienaga or the adjoining hardwood forest have a wider geographical distribution, with some found as far away as the Atlantic seacoast. Plants such as woolly sedge, blue joint grass, marsh bluegrass, manna grass, smooth horsetail, canary grass, and American brooklime are found all across the northern half of the United States and into adjacent Canada.

This phenomenon that plants living in wetlands generally have a broader geographical range than those living under drier conditions can be observed throughout the country. One possible explanation is that the seeds of wetland plants are ingested by waterfowl or cling to the birds' feathers and muddy feet. Because of the great mobility of waterfowl species, the seeds are disseminated over a wide territory.

Tajique Canyon

During the middle two weeks of October, an orange-red glow brightens Tajique Canyon, in the heart of New Mexico's Manzano Mountains. The color radiates from the leaves of the bigtooth maple, a tree that ranges from Idaho and Wyoming southward into Mexico but occupies only a few canyons in New Mexico. Local residents recall that many more maples once grew in Tajique Canyon, but a severe drought in the 1930s killed nearly half of the stand.

Tajique Canyon lies between 9,509-foot Mosca Peak to the north and Capilla Peak, slightly lower, to the south. A dirt road leading westward from Tajique, a village first settled in prehistoric times, meanders for seven miles up the canyon. Bigtooth maples, which are a kind of sugar maple, become evident three-fourths of the way up the road but are most concentrated in the narrow end of the canyon, accessible only by a trail. Other trees—ponderosa pine, spruce, Rocky Mountain juniper, Gambel's oak, and the pink-flowered New Mexico locust—add variety to the forest.

The sugar maples of the United States are an enigmatic group. The trees grow from Maine to eastern Idaho and from Florida to western Colorado and western Arizona. They are also found in northern Mexico. All have a similar leaf shape, put forth small clusters of flowers after the leaves have expanded, and produce a sugary sap that can be tapped for maple syrup. Yet anyone who has inspected sugar maples throughout the country can see that the leaves of Vermont sugar maples do not look quite like the leaves of big-tooth sugar maples in Tajique Canyon or of southern sugar maples in the Florida panhandle or of those in Wisconsin or southern Illinois.

The leaf differences have led botanists to argue over whether the sugar maples of the United States all belong to one highly variable species or to several different but closely related species. Botanists generally prefer to distinguish between species of flowering plants on the basis of reproductive characteristics (flowers and fruits) rather than of leaf characteristics. Flowers and often fruits tend to be stable, showing little variation in response to environmental conditions. Leaves, on the other hand, may vary considerably depending on whether they are in the sun or the shade, whether they are on main branches or on sprouts, and sometimes whether they are on limbs that have flowers or limbs that do not have flowers.

Charles Sprague Sargent, the foremost student of American trees at the turn of the 20th century, and a botanist who had little trouble distinguishing the great number of American oaks, found it impossible to draw lines of distinction between all the different forms of sugar maple in the United States. Consequently he called them all one species, *Acer saccharum*, commenting that the sugar maple is the most variable of American tree species. Many botanists have agreed, pointing out that the flowers and winged fruits formed by all the sugar maples are virtually indistinguishable. Others, however, have felt that the differences in the leaves are too great to be ignored.

In recent years, new techniques have enabled scientists to analyze and distinguish plants chemically and genetically. As a result of new evidence, botanists are now leaning toward recognition of five closely related species: sugar maple of the Northeast; black maple of the North; southern sugar maple and chalk maple, both of the Southeast; and the bigtooth maple of the West. The genetic differences among these five species probably contribute to some of the leaf differences that have been observed, but other factors are also involved, including geographic location, climate, soil type, and underlying rock substrate.

Whether the trees in Tajique Canyon are bigtooth maples or merely variations on a more general theme, they are a vivid sight during mid-October. The phenomenon of fall leaf coloration is tied in with the aging and shedding of leaves.

In temperate regions where there is a cold season, most species of trees are deciduous. The shedding of leaves prevents damage caused by excessive drying. During the growing season, when water is usually plentiful, a tree takes in enough water to offset the loss of water from its leaves through transpiration. But in winter, water is often tied up in the form of ice. In addition, if living leaves were to remain on the tree during winter, the leaf cells would be damaged or destroyed by the constant freezing and thawing of water within them. (In coniferous trees, whose slender needles are adapted to wintry weather, and in most tropical trees, leaves do not fall annually during a particular season.)

The green color of most spring and summer leaves is due to the presence of great quantities of chlorophyll, the chemical compound that plays a major role in photosynthesis. Small quantities of other pigments—orange and yellow carotenes and pale yellow and brown xanthophylls—are also present in certain leaves but are masked by chlorophyll.

All through spring and summer, most of the leaf cells manufacture sugars. Some sugars are used immediately for needed energy, whereas others are transported to other parts of the plant and stored in the form of starch and other compounds. Still other sugars accumulate in the leaves, where they eventually cause a breakdown of the chlorophyll. As the green pigment disappears, the oranges, yellows, and browns of the carotenes and xanthophylls become apparent. The accumulation of sugars in some leaf cells may also result in the production of bluish and reddish pigments, called betacyanins and anthocyanins.

Some trees, such as aspens, which turn light yellow, display a uniform color because of the dominance of a single pigment. The leaves of ashes often turn a deep maroon, sumacs radiate with red. Sweet gums, instead, offer a variety of colors, even on the same tree. Other trees, including the orange sugar maples, vary considerably in color from place to place or from tree to tree.

As the leaves age further, any remaining sugars are conducted to various food storage organs of the plant, such as the roots. The leaves are then shed through a process called abscission. When a maple leaf is young, hormones inhibit this process. But after the heat of summer, hormonal changes give rise to new regions of leaf cells. Several layers of cells at the base of each leaf stalk become filled with a fatty substance known as suberin, the same substance that gives cork cells their spongy texture. Just above these suberized cells, a separation layer forms as other cells swell with a gelatinous material. At the end of summer, as the cement that holds plant cells together breaks down, these enlarged cells become especially liable to separate. Eventually wind and rain knock the leaf off the plant where the separation layer has developed. The exposed corky cells now serve as a kind of scar tissue, protecting the tree from water loss and the introduction of disease during the long winter months.

Paxton Cone

Volcanoes have been an active force in northwestern New Mexico for the past four million years, beginning with the first violent eruption of Mount Taylor. Some 590 square miles of lava between the Zuni Mountains and Acoma now provide a museum of volcanic phenomena. This arid, inhospitable area is known as El Malpais, "the badlands." Part of the land is private, part is managed by the National Park Service, and part is managed by the Bureau of Land Management. A small area falls within Cibola National Forest.

Paxton Cone, on the national forest land, was created between 10,000 and 40,000 years ago when an eruption sent a river of lava northeastward down Zuni Canyon. Lying about 30 miles southwest of the present-day community of Grants, the cone built up from cinders that fell around the eruption orifice. The lava that flowed northeast was thick and tarlike; it solidified, leaving very rough, sharp surfaces and an intricate network of fissures. This type of lava is called aa (the word is Hawaiian).

Cinder cones are only one of four volcano types found in the Malpais area. The most violent, exemplified by long-extinct Mount Taylor, is the stratovolcano, which ejects material into the upper atmosphere. When it last erupted, Mount Taylor sent tons of lava, cinders, ash, and steam into the air as its crater walls fell inward to form a caldera. Less violent are shield volcanoes, broad, flat volcanoes that often release their energy through several orifices. Shield volcanoes usually can be recognized by multiple craters at the top. Finally, basalt cones, with wide, steep-sided craters, erupt rapidly and send out a rather thin-textured lava that cools to a smooth or somewhat ropy surface. This type of lava, referred to as pahoehoe, is the most common in El Malpais.

At higher elevations, where conditions are relatively cool and moist, the Malpais area is forested with well-developed coniferous trees. Douglas firs and ponderosa pines are found at elevations between 7,000 and 8,900 feet, along with a lower layer composed primarily of Rocky Mountain juniper. Douglas firs, which require more moisture, are found mainly on northern slopes and on rough lava where rainwater tends to accumulate in the fissures. Quaking aspens, which also need a lot of water, can also be found in these locations.

Douglas firs germinate poorly in the lava because of the heated surface of the rock. Botanist Alton Lindsay has found that during summer the surface temperature of the lava rises as high as 129 degrees F. According to Lindsay, the roots of Douglas firs get under the surface crust of the lava and grow along small tunnels that are warm and moist, but contain no soil. As the roots get older, they may break through the thin lava crust and be partly exposed. The growth of many of these trees is stunted due to lack of nutrients and water, and they are often bent eastward in response to the strong prevailing winds. Lindsay, who studied the vegetation patterns on El Malpais for years, found one mature, cone-bearing Douglas fir that was only 16 inches high.

At about 7,000 feet and below, Douglas firs drop out and the plant community is dominated by ponderosa pines, with a variety of shrubs, wildflowers, and grasses often creating an understory. Ponderosa pines have thicker needles than the Douglas firs, and their roots penetrate more deeply, keeping them well supplied with water. And because their very large seeds produce sturdy seedlings that send out roots promptly and deeply, they can germinate in spite of the hot lava surface.

The Douglas fir zone and the ponderosa pine zone extend to lower altitudes in El Malpais than in nearby areas free of lava. As a possible explanation for this, Lindsay suggests that the dark lava becomes hotter than nonlava

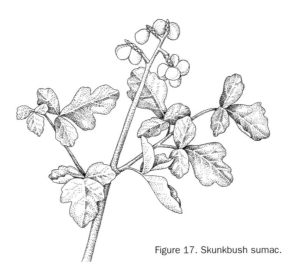

Figure 17. Skunkbush sumac.

rock, stimulating an upward convection of heated air that causes an extra measure of rain to fall on the lava. Rainwater accumulates in the fractured lava long enough for plants growing there to replenish their supply.

Here and there in El Malpais are sinkholes in which water accumulates, draining down from the Zuni Mountains or emerging from natural springs. These oases are home to duckweeds, sago pondweed, and watercress, surrounded by a border of cattails, softstem and three-square bulrushes, reed grass, and swamp milkweed. But at the lowest altitudes, between 6,200 and 7,000 feet, water is usually scarce. Plants that can make it here include piñons, one-seeded juniper, banana yucca, and cacti. Broad-leaved shrubs, such as Apache plume, skunkbush sumac (fig. 17), New Mexico privet, and a couple of gnarly oaks, grow in lava-free zones or where shallow soil has slowly built up in lava fissures. The broad-leaved plants often have some mechanisms to prevent desiccation, such as leaves that are extremely small, succulent, or covered with hairs.

In many places, the aa supports only gray, yellow, or orange lichens, which cement themselves to the black, craggy surface of the lava. Requiring few nutrients for their minimal growth and effectively conserving the moisture in their tissues, the lichens may remain glued to the lava for hundreds of years. Lava does not cover all of the Malpais area, however. Islands of deeper soil with richer vegetation, called kipukas, dot the landscape. Today's kipukas probably resemble the region as it was prior to volcanic activity.

Gila National Forest

SIZE AND LOCATION: Approximately 3.3 million acres in southwestern New Mexico. Major access routes are U.S. Highways 60 and 180 and State Routes 12, 15, 32, 35, 59, 78, 90, 152, 159, 163, 174, and 435. District Ranger Stations: Glenwood, Quemado, Reserve, and Silver City. Forest Supervisor's Office: 3005 E. Camino del Bosque, Silver City, NM 88061, www.fs.fed.us/r3/gila.

SPECIAL FACILITIES: Boat launch areas.

SPECIAL ATTRACTIONS: Gila River Bird Habitat Area; Catwalk; Indian ruins; Inner Loop National Scenic Byway.

WILDERNESS AREAS: Gila (558,065 acres); Aldo Leopold (202,016 acres); Blue Range (29,304 acres).

The Gila National Forest has the distinction of being the first national forest in the country to have a wilderness area when, on June 3, 1924, Gila Wilderness was established at the urging of naturalist and author Aldo Leopold. Because of Leopold's enthusiasm for having wilderness areas set aside, a second wilderness in the Gila National Forest has been named in his honor. Along with the smaller Blue Range Wilderness, nearly 25 percent of the Gila National Forest is in wilderness areas. The three wildernesses are virtually side by side from west to east, with the Blue Range Wilderness abutting Arizona, the Gila Wilderness the middle one, and the Aldo Leopold Wilderness the easternmost.

The Blue Range Wilderness is adjacent to Arizona's Blue Range Primitive Area, which is on the border of the Gila National Forest. Although this very rough and rocky area is mostly dry, limited forests of spruce and fir can be found at the highest elevations. Cacti and yuccas can be found at the lowest elevations, whereas lower forested slopes have open stands of piñon pines and junipers, including three kinds of each. Above the piñon–juniper zone are forests of ponderosa pines, with aspens growing in cooler ravines. Canyons in this wilderness are steep and rock-walled and generally very difficult to hike in. To reach the wilderness, take U.S. Highway 180 nearly 100 miles north of Silver City to the Cottonwood Campground. One mile north of the campground is a dirt forest road to the west that follows the northern boundary of the wilderness all the way to Arizona. At about midway on this road is Pueblo Park Campground, with areas of past Indian occupation in the vicinity. A trail from the campground enters the wilderness and passes Chimney Rock Canyon before coming to Pueblo Creek. A branch trail to the west of Pueblo Creek traverses Tige Canyon. The Brushy Mountains form the eastern boundary of the wilderness.

East of the Blue Range Wilderness is the vast Mogollon Plateau that includes the Mogollon Mountain Range and lesser mountains such as Saliz, San Francisco, Tularosa, Kelly, Mangas, and Elk, all in the Gila National Forest. While hiking in the Mogollons, examination of rocky crevices may reveal some very rare species of plants, including Wright's catchfly, a handsome wildflower with white or pink-tinged petals that are cleft at their tip; Mogollon whitlow grass, a dwarf mustard with tiny white flowers; and Hess's fleabane and rock fleabane, two species that have heads like little asters. Around a few natural springs in the mountains is the Mogollon clover, whose red-purple heads are borne on long stems.

If you continue your drive on U.S. Highway 180 north of the Blue Range Wilderness, you come to the small settlement of Luna after crossing the San Francisco Mountains. At Luna the highway heads westward to Arizona, passing the Head of the Ditch Campground with Indian ruins in the vicinity. To explore the extreme northwestern corner of the Gila National Forest, several gravel and dirt roads can take you north of Luna. Among features in this very remote part of the national forest are Hulsey Lake, Jim Smith Peak (9,275 feet), and several creeks and natural springs.

The Gila Wilderness is the largest in New Mexico and one of the largest in the national forest system. Its southern boundary is about 12 miles north of Silver City and is about 35 miles west to east. The Mogollon Mountains are aligned northwest to southeast in the wilderness, and there are numerous flat-topped mesas, rolling hills, and incredibly deep canyons. The Gila River, one of the few large free-flowing rivers that has never been dammed,

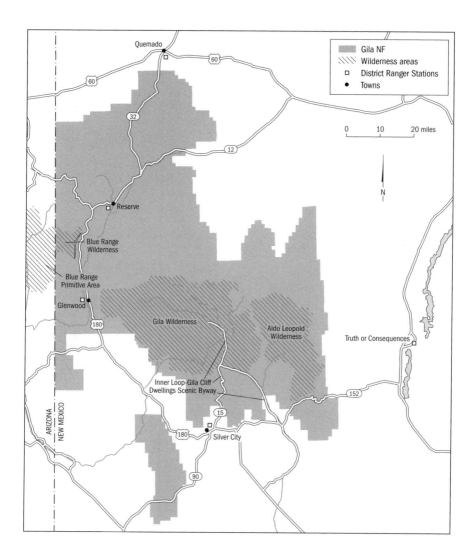

penetrates the wilderness and is often lined by sheer rock cliffs. The Gila River actually consists of three forks—West, Middle, and East—and there are several hot springs, but most of them are very difficult to get to. One of the hot springs reportedly emits water with a temperature of 130 degrees F. Signs of past Indian life are found throughout the forest, including significant well-preserved cliff dwellings. The southeastern corner of the wilderness is separated from the remainder of the wilderness by State Route 15, which goes to the Gila Cliff Dwellings National Monument. The visitor center at the monument is operated jointly by the National Park Service and the U.S. Forest Service. State Route 15 is the upper end of the Inner Loop National Scenic Byway that begins in Silver City as State Route 35. Just outside the southwestern edge of the Gila Wilderness, about 15 miles northeast of the community of Cliff, is the Turkey Creek Research Natural Area. This is a special area where two perennial streams have pristine riparian communities along them. The terrain is rough, with steep, rocky slopes leading into the canyon bottom. In the canyon you see Wright's sycamore, narrowleaf cottonwood, Arizona alder, Arizona walnut, velvet ash, box elder, and chokecherry. Less common

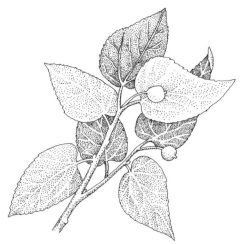

Figure 18. Netleaf hackberry.

trees include Chihuahua pine, silverleaf oak, netleaf hackberry (fig.18), New Mexico locust, and white fir. To get to this area requires about a five-mile hike, most of it along an old jeep road, through Brushy Canyon from the end of Forest Road 155.

The Black Range is the major topographic feature of the Aldo Leopold Wilderness, and the terrain is exceedingly rough with deep canyons, steep, rocky cliffs, and dense forests. Only a strip of land less than two miles wide separates the Gila Wilderness from the Aldo Leopold Wilderness, and this narrow corridor is traversed by Forest Road 150. The Continental Divide crosses the wilderness, paralleled by the Continental Divide National Recreation Trail. Reeds Peak, at 10,015 feet, is the highest in the wilderness and is situated on the Continental Divide. Numerous springs are dotted through the wilderness and there are several sparkling streams but few lakes. Diamond Creek is home to the endangered Gila trout.

Despite all of the roadless areas in the national forest, much of the Gila can be observed by driving the numerous roads that cross the mountains. The Inner Loop–Gila Cliff Dwellings Scenic Byway is a scenic and historic 110-mile paved route that begins and ends in Silver City. Leave Silver City on State Route 15, heading north toward the Gila National Forest. Just before reaching the national forest boundary is the tiny town of Pinos Altos, whose residents try to keep the village appearing the way it did decades ago, with dirt streets and wooden sidewalks. Immediately north of Pinos Altos, the Continental Divide National Scenic Trail crosses the scenic byway. The scenic byway then enters the national forest and becomes a very narrow, winding road through a dense ponderosa pine forest. A pullout and interpretive sign show and explain an arrastra left over from gold mining days. An arrastra is where gold-bearing ore was crushed by a boom that was dropped from mules. Where the scenic byway makes one of its sharpest bends, there is a plaque in memory of mountain man and hunter Ben Lilly. A short trail from the plaque leads to a super view of the surrounding forest. The scenic byway then follows Cherry Creek lined by rock columns in Cherry Creek Canyon. Cherry Creek and McMillan Campgrounds are nestled along the creek in picturesque settings. Less than two miles beyond McMillan Campground are two trails. The trail to the southeast climbs to 9,001-foot Signal Peak, an old Forest Service lookout that was the site of a heliograph station many decades ago.

After several twists and turns, the scenic byway comes to a dirt side road that goes for about five miles through Sheep Corral Canyon. At Lookout Point there are extensive views of the mountains to the west. Although you are at an elevation of about 7,000 feet, the vegetation around you consists of piñon pines and junipers and even some cacti. The scenic byway straightens out a bit and descends northward to an intersection with State Route 35. Stay on State Route 15 and proceed northward in a narrow corridor separating two lobes of the Gila Wilderness. The scenic byway now follows Copperas Creek, climbing past Copperas Peak on your left to a superb overlook that provides magnificent views into the wilderness. A trail from here also enters the eastern lobe of the wilderness and eventually climbs around to the south where it terminates at State Route 35 just east of Lake Roberts. More campgrounds sit along the byway just before the road reaches the privately owned Gila Hot Springs. The scenic byway then ends at the remarkable Gila Cliff Dwellings where you should stop at the visitor center before starting the mile-long loop trail to the ancient dwellings. The dwellings stand 175 feet above the valley floor. The rooms were occupied by Mogollon Indians during the thirteenth and fourteenth centuries.

After learning about the cliff dwellings, retrace your route on State Route 15 southward to its junction with State Route 35 and follow the latter road eastward to continue on the Inner Loop Scenic Byway. Lake Roberts is a 72-acre lake that is a fisherman's paradise. Self-guided nature trails start at the Upper End Campground, one of them going to a cienaga, or marsh, where a good variety of wetland plants grows, and the other is a trail past ancient pictographs. From the Vista Point, there is a trail to prehistoric Mimbres Indian ruins that have remained since the Mimbres occupied the area between A.D. 200 and 1150. Trees you may see while hiking in the Lake Roberts area are piñon pine, ponderosa pine, southwestern white pine, Chihuahua pine, Gambel's, gray, and silverleaf oaks, Arizona walnut, velvet ash, Douglas fir, three kinds of junipers and, in wetter areas, Arizona sycamore, Fremont's cottonwood, narrowleaf cottonwood, and a complex of wetland species.

Common shrubs are Apache plume, cliff rose, Fendler bush, mountain mahogany, squawbush, wild olive, and Wright's silk tassel. Be on the lookout for poison ivy. At least seven different kinds of cacti have been found in the Lake Roberts area, as well as numerous wildflowers.

The scenic byway continues in the national forest along Sapillo Creek. About eight miles east of Lake Roberts, Forest Road 150 follows the narrow corridor that separates the Gila and Aldo Leopold Wildernesses. This dirt road is a great one, but not for the fainthearted. For a few miles it stays on the Continental Divide, paralleling the national trail. Between the two wildernesses, there are three secluded campgrounds from which you may hike into the wilderness areas. The forest road continues to state-owned Wall Lake where the road improves as it goes to the Forest Service's Beaverhead Work Center on State Route 59.

The scenic byway leaves the national forest and intersects with State Route 152, which returns to Silver City. East of town is the site of old Fort Bayard, once a cavalry post and now a hospital. From near here are the Sawmill and Wood Haul Wagon National Recreation Trails, both winding northward back into the Gila National Forest.

A part of the Mimbres Mountains south of the Aldo Leopold Wilderness may be explored from State Route 152, which crosses the Gila National Forest between San Lorenzo and Hillsboro. This is a highly scenic road that enters Ancheta Canyon as soon as it leaves San Lorenzo. The Upper Gallinas and Iron Creek Campgrounds are on the route as it climbs to the Continental Divide where there is a fine viewpoint at Emory Pass. From the pass, a one-mile hiking trail runs northward into the Aldo Leopold Wilderness. An equally interesting trail from the pass to the south is not in the wilderness. This trail forks after about one mile, the left fork going into Silver Creek Canyon and the right fork, which is much longer, passing Sawyers Peak and Seven Brothers Mountain. East of Emory Pass, State Route 152 becomes extremely crooked as it descends from the Continental Divide. Just before the highway leaves the national forest, it comes to the ghost town of Kingston. This town sprang up in 1882 when a rich lode of silver was discovered nearby. By the time the Solitaire Mine was in full operation, Kingston boasted of a population of 1,800 people. Although the town had a three-story hotel, a church, and reputedly 22 saloons, it dwindled rapidly after the silver had been extracted. Only a handful of residents live in Kingston today.

Two trips that are musts for exploring the Gila National Forest begin near the small community of Glenwood on U.S. Highway 180, about 60 miles north of Silver City. State Route 174 from Glenwood is only five miles long, ending at Whitewater Canyon and the Catwalk, the most visited area in the Gila National Forest. The highway follows Whitewater Creek to a picnic area.

Gold and silver were discovered around 1889 in the mountains above White-water Canyon where the picnic area is now located. In the canyon, John T. Graham built a mill in 1893 to process the ore, and the small town of Graham (later changed to Whitewater) developed here. The ore deposits were located about four miles upstream from the little town and mill, and the narrow, rocky canyon between the mill and town prevented passage between the two. Plenty of water was available upstream where the mine was located, but by the time the stream reached the mill and little town below, it was often reduced to a trickle or even was dry. For water to get from the mine to town, a four-inch-diameter metal pipe was built in the canyon. Because the pipeline needed frequent repairs, workers had to walk the narrow line, which became known as the Catwalk. After the mill was closed in 1913, the pipeline fell into disrepair. In 1935, the Civilian Conservation Corps rebuilt the Catwalk out of wood as a recreational attraction in the national forest. In 1961, the Catwalk was replaced by a metal structure that hangs from the spectacular cliffs above the bouldery Whitewater Creek.

A more lengthy scenic drive over mostly gravel roads is the Outer Loop Scenic Drive. Begin this adventure on State Route 159 north of Glenwood. The road climbs to the once-thriving mining town of Mogollon, surrounded by the remains of old mines at an elevation of 6,600 feet. Gold and silver were mined as early as 1875. By 1904, 40 percent of the gold and 25 percent of the silver mined in New Mexico came from the deposits around Mogollon. The town now caters to tourists and those interested in mining history. From Mogollon, the Outer Loop swings southward to the northern edge of the Gila Wilderness. All the way to the Gilita Campground are several trailheads for hikes into the wilderness. The road then loops around to lovely Snow Lake, at 7,400 feet, one of the highest lakes in the national forest. Fishing is good here, and the surrounding forest provides an opportunity to study the species of trees in the region.

East of Snow Lake the road becomes dirt as it continues to stay a short distance north of the Gila Wilderness. After the Outer Loop Drive climbs to Black Mountain Mesa, it becomes State Route 59 and is gravel to the national forest's Beaverhead Work Center. Along this stretch is the Wolf Hollow Trailhead where hikers may take one of the most popular trails into the wilderness. After passing through Wolf Hollow, the hiking trail climbs to the Black Mountain Lookout at the edge of the wilderness.

From the Beaverhead Work Center, you may either take Forest Road 150 southward between the Gila and Aldo Leopold Wildernesses, or you can proceed eastward on the now-paved State Route 59. This highway winds, twists, and climbs for nearly 30 miles to the eastern edge of the national forest, at one time climbing over the Continental Divide just beyond Burnt Canyon

Flat. From Beaverhead to the forest boundary, the Outer Loop follows Indian Creek and then Poverty Creek before going through Wildhorse Canyon.

Between Glenwood and Luna, State Route 12 to the east is the best access for points of interest in the northern part of the Gila National Forest. At the village of Reserve, where there is a district ranger station, there are several options. Paved State Route 435 south of Reserve follows the San Francisco River for a few miles before turning eastward up Willow Spring Canyon. The highway follows a series of switchbacks to the Sign Camp Ridge where the route becomes gravel. Side roads to the southwest go to Devil's Park and Devil's Den and to the east through Patch Canyon to Six-shooter Saddle. If you stay on the main gravel road, it passes several side canyons and goes through others, eventually coming to the Negrito Work Center. From here you may take a gravel road southward to Negrito Mountain past Adam Lake and to the Gilita Campground on the Outer Loop Road, or you may take the dirt road eastward that ends in the Elk Mountains, or you may take the gravel road northward that seems to be endless as it follows the western side of the Continental Divide. This very interesting road passes several openings such as Turkey Park and Collins Park, several springs, small lakes, and peaks as it crosses the Tularosa Mountains. The road eventually swings to the northwest and joins State Route 12 at the primitive Apache Creek Campground.

East of Reserve is a forest road that goes into the heart of the Tularosa Mountains, ending a short distance north of Buzzard and Eagle Peaks. North of Reserve, State Route 12 continues northeast to the Apache Creek Campground (see above) where there is a road junction. State Route 12 continues eastward, crossing the Continental Divide just south of the main Mangas Mountain Range where there is a primitive campground at Valle Tio Vinces Springs. The Continental Divide National Recreation Trail crosses the campground. Three miles to the east along the trail is 9,961-foot Mangas Peak, after which the trail drops into Mangas Canyon and exits the national forest. Gravel roads to the west of the campground climb onto Slaughter Mesa and into the Gallo Mountains and to the north come to 131-acre Quemado Lake, where there are three campgrounds and a boat ramp. El Caso Peak looms over the eastern end of the lake.

From the road junction at Apache Creek Campground, State Route 32 branches off of State Route 12 to the north. This highway climbs into the Gallo Mountains and eventually comes to State Route 103, which in four miles reaches Quemado Lake. Where State Route 103 intersects State Route 32 is an interesting rock formation known as Castle Rock.

One disjunct unit of the Gila National Forest is west of Silver City and includes the Big Burro Mountain Range, covering an area 30 miles from north to south and about 11 miles from east to west. The Continental Divide cuts across the eastern edge of the range. This area has limited access, and only a few points of interest can be readily reached. At the northern end of the mountains is the Gila River Bird Habitat Area. At the southwestern corner of this preserve is the Gila River Research Natural Area. Both of these areas are of interest to nature lovers. The birding area is along the Gila River and has been further enhanced by the Forest Service bank stabilization project, which includes planting riparian vegetation and creating marshes. Bird watchers have recorded 337 species of birds from the area, including 166 species known to breed in the Gila National Forest. One of the nesting species is the southwestern willow flycatcher, one of the nation's endangered species. To reach the Gila River Bird Habitat Area, drive north of Cliff on U.S. Highway 180 and turn on the road to Bill Evans Lake. After 3.5 miles, the gravel road to the right goes to the northern end of the area.

The Gila River Research Natural Area, which contains the 600-foot-deep Gila River Canyon, was initially set aside to protect hawks, falcons, herons, songbirds, and native fish. The native fish in the Gila River here are spikedace, loach minnow, longfin dace, Gila coarse-scaled sucker, and Gila mountain sucker. Other rare inhabitants of the natural area are the Gila monster and Arizona coral snake. The river is lined with cottonwoods, sycamores, ashes, and hackberries. Below 4,000 feet elevation are 400 acres of desert scrub. The desert scrub includes several species of yucca, sotol, mesquite, shrub live oak, turpentine bush, snakeweed, wolfberry, oreganillo, and several prickly species such as whitethorn acacia, wait-a-minute bush, ocotillo, and prickly pear cacti. In dry washes are such plants as alligator juniper, burrobrush, rock sage, hedge nettle, seep willow, and several large grasses, including bull grass muhly, deer grass muhly, and bristle grass.

Lincoln National Forest

SIZE AND LOCATION: 1.1 million acres in south-central and southeastern New Mexico. Major access routes are U.S. Highways 54, 70, 82, and 380 and State Routes 24, 37, 48, 130, 137, 246, 248, 349, 388, and 563 (Sunspot National Scenic Byway). District Ranger Stations: Carlsbad, Cloudcroft, and Ruidoso. Forest Supervisor's Office: Federal Building, 11th and New York, Alamogordo, NM 88310, www.fs.fed.us/r3/lincoln.

SPECIAL FACILITIES: Winter sports areas.

SPECIAL ATTRACTIONS: Smokey Bear Museum; Sitting Bull Falls; Billy the Kid National Scenic Byway; Sunspot National Scenic Byway.

Three distinct units make up New Mexico's Lincoln National Forest, and the elevations within the national forest range from low desert to mountain peaks, one of which tops out at 12,003 feet. As a result, a wide range of vegetation communities may be observed; the Lincoln is one of the few national forests that encompasses part of the Chihuahuan Desert.

The southeastern district abuts Carlsbad Caverns National Park and Guadalupe Mountains National Park and extends to the Texas border. It consists primarily of a part of the Guadalupe Mountain Range. The middle district lies east of Alamogordo, with the Sacramento Mountains at the northern end of this unit. The northern district, east of Carrizozo and north of Ruidoso, contains the massive Capitan Mountain, part of the Sacramento Mountains, and the lesser Jicarilla Mountains.

The Guadalupe Mountains, extending from west of the Pecos River in Texas into New Mexico, were once under a great sea. A massive limestone deposit formed the Capitan Reef. Because of the limestone, the reef developed an extensive cavern system that today may be seen in its grandeur at Carlsbad Caverns National Park and, to a much smaller extent, in Cottonwood Cave in the Lincoln National Forest. The Guadalupe Mountains today rise above the Chihuahuan Desert. This region was occupied by Indians at least 12,000 years ago, and the Mescalero Apaches still lived there when the first white settlers arrived. Although the area at the base of the mountains received limited rainfall, several natural springs provided water so that ranching became an important activity of the earliest settlers. To ensure the protection of these watersheds of the upper Pecos, President Theodore Roosevelt created the Guadalupe National Forest in 1907. The following year, he combined the Guadalupe and Sacramento National Forests into one, calling it the Alamo National Forest. In 1917, President Woodrow Wilson added more land to the national forest and changed the name of all of these parcels to Lincoln National Forest.

The Guadalupe District, the southernmost in the national forest, extends for 41 miles north of the Texas border. Around the base of the mountains at about 3,500 feet is the Chihuahuan Desert, and typical Chihuahuan plants such as creosote bush, lechuguilla, mescal, agave, yucca, and sotol are readily abundant. Rare plants occur here, as well, including the southwestern barrel cactus and spoonleaf rabbitbrush. Between 5,000 and 6,500 feet, the vegetation in this elevation zone of the Guadalupe Mountains consists of piñon pine, one-seeded juniper, mountain mahogany, prickly pear cacti, and blue grama. The rare threadleaf wild carrot occurs at this elevation. Above the

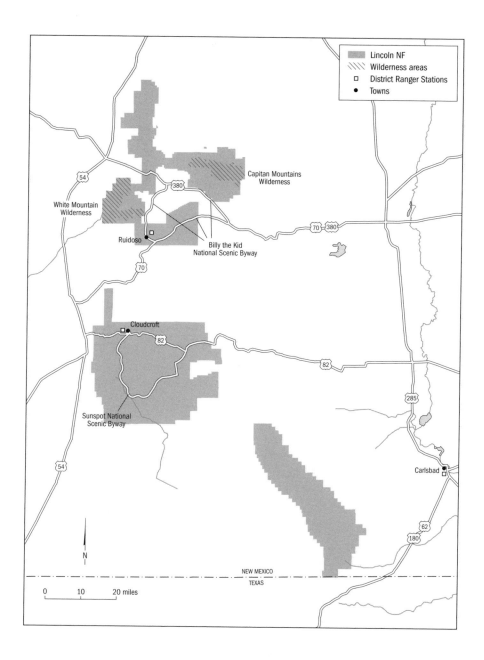

Lincoln NF
Wilderness areas
□ District Ranger Stations
● Towns

Capitan Mountains Wilderness

White Mountain Wilderness

Ruidoso

Billy the Kid National Scenic Byway

Cloudcroft

Sunspot National Scenic Byway

Carlsbad

NEW MEXICO
TEXAS

0   10   20 miles

N

piñon–juniper community in the uppermost forests of the mountains are ponderosa pine, Douglas fir, and limber pine. Rocky canyons permeate the Guadalupe Mountains, and bigtooth maple, chokecherry, walnut, hop hornbeam, and Texas madrone are the woody species found most commonly here. Several rare species of plants live in the limestone crevices, including Hershey's cliff daisy, five-flowered rock daisy, Wright's spider lily, McKittrick pennyroyal, supreme sage with its pink petals dotted with red, Guadalupe beardtongue, Lee's pincushion cactus, and curl-leaf needlegrass. In a few seeps at the base of limestone cliffs may be found Chaplin's columbine, an attractive and delicate yellow-flowered beauty. Geronimo, the Apache chief, claimed that the richest gold deposits in the world were in the Guadalupe Mountains, but apparently only one old prospector ever found gold, and he never divulged its location.

State Route 137 enters the Guadalupe District from the eastern side. After entering the national forest, The Pinnacle is an attractive rock formation immediately to the south. The highway crosses the entire district, eventually climbing over the Guadalupe Mountains before leaving the national forest. However, about six miles west of the Guadalupe Work Center, one of the most scenic roads anywhere in the Lincoln National Forest heads south. This is Forest Road 540, which has several outstanding vistas, including Five Points Vista. After this vista, the road finally peters out near Devil's Den Canyon. Beyond that, the road is strictly for four-wheel-drive vehicles until it also ends just above the Texas border and North McKittrick Canyon. This canyon is on Forest Service land, but it then crosses into Guadalupe Mountain National Park and joins the main McKittrick Canyon in the park. In these two canyons some of the rarest plants of the region occur. Animals that you could encounter in this area are black bear, mule deer, mountain lion, and ringtails.

Fewer than four miles west of the Forest Road 540 turnoff, a lengthy forest road known as Rim Road goes northward along the entire length of the Guadalupe Mountains. To the west, the mountains drop 2,000 feet to Pecos Valley, whereas to the east, the mountains slope more gently. The Rim Road passes several scenic canyons and picturesque springs, but there are no developed recreation areas. Great vistas of Dry Canyon and the Brokeoff Mountains may be enjoyed.

The southeastern corner of the Guadalupe District is extremely rough and hard to reach. A very rough four-wheel-drive road off of Forest Road 540 climbs over Guadalupe Ridge to Dark Canyon, the site of a Forest Service lookout tower. In Dark Canyon are Indian paintings on some of the rock walls. In this area are several caves, including Cottonwood Cave and Hidden Cave. These may be visited only when accompanied by a forest ranger. Contact the district ranger in Carlsbad. Cottonwood Cave has stalagmites, stalactites, soda straws, gypsum draperies, and cave pearls.

The most visited attraction in the Guadalupe District is Sitting Bull Falls. Forest Road 276 westward off of State Route 137 leads to this attractive area where there are picnic facilities.

The large district of the Lincoln National Forest whose western edge is adjacent to the city of Alamogordo is flanked on its western side by the Sacramento Mountains, which curve around to the south. The eastern side of the district has less mountainous terrain, but there are still some low peaks. The vast Tularosa Valley is west of the national forest and extends southward for 100 miles to El Paso, Texas. U.S. Highway 82 leads eastward off of U.S. Highways 54 and 70 fewer than two miles north of Alamogordo. The highway enters the national forest through Dry Canyon and follows an old railroad route that hauled timber between Alamogordo and Cloudcroft. Called the Cloud-climbing Railroad, it had 58 trestles, the largest of which still stands a short distance outside of Cloudcroft. The railroad was in existence from 1890 to 1945. The 2.5-mile-long Osha Trail begins at the trestle.

A side road to the north off of U.S. Highway 82 goes through Fresnal Canyon and connects with a road just east of La Luz. This road heads eastward through La Luz Canyon and eventually circles its way around to Cloudcroft on the crest of the mountain. By staying on U.S. Highway 82, however, the highway goes through a tunnel where there is a parking pullout for those wishing for a great view. A hiking trail from the tunnel also goes down to a stream. After passing through a couple of small settlements, the highway goes through the eastern part of Fresnal Canyon and then climbs by means of hairpin turns to Cloudcroft. This town, founded in 1899, is a popular winter sports destination, although there is no lack of summer activities here as well. Surrounding Cloudcroft are several campgrounds, with hiking trails available at the Apache Campground and Sleepy Grass Campground. A trail for the blind is available at La Pasada Encantada on the road through Apache Canyon. If you hike in the mountains above 9,000 feet, you encounter forests of Douglas fir and white fir, with several attractive high mountain wildflowers. At somewhat lower elevations you encounter forests of ponderosa pine, Engelmann spruce, New Mexico locust, Rocky Mountain maple, and chokecherry, with many kinds of wildflowers including Indian paintbrushes, beardtongues, lupines, wild geraniums, and various members of the aster family.

Cloudcroft is the hub for several driving tours. State Route 563 to the south is the Sunspot National Scenic Byway and is a beautiful 15-mile drive along the crest of the Sacramento Mountains. This is a two-lane, paved highway with a number of pullouts. The entire route is above 8,300 feet. At the Slide Group Campground you may access the Rim Hiking Trail that follows the western side of the byway for several miles. At Nelson Vista, there is a superb view into Nelson Canyon as well as a one-fourth-mile interpretive trail. Two miles south of Nelson Vista is a side road to scenic Karr Canyon where there are picnic facilities. Just past the road to Karr Canyon is a fine road to the top of Alamo Peak. From here you can look across the Tularosa Basin to the San Andres Mountains.

A good side road to the east follows Rio Penasco. This road connects with State Route 130 12 miles to the east. About midway is Bluff Springs where there is a nice trail and picnic facility in this scenic area.

Back on the scenic byway, the road climbs to Cathey Vista at an elevation above 9,000 feet. The vista has a campground here and a short interpretive nature trail. At the end of the scenic byway are the Apache Point Observatory and the National Solar Observatory at Sunspot. You may take a self-guided tour at the solar observatory, and there is usually a guided tour on Saturday afternoon from May through October.

Although the Sunspot National Scenic Byway ends at the solar observatory, a gravel road continues southward along the Sacramento River through pretty scenery. This road passes Sacramento Lake before leaving the Lincoln National Forest after about eight miles. For a very exciting drive, take the West Side Road that leads west off of the Sacramento River Road about three miles south of Sacramento Lake. This extremely curvy dirt road passes Little Cherry Spring and Wright Spring before turning northward and making its way all the way back to U.S. Highway 82 just east of Tunnel Vista. This back road permits access to Lawrence Canyon, Mule Peak, Potato Knob, Gordon Canyon, Cherry Canyon, and Goat Ranch Spring.

Another highway from Cloudcroft is State Route 130, which goes southeast down Cox Canyon until it reaches Rio Penasco. The highway then follows the river east to Mayhill where it joins U.S. Highway 82 and continues to the eastern edge of the national forest.

South of Alamogordo about 10 miles on U.S. Highway 54 is a road east to Oliver Lee State Park. From the park, which is at the western edge of the national forest, the four-mile-long Dog Canyon National Recreation Trail is a popular hiking trail that enters the national forest and winds its way toward the Apache Point Observatory, climbing 3,130 feet in elevation. An old stone cowboy cabin is along the way. Evidence exists that the Dry Canyon Trail was used to travel from the Tularosa Basin to the Sacramento Mountains and back since the time of the first prehistoric people. The Apaches used the canyon for protection from the U.S. Cavalry, and with the coming of cattle ranching, cowboys moved their cattle through the canyon.

The northern district of the Lincoln National Forest lies east of Carrizozo and is home to the northern extremities of the Sacramento Mountain Range. The most northern part of the range is known as the Jicarilla Mountains, and south of these are the Capitan Mountains and the Sierra Blancas. Several dirt roads are in the Jacarillas, but there are no developed recreation areas.

Rising to a height of 7,900 feet, the Jacarillas consist of quartz, with some of the best examples visible on Jacks Peak at the extreme northeastern corner of the national forest. For an interesting drive through historic areas, take the road that leads southeastward of the small mining community of Ancho. In about three miles, this road enters the national forest, and in another mile it comes to an intersection. The road to the left brings you to Jacks Peak, the summit of which is at 7,553 feet. Evidence of old mines exists on the west side of the peak where magnetite and hematite were mined in the past. The right fork at the intersection takes you on a road that traverses the Jicarilla Mountains for nearly 20 miles, passing several old mining sites and the old townsite of Jicarilla where there is a historic cemetery. In 1850, placer gold was found in this vicinity of Jicarilla, and mining was profitable for a few years, although the lack of water in the area made mining difficult. The forest road passes to the east of Ancho Peak and comes to a T. The road to the left soon leaves the national forest and joins State Route 146 a few miles above the small town of Capitan. The road to the right exits the forest at the ghost town of White Oaks before becoming State Route 349 and heading to Carrizozo. Still standing in White Oaks is the faded remains of Hoyle's Castle, a two-story brick Victorian mansion that at one time had stained glass windows, a fireplace, and impressive woodwork. The town grew up after the North Homestead Lode was discovered in 1879. In 25 years, three million dollars worth of gold and silver had been mined. At one time, one of White Oaks' 4,000 inhabitants was Billy the Kid, and the first governor of New Mexico, W. C. McDonald, was born here. When the route of the El Paso and White Oaks Railroad was chosen to go through Capitan instead of White Oaks, the once prosperous village soon became a ghost town.

Patos Mountain to the east of White Oaks is crossed by the Barber Springs Trail. South of White Oaks is Carrizo Mountain whose peak may be reached via a 5.5-mile trail from White Oaks.

The highest peak in the Lincoln National Forest is in the Sierra Blanca Range, although the peak itself, at 12,012 feet, is just outside the forest boundary in the Mescalero Apache Indian Reservation. This is a huge volcanic mountain composed of tuff and boulders. In the foothills are pictographs and pueblo ruins. Ski Apache is a popular ski area just within the southern boundary of the forest, and it is the southernmost ski area in the country. State Route 532 north of Ruidoso climbs to the ski area, and there is a marvelous view of Sierra Blanca, the Monjeur Lookout, and the Capitan Mountains from the Windy Point Vista along the highway. Three high mountain campgrounds are in the area. The entire western side of the Sierra Blanca Range is in the White Mountain Wilderness, named for the continuous snow cover at the higher elevations. The western edge of the wilderness is rough, rocky, and picturesque, with the eastern portion somewhat more gentle. The trail through Argentina Canyon to Argentina Peak is strenuous, but very scenic. Another trail is at the farthest southwestern corner of the wilderness, beginning at the Three Rivers Campground and making its way up Three Rivers Canyon and into the heart of the wilderness.

U.S. Highway 70 cuts across the southern edge of this part of the national forest, following Rio Ruidoso from the town of Ruidoso to Hondo. This is part of the Billy the Kid National Scenic Byway. From Hondo, the scenic byway follows U.S. Highway 380 back into the national forest, going through the historic town of Lincoln at the southern edge of the Capitan Mountains.

The Capitan Mountains lie in an east-to-west direction east of Carrizozo and north of U.S. Highway 380. This range is 23 miles long and about eight miles wide and was just a typical mountain range until May 1950 when it became instantly famous during a windy day when a forest fire swept through the mountains northeast of Capitan. During the fire, a small, charred bear cub that was found clinging to a tree was rescued and eventually flown to a veterinarian in Santa Fe. After its recovery, this bear, who became Smokey Bear, was flown to the National Zoo in Washington, D.C., where it was on view to millions of Americas until it died. Smokey Bear is buried in the Capitan Mountains, and a museum commemorating this famous animal is in Capitan. Much of the Capitan Mountains are in the Capitan Mountains Wilderness. The northern edge of the wilderness is dissected by several narrow rocky canyons, although the crest is fairly flat with mountain meadows alternating with forests of Douglas fir, corkbark fir, Engelmann spruce, and ponderosa pine. On the lower slopes are the drier piñon–juniper woodlands. A road off of State Route 246 reaches the northeastern edge of the wilderness, terminating at Boy Scout Mountain. From here, the Capitan Peak Trail passes to the west of Chimney Rock and eventually to Capitan Peak, at 10,083 feet. From Capitan Peak the excellent Summit Trail follows the crest of the mountains to Summit Peak, at 10,179 feet. Between the two peaks, the trail goes through Pierce Canyon Pass. Just east of the pass is the Seven Cabins Canyon Trail to the north and the Pierce Canyon Trail to the south.

Santa Fe National Forest

SIZE AND LOCATON: 1,568,820 acres in north-central New Mexico, including the southern Sangre de Cristo Range. Major access routes are Interstate 25, U.S. Highways 84 and 85, and State Routes 44, 50, 63, 96, 112, 126, 290, 501, and 565. District Ranger Stations: Coyote, Cuba, Espanola, Jemez Springs,

Las Vegas, and Pecos. Forest Supervisor's Office: 1200 St. Francis Drive, Santa Fe, NM 87504, www.fs.fed.us/r3/sfe.

SPECIAL FACILITIES: Winter sports areas.

SPECIAL ATTRACTIONS: Rio Chama National Wild and Scenic River; Santa Fe National Scenic Byway; Cañada Bonita.

WILDERNESS AREAS: San Pedro Parks (41,132 acres); Dome (5,200 acres); Pecos (112,222 acres, partly in the Carson National Forest); Chama River Canyon (50,300 acres, with a small part in the Carson National Forest).

East of Santa Fe, the southern end of the Sangre de Cristo Mountains rises to enormous heights, with Truchas Peak at 13,103 feet, Santa Fe Baldy at 12,623 feet, Lake Peak at 12,409 feet, and Penitente Peak at 12,402 feet, among the highest. Many of them can be seen as you drive U.S. Highways 84 and 85 between Santa Fe and Las Vegas, New Mexico. Southwest of the Sangre de Cristo Range and southeast of Santa Fe is the large and impressive Glorieta Mesa. The western side of the Santa Fe National Forest, lying west of U.S. Highways 64 and 84, contains the San Pedro and Sierra de los Valles (Jemez) Mountains and Sierra Nacimiento. The Rio Cebolla, Rio Puerco, Rio Chama, Rio Gallina, and Jemez River cross this vast forested region.

The Santa Fe National Scenic Byway is a 15-mile route that climbs from Santa Fe to the Santa Fe Ski Basin, going from an elevation of 7,000 feet to one of 10,500 feet. The byway starts up scenic Tesuque Canyon on State Route 475. At first the surrounding forest consists of piñon pines, junipers, and Gambel oaks. After reaching the Little Tesuque and Black Canyon Campgrounds, the road swings abruptly to the north. A pleasant two-mile hiking trail loops from the latter campground. As the road climbs into the Sangre de Cristo Mountains, the exposed rocky canyon walls have a reddish hue. After passing Hyde Park State Park, the scenic byway reenters the Santa Fe National Forest. The Borrego Trail, which has its beginning in the state park, also enters the national forest. In the meantime, the scenic byway begins twisting its way up the mountain. Along the way to the top are several campgrounds and picnic areas, including Big Tesuque Campground, which is nestled beneath aspens along a creek of the same name. Aspen Vista Picnic Area, where you may look down upon Santa Fe, and Aspen Basin Campground, at 10,300 feet, are beyond Big Tesuque. A nice nature trail at the picnic area is available as well as a longer trail up to Tesuque Peak, which tops out at an elevation of 12,040 feet. The Winsor Hiking Trail is at Aspen Basin and is a good trail to enter the Pecos Wilderness. To the west of Aspen Vista is a rough forest road down into Packers Canyon and eventually to the village of Tesuque on U.S. Highway 285 a short distance north of Santa Fe. As you follow the Winsor Trail northward and then eastward, explore some of the other trails that intersect it or that branch off from it. One branch trail follows Rio Nambe; another goes past Santa Fe Baldy to lovely Katherine Lake.

Although not a designated scenic byway, another driving tour is equally scenic and provides other great accesses into the national forest and the Pecos Wilderness. After leaving Santa Fe, Interstate 25 ascends to Glorieta Pass. From the pass, a jeep trail to the north goes part way up Glorieta Canyon. After this road peters out, you may hike another 2.5 miles to 10,199-foot Glorieta Baldy where there is a primitive campground. The campground may also be reached by a four-wheel-drive road from the east (see below). Although Interstate 25 continues past Glorieta Pass on its way to Las

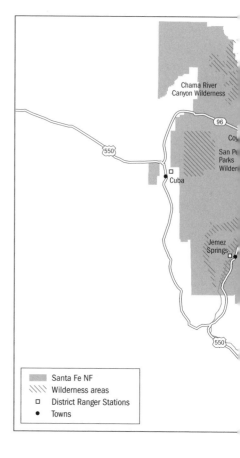

Santa Fe NF
Wilderness areas
District Ranger Stations
Towns

Vegas, New Mexico, State Route 50 heads due east from the pass to the town of Pecos. At Pecos, State Route 63 goes north into the Santa Fe National Forest all the way to the southern end of the Pecos Wilderness where it dead-ends. This scenic road passes Lisbon Spring Fish Hatchery, three campgrounds, and the tiny settlement of Tres Lagunas. At the Dalton Creek Campground, a four-wheel-drive road cuts westward across the mountain to Glorieta Baldy and its primitive campground. The highway to Tres Lagunas closely follows the Pecos River. One-and-a-half miles north of Tres Lagunas, State Route 63 follows Holy Ghost Creek to a campground at the edge of the wilderness. When the road begins its course along the creek, however, a dirt road branches off of State Route 63 to the north into a beautiful area. At the upper end of this road, at the border of the Pecos Wilderness, are a series of inviting campgrounds. From each of these campgrounds there are hiking trails into the wilderness, and there is also a nature trail at Jack's Creek Campground.

Just outside the northwest corner of the Pecos Wilderness is Borrego Mesa

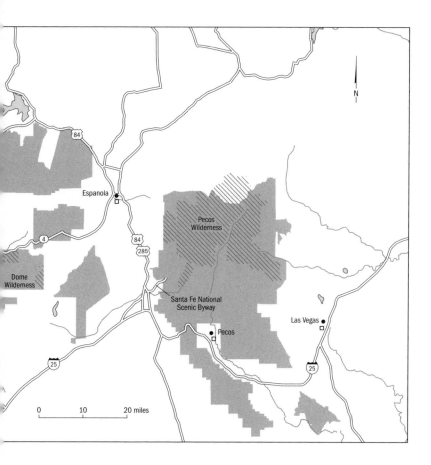

where there is a campground and the trailhead for the Rio Medio Trail that follows the river across the northern part of the wilderness. This trail circles around the southern end of Truchas Peak and then along the eastern side of the peak to the picturesque Truchas Lake. At 13,103 feet, Truchas Peak is the highest in the Santa Fe National Forest.

West of Espanola and surrounding Los Alamos is an even larger part of the Santa Fe National Forest. Vegetation ranges from near-desert in some of the foothills to piñon–juniper woodlands on the lower mountain slopes to ponderosa pine forests to Douglas fir, spruce, and fir forests at the highest elevations. Fewer than three miles west of Los Alamos is Cañada Bonito, a pristine high mountain grassland on a steep southwest-facing slope between elevations of 9,000 and 9,860 feet. Forests sit above and below this remarkable grassland.

South of Cañada Bonito is the Pajarito Plateau at the eastern edge of the Jemez Mountains. This plateau was occupied by prehistoric people nearly 10,000 years ago. By the late 1300s, Pueblo Indians built villages and farmed nearby canyon bottoms and mesas. Results of past occupation of this land may be seen in the national forest and at nearby Bandelier National Monument. Pajarito Canyon has a broad, flat bottom, although it narrows in places. Wet meadows occur on some of the upper slopes at elevations between 8,500 and 9,000 feet. Below 6,400 feet, the vegetation is mostly grasses such as blue grama, side-oats grama, and false buffalo grass beneath a scattering of one-seeded juniper, Gambel's oak, and several species of shrubs. Some of the shrubs are four species of sagebrush, mountain mahogany, two kinds of rabbit brush, skunkbush, Apache plume, and Fendler bush. Here and there are wildflowers, including Rocky Mountain nodding onion, sego lily, Indian paintbrush, golden aster, Rocky Mountain beeflower, false pennyroyal, scarlet trumpet, lupine, plains blackfoot, and several species of beardtongues and evening primroses.

After visiting Cañada Bonito, drive State Route 501 southward from Los Alamos to the junction with State Route 4 at the northern edge of Bandelier National Monument. If you follow State Route 4 westward, you enter the Jemez Mountains and the Santa Fe National Forest. A main feature of the Jemez Mountain Range is a large volcanic depression 15 miles in diameter, which is a caldera that was formed when the land surface collapsed during volcanic activity. The depression is known as Valle Grande. Recent volcanic eruptions created the Pajarito Plateau. Much of this area is now forested, although ancient lake beds are occupied by lush meadows. On the southeastern side of Valle Grande is a small mountain ridge called San Miguel Ridge. The high point of the ridge is St. Peter's Dome. The dome, at 8,464 feet, is the centerpiece of the small Dome Wilderness. The St. Peter's Dome Road is a four-wheel-drive road that leads to the western edge of the wilderness.

From the East Fork and Jemez Falls Campgrounds are trails to a lovely waterfall along the East Fork of the Jemez River. About 1.5 miles southwest of the Jemez Falls Campground is a pristine area of ponderosa pine and southwestern white pine, and on north-facing slopes, Douglas fir and white fir. This area has been designated the Monument Canyon Research Natural Area. The natural area preserves not only a fine forest but 60-foot-tall spires and columns of a rock type known as Bandelier tuff. The 640-acre area is easily reached by a forest road. Rocky outcrops in the natural area support some Rocky Mountain juniper, mountain mahogany, New Mexico locust, Gambel's oak, and the shrubby Apache plume, oceanspray (pl. 30), wax currant, and mock orange. Tassel-eared squirrels are usually seen scurrying around the area. A study of birds in the region has revealed the band-tailed pigeon, flammulated owl, saw-whet owl, broad-tailed hummingbird, white-throated swift, Grace's warbler, and gray-headed junco. Nearby, and maybe in the natural area, is the Jemez Mountain salamander.

Along the East Fork of the Jemez River and near the falls are Fremont's cottonwood, New Mexico olive, netleaf hackberry, canyon grape, Virginia creeper, spectacle pod, willow-weed, monkey flower, heart-leaved buttercup, and helleborine, the last a type of wild orchid.

After leaving Jemez Falls, State Route 4 swings northward to Redondo Campground where there is a nature trail adjacent to the nice camping area. After three more miles on State Route 4, there is a junction with State Route 126. Take this side road that ends at Fenton Lake near the Rio Cebolla. North of Fenton Lake, via a dirt road that eventuallly goes to the secluded Seven Springs Campground, is a wonderful example of a high mountain marsh.

From the intersection with State Route 126, State Route 4 turns abruptly southward, following the Jemez River to the small community of Jemez Springs. The road passes a fantastic rock formation called Battleship Rock, where there are picnic facilities. You may hike a trail that connects the Battleship Rock area with Jemez Falls. After leaving Jemez Springs, State Route 4 enters Cañon de San Diego before coming to the town of Jemez. The very scenic State Route 485 branches off of the highway, following Rio Guadalupe into the Sierra Nacimiento area.

To access the large northwestern section of the Santa Fe National Forest, the best approach is from the north via State Route 96. Paved roads enter the national forest out of both Youngsville and Coyote, but no developed Forest Service areas exist along the routes. Nearly five miles west of Coyote, however, there is a forest road that approaches the eastern side of San Pedro Parks Wilderness. Several hiking trails into the wilderness are along this road. A side road southward to the Rio Puerco Campground continues to an interesting rock formation known as Teakettle Rock and to the natural Jarosa Spring.

Farther west on State Route 96, just east of the town of Gallina, a delightful but very curvy forest road goes all the way to the Chama River Canyon Wilderness. This road climbs over Mesa Burule and Mesa Alta and past numerous small natural springs. The remarkable Chama River Canyon cuts across the wilderness.

Beyond Gallina, State Route 96 comes to a T with State Route 112. This latter road circles around to Dead Man Peak at the northwestern corner of the Chama River Canyon Wilderness.

Cañada Bonito

An all-weather road winds westward for six miles from Los Alamos to a popular ski area in north-central New Mexico's Jemez Mountains. From the ski area, abandoned Forest Road 282 now serves as a mountain trail into a verdant section of the Santa Fe National Forest. After about a mile, the trail leads out of the dense forest and into a clearing of grassland, which extends from an elevation of nearly 9,000 feet to a 9,860-foot ridge crest on a slope that has as much as a 40 percent grade in places.

At its base, the grassland merges imperceptibly into the forest, whereas at the crest a sharp transition to a dense forest of Engelmann spruce occurs. Some woody plants encroach along the edges of the grassland, but only a few have succeeded in establishing themselves within it. Biologists consider the grassland unique in the Santa Fe National Forest because it has not been subjected to grazing by domestic livestock since 1943, when it was removed from public access with the establishment of the nearby nuclear laboratory at Los Alamos. Other isolated mountain grasslands in the area are heavily grazed and have a mowed appearance, with numerous gullies eroded into the terrain.

In the generally semiarid region, the Jemez Mountains stand as an island of cool, moist forests. The area receives little precipitation between April and June, but frequent summer thunderstorms and winter snows, which pile up to 10 feet or more, provide adequate moisture for the rest of the year. Several grassy slopes, including Cañada Bonito grassland, have developed, usually on predominantly southern exposures. Geographer Craig Allen refers to these areas as montane grasslands, reserving the term meadow for usually moist, low-lying level areas.

Although the high elevation limits the growing season to about 100 days, the vegetation is dense. The grasses, which grow nearly waist high, are mostly Thurber fescue and Parry danthonia, but botanists have recorded 15 other kinds, including two other fescues, two June grasses, bluegrasses, bromes, bent grasses, and even Virginia wild rye, which is widespread in the United States. Conspicuous flowers—Indian paintbrushes, western wallflower, orange sneezeweed, cinquefoils, bluebells, scarlet gilia—add color to the landscape.

Adjacent to the grassland, below 9,000 feet, is a mixed conifer forest dominated by a few large Douglas firs and an abundance of smaller white firs and Rocky Mountain maples (pl. 31). Aspen and Engelmann spruce are also present, but they are relatively small and show little evidence of recent reproduction. On the ridge crest and down the even steeper north-facing slope, Engelmann spruces form a closed forest, with a scattering of corkbark fir, white fir, Douglas fir, quaking aspen, and Rocky Mountain maple.

Beneath the vegetation of the Cañada Bonito grassland, a thick, nearly continuous blanket of sod and litter has developed, bare only where pocket gophers have been at work. Allen reports that the fine loamy mixed soils beneath the grasses contain at least 1 percent organic matter to a depth of as much as three feet. In contrast, the soils beneath the mature Engelmann spruce and mixed conifer forests are comparatively thin and stony, extending only inches below the surface, except in areas where the trees have recently encroached on the grassland. In both grasslands and forests, soils develop from the breakdown of underlying rocks controlled by topographic and weather conditions and the effects of living organisms. Grasses, with their high production of living tissue and their propensity to add it to the soil, aid in the development of deep soils. In addition, the pocket gophers churn the soil and continue to bury the organic matter into deeper layers.

In 1984 Allen counted 171 trees that had invaded the Cañada Bonito grassland. The age of these trees indicates that the invasion started in 1920, with most of the trees becoming established between 1928 and 1953. Only four of the trees—two Douglas firs and two ponderosa pines—are large and earlier in origin. These large trees have probably been the seed sources for the recent invasion.

Allen postulates that Cañada Bonito arose following a catastrophic crown fire in what was a spruce–fir forest. Because forests of Engelmann spruces and corkbark firs do not recover after a fire, the burned-out forest may have been invaded by herbs. The steep, south-facing slopes provided a dry habitat more suitable for the growth of grasses than trees, whose reproduction was hampered by the early dry season and the lack of shade. Engelmann spruce and corkbark fir would not have regenerated in conditions such as these, and the more drought-tolerant Douglas fir and ponderosa pine would not have been able to do much better. In addition, there may not have been enough ponderosa pine and Douglas fir in adjacent forests to provide an adequate source of seeds. Furthermore, if grasslands persist long enough, they develop a thick cover of sod and litter, and with the thicker cover, establishment of conifer seedlings becomes even more difficult.

The most powerful force that deters woody invasion of a grassland is fire, as suggested by Wisconsin ecologist John T. Curtis in 1959. Because grasslands occur on dry slopes, fire could spread rapidly. The abrupt change at the ridge crest from grassland to spruce forest might also be explained by fire. Ground fires burn upslope with ease, but they usually peter out at the ridge-top because of the difficulty in burning downhill on moist slopes.

Fires once occurred with great regularity in the grasslands and forests because of the annual buildup of dry plant litter, which provided a ready source of fuel. The intensity of grassland fires killed any tree seedling that might have germinated. Before settlement by modern inhabitants of the West, Indians may have purposely set fire to the grasslands to improve hunting or perhaps for other reasons. Lightning strikes undoubtedly set many fires (weather records kept at nearby Los Alamos show that of the 62 days of July and August, an average of 46 days has thunderstorms). When shepherds and goatherds began to occupy the Jemez Mountains during the 1800s, they too burned the grasslands for pasture until the end of the second decade of the twentieth century. According to Richard J. Vogl, a biologist at the University of Arizona, other settlers burned the grasslands to clear the area for agriculture, reduce undergrowth for better visibility, and promote heavy vegetation growth for ground-dwelling animals.

In about 1920, however, the U.S. Forest Service, convinced that fire was a force of destruction, began a policy of fire suppression. This suppression coincided with an invasion of tongues of forest into the grasslands, such as the invasion documented at Cañada Bonito by Craig Allen. As long as this policy of fire suppression continued, the forest advanced slowly into the grasslands.

The policy also has permitted dangerously large quantities of dry plant material to accumulate in the grasslands and the forests. Natural fires allow this dry plant material to be recycled into the soil in the form of ash. Now, however, it has accumulated to such an extent that devastating forest fires break out every year across the nation.

In the last several years, as ecologists have begun to realize that fires can be beneficial, the Forest Service has been amenable to altering its policy. In fact, burning a grassland to maintain its integrity has become a standard management technique.

Because Cañada Bonito is one of the few remaining Thurber fescue grasslands that have not been subjected to recent livestock grazing, it has been designated as a Research Natural Area by the Forest Service to permit the study of this ecosystem, to maintain its genetic pools, and to preserve this grassy island in the heart of a coniferous forest.

Rio Cebolla Marsh

Originating some 25 miles west of Los Alamos, a clear mountain stream known as Rio Cebolla descends to the southwest, passing through Seven Springs Campground, a secluded section of the Santa Fe National Forest. Four miles beyond the campground, on its way to merging with the Rio de las Vacas, the stream flows out of forested terrain into a broad, half-mile basin to form an extensive marsh. Mountain ranges loom in every direction—the Jemez Mountains to the east, the San Pedro Mountains to the north, and the Sierra Nacimiento to the west and southwest. The marsh itself, one of the few wetlands in the region, is 7,730 feet above sea level.

State Highway 126 cuts across the southern end of the marsh, and the roadway serves to hold back some of the water. To the north, the marsh water is knee-deep in places, whereas south of the road, mucky soils prevail, with little standing water. Most of the marsh on either side of the road consists of herbaceous plants, with only a scattering of woody species (if many trees and shrubs grew there, the area would qualify as a swamp rather than a marsh). What woody plants there are—willows, alders, and viburnums—are all less than 15 feet tall.

Some plants with very showy flowers grow in the northern part of the marsh, including two species of yellow groundsels, a white bittercress of the mustard family, the white prairie mallow, and goldenglow, which in bloom resembles a sunflower. Among other botanical jewels is an orchid with small greenish white flowers. Large tussocks, or clumps, of reed canary grass and aquatic sedge are prominent where the water is deepest. Agile visitors may step or hop from tussock to tussock, but the best way to survey the vegetation is to don a pair of waders and slosh through the water.

South of the roadway, the mucky soil is very soft, and it is a good idea to test your weight before striding across the area. The most conspicuous plant by early summer is a six-foot-tall, thick-stemmed member of the lily family called corn plant. It has broad, heavily veined leaves, but its greenish white flowers are tiny. Blue Missouri iris contrasts with yellow-flowered cinquefoil and buttercup. Wide-leaved and narrow-leaved cattails are also common.

Visiting the Rio Cebolla Marsh in July 1990, I identified 71 different kinds of plants in flower. Other species had finished blooming or had yet to bloom, making positive identification of them difficult or even technically impossible. In reviewing the plant list, some interesting points stand out. One is that nearly half of the species are also found in Tuttle Marsh, Michigan, a marsh several hundred miles to the northeast that is more than 6,000 feet lower in elevation. This finding is consistent with other studies that indicate that many wetland or aquatic plants have relatively broad geographical ranges. This may result, at least partly, from the dissemination of wetland plant seeds by waterfowl, which may carry the seeds on their feet or pass them through their digestive tract.

A second observation, also typical of wetland studies throughout the country, is that 40 percent of the plant life in the Rio Cebolla Marsh consists of 16 sedge species, eight grasses, and five rushes. The prevalence of these types of plants may also be related to waterfowl, which eat their fruits and seeds.

The distinctions between sedges, grasses, and rushes are usually unimportant to the average person, troublesome to biologists, and a scientific puzzlement for many professional botanists. All three groups of plants generally have long, narrow, grasslike leaves; all three have tiny, nonshowy flowers, so that differences are not readily detected with the naked eye; and all do best in wet or at least damp places. They are different enough, however, to be classified into separate families, with the rushes considered to be more closely related to members of the lily family than to either sedges or grasses.

Some general characteristics can be useful clues in distinguishing the three groups. The stems of sedges are often triangular and almost always solid, whereas the stems of grasses are usually round and hollow, and the stems of rushes are mostly round and solid. The leaves of sedges are normally flat and arranged in three distinct rows on the stem. The leaves of grasses are normally flat or curved under the edges and arranged on two sides of the stem. Rush leaves may be flat or cylindrical and hollow, and as in grasses, they are arranged along two sides of the stem.

Three-fourths of the rush family, including most species in the United States, belong to the genus *Juncus*. The male parts of the flowers, which produce the pollen, and the female parts, which produce the eggs and ultimately the seeds, are surrounded by six tiny, usually greenish or brownish structures that botanists consider to be equivalent to petals and sepals. A similar pattern is found in most members of the lily family, except that the petals and sepals are large and colorful, forming flowers attractive to pollinating insects. Rushes are pollinated by the wind, and their small, drab flowers are therefore no disadvantage. Their fruits (the developed ovaries and the seeds they contain) consist of small, dry capsules encasing many tiny seeds.

The 4,000 species of sedges and 400 of rushes found in the world are exploited for numerous purposes. Leaves and stems are used to make baskets, chair bottoms, and other objects. In Europe, one kind of rush serves as a forage plant for sheep, and in the United States, one sedge, with the misnomer of nutrush, produces small underground tubers, which pigs dig up and eat. Papyrus of antiquity was made from sedge stems that were split into thin strips and then pressed together when wet.

Grasses and sedges are also wind pollinated and have similarly drab flowers, usually greenish or brownish. But the male and female parts are not surrounded by six structures. In grasses, a pair of tiny structures usually encloses the male and female flower parts, at least for a while; in sedges, there is only a single accompanying structure. The seeds are borne singly and never grouped inside capsules. In sedges, the developed ovary provides a thin coat over the seed and can be rubbed off, whereas in grasses, the fruit (called a grain) consists of a seed fused with its surrounding ovary.

The world's 10,000 grasses, some of which grow in relatively dry habitats, are even more important economically. Bamboo, a member of the grass family, is a major raw material in the Far East. Wheat, corn, oats, barley, rice, rye, sorghum, and sugarcane —all domesticated grasses—provide staple foods on every continent. Many other grasses, such as timothy, serve as forage for animals. Fibers, paper, oils, adhesives, plastics, alcoholic beverages, packing materials, insulation, and thatching are among the many products derived from grasses.

Grasses are preeminent as human food because of the rich supply of starch found in their grains, compared with that generally found in the seeds of sedges and rushes. Grasses also excel as forage plants for grazing animals because their leaves and stems begin to grow back immediately after being nibbled, whereas those of sedges and rushes do not grow again.

NATIONAL FORESTS IN
SOUTH DAKOTA

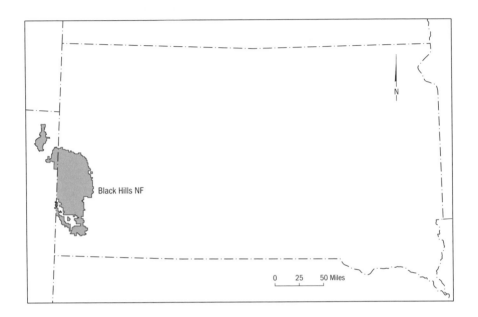

Black Hills NF

N

0   25   50 Miles

The Black Hills National Forest is the only national forest whose major acreage is in the state of South Dakota. The Custer National Forest, mostly in Montana, is partly in South Dakota. These forests are in Region 2 of the U.S. Forest Service.

Black Hills National Forest

SIZE AND LOCATION: 1,233,000 acres in western South Dakota and northeastern Wyoming. Major access routes are Interstate 90, U.S. Highways 14A (Spearfish Canyon National Scenic Byway), 16, 16A, 18, 85, and 385, and South Dakota State Routes 40, 87, 89, and 244 and Wyoming 24 and 585. District Ranger Stations: Custer, Deadwood, Hill City, Rapid City, and Spearfish in South Dakota and Newcastle and Sundance in Wyoming. Forest Supervisor's Office: Route 2, Custer, SD 57730, www.fs.fed.us/r2/blackhills.

SPECIAL FACILITIES: Boat ramps; winter sports areas; swimming beaches.

SPECIAL ATTRACTIONS: Peter Norbeck National Scenic Byway; Spearfish Canyon National Scenic Byway.

WILDERNESS AREA: Black Elk (13,426 acres).

The ponderosa pines that clothe the hills in western South Dakota and northeastern Wyoming look very dark from a distance, and because of this, the region became known as the Black Hills. Visitors to the area to see Mount Rushmore or the Crazy Horse carving pass through a part of the Black Hills National Forest. Within the national forest is Harney Peak, at 7,242 feet, the highest point in the United States east of the Rocky Mountains. In northeastern Wyoming, the Black Hills National Forest also includes most of the Bear Lodge Mountains.

The flora and fauna of the Black Hills have piqued the interest of naturalists for decades. For one thing, the Black Hills are near the easternmost edge of the range of ponderosa pine. A 90-acre grove of lodgepole pine also stands more than 200 miles west of the nearest stand in the Big Horn Mountains of central Wyoming. Limber pine, also isolated from the Big Horn Mountains, grows in limited numbers on the granite spires in the Black Hills. Persons from the eastern United States also recognize some plants that have found niches in the Black Hills, including maidenhair fern, maidenhair spleenwort, and wintergreen. The rugged terrain around Harney Peak is home to mountain goats, bighorn sheep, and mountain lions, and it was the last home in the world for the Great Plains lobo wolf before it was exterminated by hunters in 1926.

Although the granite spires are probably the most spectacular land forms in South Dakota's Black Hills, there are scenic limestone canyons with picturesque waterfalls. The Bear Lodge Mountains in northeastern Wyoming have rugged, scenic topography as well, particularly in the vicinity of the Warren Peaks.

Seven historic fire lookout towers are scattered throughout the national forest, all of them on the National Register of Historic Places. Gold was discovered and mined in the Black Hills, and the remains of several mining operations may still be seen.

Two of the best ways to get acquainted with the South Dakota part of the Black Hills National Forest is to drive the Spearfish Canyon (pl. 32) and Peter Norbeck National Scenic Byways. The Spearfish Canyon National Scenic Byway is also U.S. Highway 14A. This marvelous route is over an old railroad grade that follows Spearfish Creek into the heart of the Black Hills. The railroad was abandoned after destructive flooding in 1933. Although only 20 miles long, the Spearfish Canyon National Scenic Byway provides for a lifetime of memories. Begin the scenic byway just outside the southern city limits of Spearfish and start an ascent into the canyon. You may stop your vehicle and view the breathtaking scenery at several pullouts along the byway. The first major stop is across from Bridal Veil Falls, with scenic Spearfish Peak behind it. Within the canyon, gold was discovered, and the remains of gold mining may be seen in the form of slag piles, old mining skids, and the occasional remains of a water flume. Here and there, fishermen line Spearfish Creek, hoping for trout that was originally imported from Colorado more than 100 years ago. In the depths of Spearfish Canyon, the limestone cliffs may rise 1,200 feet above the highway. Near the creek and at lower elevations are deciduous trees such as aspens, cottonwoods, green ashes, black walnuts, and oaks, whereas ponderosa pines and white firs grow on the upper parts of the cliffs. At two of the pullouts are picnicking facilities.

A little more than halfway along the scenic byway is the old mining community of Savoy where there are lodging and eating establishments. At Savoy you may take a side trip on the Iron Creek Forest Road, which has numerous small waterfalls and catch pools alongside it. A short trek to Roughlock Falls rewards the hiker with several cascades swirling into the creek below and typical summer- and spring-blooming Black Hills wildflowers. Two major Forest Service hiking trails may be picked up from the Rod and Gun Campground situated along Iron Creek. Little Spearfish Trail makes a loop, eventually following Little Spearfish Creek for a short way. Roughlock Trail goes along the north side of Little Spearfish Canyon through ponderosa pine forests and occasional grassy openings. Near the Rod and Gun Campground is a meadow where winter scenes from the motion picture *Dances with Wolves* were filmed.

Several forest roads branch off the Iron Creek Forest Road. Forest Road 105 to the northwest proceeds through Schoolhouse Gulch and past abandoned mines until the road eventually peters out. Forest Road 222 to the west abruptly turns northward and climbs to the historic Cement Ridge Lookout.

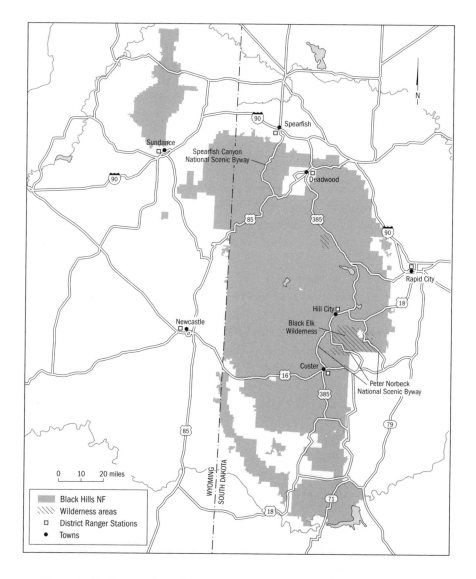

From the lookout at 6,700 feet you have great views to the west of the Bear Lodge Mountains and to the north of a large segment of the Black Hills. Forest Road 117 southward along Little Spearfish Creek joins U.S. Highway 85 in about 10 miles.

From the community of Savoy, the Spearfish Canyon National Scenic Byway continues southeastward along Spearfish Creek. It joins U.S. Highway 85 at Cheyenne Crossing, where it ends.

U.S. Highway 85 eastward from Cheyenne Crossing goes to the towns of Lead, Deadwood, and Sturgis and is mostly not through Forest Service land. About three miles east of Cheyenne Crossing, however, Forest Road 194 heads northward and in four miles climbs to Terry Peak where there is a lookout tower and a ski area.

To the south of Cheyenne Crossing, U.S. Highway 85 remains in the Black Hills National Forest for several miles, eventually reaching O'Neil Pass where there is good snowmobiling during winter. From the Hardy Guard Station is scenic Forest Road 875 through the Grand Canyon and past several abandoned mines.

The Peter Norbeck National Scenic Byway, which consists of a short loop within a longer outer loop, not only traverses much of the Black Hills National Forest but also makes its way through Custer State Park and Mount Rushmore National Monument. At the eastern edge of the community of Custer, State Route 89 to the north is the beginning of the scenic byway. For six miles, this stretch of the highway, which is part of both the inner and outer loops, goes through dense forests of ponderosa pines punctuated by grassy meadows. At picturesque Sylvan Lake (not in the national forest) is the junction of State Routes 87 and 89. State Route 87 to the east is the inner loop of the Peter Norbeck National Scenic Byway and is a very curvy road through spectacular scenery. Although the inner loop is not in the Black Hills National Forest, it is surrounded by it on either side. The scenic byway passes the Cathedral Spires and Needle Eye. South of Needle Eye are the spectacular granite spires of The Needles, which are in the national forest. North of this inner loop road is the Black Elk Wilderness, the only wilderness area in the Black Hills National Forest.

Should you elect to stay on State Route 89 and continue on the outer loop, the scenic byway is now State Route 244 and circles around the western and northern sides of the Black Elk Wilderness. Within the wilderness is rugged Harney Peak, surmounted by a historic stone lookout tower, with several trails, all rough, to the top of the peak. These trails may be accessed from both the inner and outer loops of the scenic byway. East of Harney Peak and also in the wilderness is the Upper Pine Creek Research Natural Area, a three-square-mile area of a pristine ponderosa pine community. White spruce and aspen are interspersed among the ponderosa pines, with the shrubby snowberry, hazelnut, and wild spiraea. Among the wildflowers are kinnikinnick and golden pea. A hiking trail from the William Creek Campground along State Route 244 passes the western edge of the Research Natural Area on its way to Harney Peak. The outer loop comes to Horsethief Lake in a beautiful ponderosa pine forest setting where there is a campground. The scenic byway then passes through Mount Rushmore National Monument. After leaving the national monument, the scenic byway joins U.S. Highway 16A to the south and makes a double circle through the Pigtail Bridges, a remarkable engineering feat designed by C. C. Gideon. Beyond the Pigtail Bridges, the road gradually descends to Lakota Lake where old-growth ponderosa pines surround the small artificial lake. Incidentally, there are no natural lakes in the Black Hills. The Iron Creek hiking trail originates at Lakota Lake. The scenic byway then enters Custer State Park.

Other drives in the Black Hills National Forest not designated as scenic byways are worth taking. U.S. Highway 14A west from Sturgis follows Boulder Creek through Boulder Canyon. Just off of the highway is Mount Roosevelt where there is a friendship tower dedicated to President Theodore Roosevelt and built by Seth Bullock, a Deadwood resident and friend of the president. Mr. Bullock was also the second supervisor of the Black Hills National Forest. The tower offers fine views.

A short distance west of Spearfish is a road southward off of Interstate 90 that enters Higgins Gulch and is east of Crow Peak. Crow Peak rises 1,500 feet to an elevation of 5,760 feet where spectacular views may be enjoyed if you can climb the difficult hiking trail.

U.S. Highway 385 between Deadwood and Hill City penetrates other fine areas in the Black Hills National Forest. At first, the highway parallels Strawberry Creek and then Bear Butte Creek to the junction with Nemo Road. Nemo Road is a direct route to Rapid City through the national forest. This interesting road stays near Hay Creek to Box Elder Creek. Near Box Elder Creek the first timber sale ever in a national forest took place more than 100 years ago. Because of proper forest management, this area, known as Case Number 1, has more trees than it had when the timber sale was authorized. From the Nemo Work Center along Box Elder Creek, Nemo Road continues to Rapid City, passing a picnic area beneath a formation known as Steamboat Rock. Should you take a side road north from the Nemo Work Center, you may reach Dalton Lake and a couple of interesting caves—Wonderland Cave and Bethlehem Cave.

From the junction with Nemo Road, U.S. Highway 385 passes the Custer Park Quarry, Roubaix Lake, and the Black Hills Experimental Forest Station. You may learn about forest and wildlife management in the Black Hills if you visit the station. Beyond the experimental forest is the Pactola Reservoir Recreation Area. The center of the recreation area is the Pactola Reservoir where there are several campgrounds, boat ramps, picnic areas, swimming beaches, hiking trails, and a visitor center.

Six miles south of Pactola Reservoir is the Sheridan Lake Recreation Area where campgrounds, boat ramps, and a swimming beach are available. At the eastern edge of the reservoir you may access two of the finer hiking trails in the national forest. The Centennial Trail is 111 miles long and is available for hikers, bikers, and horseback riders. The northern terminus of the trail is at Bear Butte northeast of Sturgis and outside the national forest. The southern terminus is at Wind Cave National Park. Thirty trailheads access the Centennial Trail. The Flume National Recreation Trail follows the old 1885 mining flume from the eastern edge of Sheridan Lake to the community of Rockerville. Although the air distance between the two locations is roughly five miles, the trail twists and turns for 11 miles. As you hike this very scenic and historic trail, you see part of the flume as well as other mining memorabilia. Part of the trail is in narrow canyons, some of it through ponderosa pine forests, and some of it through meadows teeming with wildflowers.

U.S. Highway 16 west of Custer crosses the Black Hills National Forest on its way to Newcastle, Wyoming. Although Jewel Cave National Monument is the major attraction along this route, national forest points of interest are Lithograph Canyon, Hell Canyon, and Tepee Canyon, three scenic gorges. A good but rugged trail in Hell Canyon follows a bench beneath picturesque limestone cliffs.

As U.S. Highway 16 nears the Wyoming border, County Road 769 leads southward to a forest road that curls back to the north and climbs to the historic Elk Mountain Lookout Tower. The Elk Mountains are a small range that straddles the South Dakota–Wyoming state lines south of U.S. Highway 16. Where the highway crosses into Wyoming, there is a forest road northward to another historic lookout tower on Summit Ridge.

West of Pactola Reservoir and Sheridan Lake is the Deerfield Lake Recreation Area. The Lake Loop is a 10-mile hiking trail that circles the recreation area.

The Mickelson Trail is a projected 110-mile trail that follows the historic Burlington Northern Railroad between Deadwood and Edgemont. When completed, the trail will cross much of the Black Hills National Forest. Those wishing to hike, horseback ride, or cross-country ski this trail must pay a user's fee.

Although nearly 130 square miles of the Black Hills National Forest are in the Black Hills of eastern Wyoming, most of the Forest Service recreation areas in Wyoming are centered around the Bear Lodge Mountains north of Sundance. Warren Peaks are in a rugged area with an historic lookout tower on one of them.

A few miles north of Warren Peaks is Cook Lake in the heart of the Bear Lodge Mountains. Cook Lake Trail is a mile-long trail that encircles the lake. From the northern end of Cook Lake is the 3.5-mile Cliff Swallow Trail that passes several beaver dams along Beaver Creek before climbing to rock ledges above the creek. The trail gets its name from the cliff swallows that nest in the limestone cliffs. In addition to cliff swallows, you may also see beaver, ospreys, Merriam's turkey, water ouzels, turkey vultures, Rocky Mountain elk, and several species of waterfowl. Wyoming Highway 24 crosses the northern part of the Bear Lodge Mountains.

Black Hills

Home to such well-known landmarks as Mount Rushmore, Devils Tower, and the Crazy Horse Memorial, the Black Hills straddle the border between South Dakota and Wyoming. The modern name for these mountains reflects one bestowed by the native Lakota, for whom this is sacred land: Paha Sapa (hills of black). The range does look dark from a distance, because 95 percent of its tree cover consists of ponderosa pine, which has deep green needles.

Some 150 million years ago, when dinosaurs roamed the earth, this region was relatively flat and the climate was tropical or subtropical. Instead of ponderosa pines, the dominant plants were cycads, which resemble tree ferns. Between 150 million and 100 million years ago, the cycads were joined by figs, sassafras, oaks, and willows, as well as such evergreen plants as sequoias and palms. Then, about 60 million years ago, the terrain began to be uplifted, blocking the eastward flow of warm air from the Pacific Ocean. As the region became cooler and drier, temperate species gradually began to replace the tropical plants, and eventually some of the evergreens were replaced by deciduous trees (most of them migrants from the East).

As the Cascade Range and the Rocky Mountains rose up in the West, beginning about 30 million years ago, the Black Hills region became cold and arid. Coniferous forests developed, and grasslands appeared in the driest areas along their periphery. Following the last ice age, which ended about 12,000 years ago, northern forest species such as white spruce and paper birch migrated southward into the region. Because several habitats—Rocky Mountain, Great Plains, northeastern deciduous forest, and northern forest—seem to meet here, the Black Hills boast an unusual collection of trees, ferns, and wildflowers. According to Wyoming botanist Robert Dorn, 30 percent of the 1,260 species now found in the Black Hills originated in the Great Plains, 25 percent (including ponderosa pine) in the Rockies, 5 percent in deciduous forests, and 1 percent in northern forests. (The remaining species generally have widespread distribution.)

The range's uninhabited areas fall mainly within Black Hills National Forest, most of which lies in South Dakota. One of the best introductions to the region is to drive U.S. Highway 14A, a national scenic byway that passes through Spearfish Canyon. Exposed on the canyon walls are various types of shale and limestone, whereas the streams in the canyon bottom are lined with lush vegetation. The byway includes a number of interpretive stops where visitors can park and learn about particular habitats. For hikers, a good way to sample a cross section of vegetation is to follow the trail that begins at the Spearfish Canyon Resort and leads to Roughlock Falls.

Beneath the pines are several distinct communities of understory plants. On exposed, rocky, south-facing slopes, particularly in the southern Black

Hills, the major species are little bluestem, yucca, sagebrush, sand lily, and various gramas and needle grasses. Juniper, Oregon grape, buffalo berry, and blue wild rye dominate at about 7,000 feet on a centrally located limestone plateau. In the northern part of the region, in relatively moist areas between 4,000 and 5,000 feet, the pines are joined by bur oak, and the understory contains chokecherry, Oregon grape, and melic grass. Plants with a western flavor—mountain mahogany, skunkbush, and American black currant—appear at the western edge of the Black Hills.

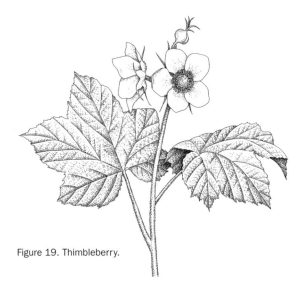

Figure 19. Thimbleberry.

On some dry, rocky exposures in the southern Black Hills, the principal tree is Rocky Mountain juniper. The understory includes several prairie grasses as well as the shrubby skunkbush.

In waterside habitats in the eastern mountains, box elder and green ash grow with American elm, eastern cottonwood, red osier dogwood, and Bebb's willow. Wildflowers here include water parsnip, fringed loosestrife, and hedge nettle. In the northeastern foothills, on relatively dry slopes, bur oak dominates above an understory of hop hornbeam, smooth sumac, coralberry, and poison ivy. Among the wildflowers are a red columbine (pl. 33), aster, figwort, wild sarsaparilla, fleabane, and avens. In the northwest, particularly in the wake of natural forest fires or controlled burns, quaking aspen and paper birch take over. Common in the understory are chokecherry, beaked hazelnut, a wild rose, red baneberry, thimbleberry (fig. 19), and bracken fern.

The foothills harbor several treeless habitats. The wetter areas support meadows containing Missouri goldenrod, false toadflax, golden glow, Indian paintbrush, Mariposa lily, death camas, and prairie smoke. Grasslands are found in drier zones. At the southern edge of the hills is a bunch grass community of little bluestem, whereas elsewhere, particularly along the western edge, grow short prairie grasses—the most prominent being various bluegrasses, buffalo grass, and wheat grass—as well as prickly pears. Rabbit brush and three kinds of sagebrushes give the lowlands to the west and south a distinctly Wild West appearance.

NATIONAL FORESTS IN TEXAS

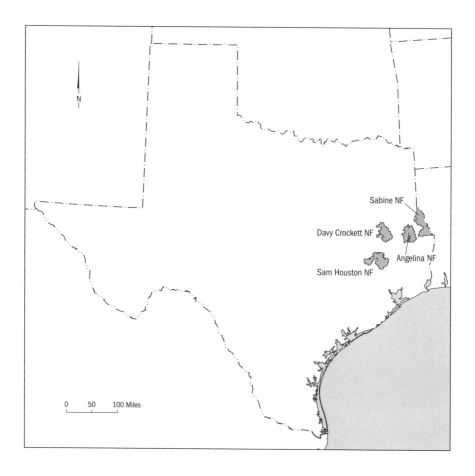

The state of Texas has four national forests, all administered from the same forest supervisor's office. The forests encompass more than 637,000 acres, all in the Pineywoods section of eastern Texas. The forests contain several lakes and reservoirs, several campgrounds, many miles of hiking trails, and five wilderness areas. Forest Supervisor's Office: 701 N. 1st Street, Lufkin, TX 75901. The Texas national forests are in Region 8 of the U.S. Forest Service.

Angelina National Forest

SIZE AND LOCATION: 153,160 acres in southeastern Texas between Lufkin and Jasper and surrounding most of the Sam Rayburn Reservoir. Major access routes are U.S. Highways 69 and 96 and State Routes 63, 103, and 147. District Ranger Station: Zavalla. Forest Supervisor's Office: 701 N. 1st Street, Lufkin, TX 75901. www.southernregion.fs.fed.us/texas.

SPECIAL FACILITIES: Boat ramps.

SPECIAL ATTRACTIONS: Aldridge Sawmill Historic Site; Stephen F. Austin Experimental Forest Interpretive Trail System; Sam Rayburn Reservoir; Boykin Springs Longleaf; Black Branch Barrens.

WILDERNESS AREAS: Turkey Hill (5,473 acres); Upland Island (13,331 acres).

The huge Sam Rayburn Reservoir divides the Angelina National Forest into two nearly equal halves, one northeast of the reservoir centered around Broaddus, and one southwest of the reservoir around Zavalla. The national forest is in the Pineywoods Vegetational Province of Texas, but the forests were cut before they became public lands in the 1930s. Present timber management is carried out following guidelines in the land management plan. The national forest has outstanding areas of longleaf pine, shortleaf pine, loblolly pine, and a good variety of hardwood species.

The national forest boasts of several different types of plant communities, many examples of them concentrated in the southern part of the forest southeast of Zavalla. Boykin Springs Recreation Area is centered around Boykin Lake where you may camp, picnic, fish, swim, and hike. The area contains a number of small hillside seeps intermixed with the pine stands. The wooded seeps contain such small trees or shrubs as sweetbay, red bay, and a kind of wax myrtle. Seeps without much woody vegetation are generally covered with grasses and sedges, with pitcher plants, sundews, and rare species of coneflowers intermixed. North of Boykin Lake is a nice forest of longleaf pine. On ridges with deep sandy soil is a community dominated by bluejack oak, shortleaf pine, post oak, blackjack oak, black hickory, and two common shrubs, American beautyberry and yaupon holly. Boykin Springs Recreation Area is connected to Bouton Lake and campground via the 5.5-mile Sawmill hiking trail. A side trail off the Sawmill trail as well as a dirt road go to Aldridge Sawmill Historic Site. The original wooden sawmill was built in 1905 and became the center of the town of Aldridge that grew to 76 buildings by 1911. When the wooden mill burned in 1911, it was replaced by a mill made from concrete. Visitors to the site today still see the concrete shells of

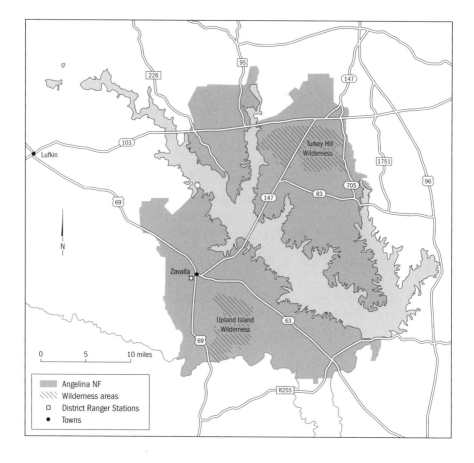

former mill buildings, several concrete foundations, the mill pond, and part of the old railroad tram. For safety and site preservation, visitors are restricted from entering buildings.

About two miles east of Boykin Springs is Shearwood Creek where the wet creek bottoms support a moist soil forest of sweetbay magnolia, red maple, swamp red bay, and swamp gum in the overstory, with gallberry, swamp cyrilla, evergreen bayberry, and maleberry in the shrub layer. Boggy depressions in the forest often contain pitcher plant, yellow-eyed grass, and several kinds of sedges. A similar sweetbay magnolia forest is also found along Trout Creek about five miles east of Shearwood Creek. State Route 63 southeast from Zavalla provides access roads to Shearwood and Trout Creeks.

A half-mile south of Post Oak Road, on either side of State Route 63, are the Black Branch Barrens, one of the best examples left of prairielike openings, called barrens, in the midst of a woodland where trees are dwarfed because of shallow, nutrient-poor soils. These barrens contain an unusual assemblage of plants, including Nuttall's rayless goldenrod.

State Farm-to-Market (FM) Road 255R branches eastward off of State Route 63 and crosses the Sam Rayburn Dam at the south end of Sam Rayburn Reservoir. A dirt road below the dam extends into a small, isolated segment of the Angelina National Forest known as McGee Bend. This area of about three square miles contains excellent examples of a beech–southern magnolia forest community, a bald cypress–tupelo gum swamp community, an overcup oak forest and, on slightly higher elevations, a community of shortleaf pine and several kinds of upland oaks. Where beech and the evergreen southern magnolia dominate, usually at the base of moist slopes, these species are joined by loblolly pine, American holly, laurel oak, white oak, and sweet gum. Bald cypress and tupelo gum occur in the wettest terrain of McGee Bend, usually associated with water hickory, water elm, Carolina ash, green ash, and overcup oak. One forested area at McGee Bend is about 90 percent overcup oak, with an intermingling of willow oak, water hickory, and a few bald cypress trees. In slightly drier habitats, shortleaf pine occurs with post, blackjack, southern red, black, and white oaks.

At the southwest corner of the Angelina National Forest is the Upland Island Wilderness, named for the "islands" of pines that were uncut because of their immaturity when the rest of the woods was logged in 1930. Several trails cut into the wilderness, some of them using abandoned jeep roads. Where Graham Creek flows into the wilderness, there is a fine forest community dominated by swamp chestnut oak, cherrybark oak, overcup oak, and sweet gum, with dwarf palmetto scattered beneath the trees.

Caney Creek Recreation Area is a few miles east of Zavalla where Caney Creek enters the Sam Rayburn Reservoir. The recreation area contains a campground, several boat ramps, and a half-mile self-guided nature trail.

That part of the Angelina National Forest east of the Sam Rayburn Reservoir includes two primitive campgrounds, three boat launch areas (Townsend, Harvey Creek, and Bayou), and the Turkey Hill Wilderness. The rolling hills of the wilderness are densely forested with pines and deciduous trees. Turkey Hill, which has an elevation of 298 feet, is at the southeastern corner of the wilderness.

The Stephen F. Austin Experimental Forest Interpretive Trail System is an isolated part of the Angelina National Forest located eight miles southwest of Nacogdoches. This is piney woods country with the Angelina River forming the southern boundary. Although the 2,560-acre site was originally designed for wildlife and timber management research, it has been expanded to include educational and recreational activities. In 1997, the U.S. Forest Service completed its first major trails designed for universal accessibility. The two trails are both barrier free and handicap accessible. The Jack Creek Loop follows a clear spring-fed stream for almost a mile, passing through a forest of pines and hardwood trees more than 100 years old. The Management Loop is a 1.5-mile trail through various forest types with information concerning the best management for the different forest types for timber and wildlife.

Boykin Springs Longleaf

When the first European settlers arrived in the southeastern United States, they found longleaf pine forests carpeting some 75,000 square miles of the Coastal Plain adjacent to the Atlantic Ocean and the Gulf of Mexico. During the last century, many of these longleaf pine stands were cut so that crops could be grown and cattle could graze; others were harvested for paper pulp and turpentine.

Clear-cut areas were frequently replanted with slash pine, a faster-growing species suitable for paper pulp. Moreover, the suppression of natural forest fires was detrimental to the longleaf pines, which depend on fires to expose bare soil for good germination of their seeds and to suppress other woody plants. Southeastern Texas, for example, once contained 1,500 square miles of longleaf pine forests, largely in a wild, undisturbed region known as the Big Thicket. By 1935, barely 30 square miles of these old-growth stands remained, and today the state's Natural Heritage Program categorizes this dwindling habitat as threatened.

Throughout much of their range, from southwestern Virginia to southeastern Texas, longleaf pine stands are parklike, with an understory sparse in other woody species but rich in grasses and wildflowers. These communities, known as longleaf pine savannas, range from being very wet to very dry. An excellent example of a dry, upland one is Boykin Springs Longleaf, which occupies 350 acres of the Angelina National Forest. Here water is scarce because it drains rapidly through the sandy soil of the region's rolling hills.

Research by botanists Edwin Bridges and Steve Orzell indicates that the oldest stands at Boykin Springs Longleaf date back more than a century. Human disturbance has been minimal, and burns by the Forest Service, carried out every three to five years during the nongrowing season, help maintain the balance of vegetation.

In most of the forest, the most common understory plants are bracken fern, poison oak, and a rich layer of grasses, including little bluestem, slender bluestem, and pineywoods dropseed. But Bridges and Orzell identified nearly 250 species of flowering plants, from such shrubs as beauty-berry, deerberry, St.-Andrew's-cross, high-bush blueberry, and the aromatic wax myrtle to such wildflowers as blue larkspur, purple coneflower, dwarf yellow milkwort, blazing star, rattlebox, and black-eyed Susan. Of particular interest is the slender gay-feather, confined only to six adjacent counties in southeastern Texas. Known only since 1959, its lavender-purple spikes bloom in July.

In the driest parts of Boykin Springs Longleaf, where the soil is sandiest and has the fewest available nutrients, the longleaf pines become more scattered and a significant midcanopy appears, composed chiefly of bluejack oak. Vegetation on the forest floor is sparse: longleaf wild buckwheat and sawtooth nerveray are common, but sizable patches are bare or covered with lichens.

At the wetter extreme, seven hillside seepage bogs are found on short, steep, south-, southeast-, or southwest-facing slopes. These bogs have formed where a permeable layer of sandy soil lies atop a more impenetrable layer. Unable to continue its downward flow, percolating water spreads out horizontally, emerging from hillsides into shallow depressions. Most of the seepage bogs at Boykin Springs Longleaf are small, but Bridges and Orzell recorded nearly 150 plant species in them, including nearly two dozen rare species.

Sphagnum frequently carpets the bog. Other typical species include two kinds of insect-trapping sundews, a pitcher plant, bogbuttons with spherical white flower heads atop slender leafless stems, nine kinds of yellow-eyed grasses (not true grasses), bog violet, obedience plant (pl. 34), rose pogonia orchid, bishop's weed, round-leaved and perfoliate-leaved bonesets. Cinnamon fern, royal fern, and Virginia chain fern grow three feet tall or more.

Bog coneflowers are confined to these and other hillside seep bogs in four southeastern Texas counties and one adjacent western Louisiana parish. Flowering vigorously in June, this three- to six-foot-tall plant bears up to a dozen flower heads; each has an elongated, purple brown center, or cone, surrounded by several lemon yellow, petal-like rays that hang downward. Large leaves as rough as sandpaper grow at the base of the plant, with progressively smaller leaves found partway up the stem. Despite the conspicuous nature of the bog coneflower, it was not described as a species until 1986.

Drummond's yellow-eyed grass has an unusual, disjunct range, growing in a few bogs from Tallahassee, Florida, to Gulfport, Mississippi, then in western Louisiana and southeastern Texas, including at Boykin Springs Longleaf. Another rare species is the yellow fringeless orchid, which until its rediscovery here, by Bridges and Orzell, was last seen in the state in 1950.

A few wetland trees and shrubs—such as sweet bay, red maple, swamp red bay, bayberry, and poison sumac—sometimes grow in the bogs. These species are most likely to intrude from an adjacent wetland forest if periodic fires are suppressed.

From a geographic and biological standpoint, the Coastal Plain of the Gulf of Mexico can be divided into the East and West Gulf Coastal Plain, with the Mississippi River being the dividing line. Biologists have tended to regard the longleaf pine savannas west of the Mississippi River as impoverished examples of those to the east. Although this may be true of the shrubs and trees, the work by Bridges and Orzell has shown that when the herbaceous plants of the savannas are analyzed, the West Gulf Coastal Plain longleaf savannas boast nearly 50 species absent from the East.

Black Branch Barrens

Barrens—natural forest openings with prairie plants—are habitats that the Nature Conservancy ranks as threatened around the world. In a dry, upland region of the Angelina National Forest, of southeastern Texas, the oak and hickory woodland contains barrens noteworthy for the rare plants that grow there. The area has been little disturbed, although it lies just three miles from the region inundated by the Sam Rayburn Reservoir.

Naturalists employ the terms barrens, glade, and prairie for forest openings with prairie plants, frequently using them interchangeably. Botanist Alice Heikens defines them more precisely, on the basis of soil and vegetation. Prairies and glades are both openings containing few or no trees and shrubs. Prairie soil has no exposed rocks, whereas a glade has rocks of various sizes scattered over the surface. Glades may be further classified by the nature of the rocks—sandstone, limestone, dolomite, and granite. Small trees may take hold in the crevices between some of the rocks. Barrens may or may not have rocks; what distinguishes them is that the prairie vegetation grows beneath a canopy of trees, which are usually stunted.

Black Branch Barrens, covering half a square mile, fits Heikens's definition of barrens because small trees are scattered across the undulating terrain. Such habitats are now rare in North America, in part because so much land has been cleared for agriculture. Elsewhere, the suppression of natural fires has disturbed the ecology, allowing forest vegetation to take over.

The blackjack and post oaks at Black Branch Barrens are stunted—most are less than 30 feet tall—because the soil is poor in nutrients and dry much of the year. Low-intensity ground fires caused by lightning strikes have also discouraged the encroachment of woody plants. Shrubs that grow along with the oaks include wild privet, parsley-leaved hawthorn, and high-bush blueberry. The trees and shrubs are spaced far enough apart to provide areas with little shade, favorable to the growth of prairie species.

Clumps of little bluestem, pineywoods dropseed, and other bunch grasses dot the ground, while the most common wildflower is Nuttall's rayless goldenrod, a plant lacking the yellow, petal-like rays of the more familiar goldenrod. Pink milkwort, baptisia, neptunia, rough buttonweed, Texas coneflower, five kinds of blazing stars, and pencil flower, all typical of dry, open areas, grow in abundance. Here and there rattlesnake master raises its spherical flower heads, belying its membership in the carrot family, which usually has flowers in umbrella-shaped clusters. Lichens are also common in the barrens.

According to Steve Orzell and Edwin Bridges, both southeastern botanists, Black Branch Barrens is enriched by a number of plants endemic, or unique, to the West Gulf Coastal Plain, as well as some species with sporadic distributions that are rare in southeastern Texas. Navasota ladies'-tresses, a small, white-flowered orchid that has been listed as federally endangered, is one of the endemics. Others are golden hedge hyssop, which is a bright, golden-yellow wildflower related to the snapdragon, and slender gay-feather, which has a few lavender-purple flowering heads arranged on a very slender spike. Nearly endemic is Texas sunnybells, a member of the lily family with small, bell-shaped flowers. Among the rare plants Orzell and Bridges have recorded at the barrens are Drummond's sandwort, least daisy, San Saba pinweed, Nuttall's milk vetch, western dandelion, Texas saxifrage, and smooth phacelia.

Botanists K. L. Marietta and E. S. Nixon have attempted to shed light on why this plant community has arisen here. They report that the topsoil is only about two feet deep but contains an accumulation of clay, so that water seeps through it slowly. The clay is montmorillonite, the same type of spongy clay that is gradually creeping down Colorado's Mesa Seco to form Slumgullion Slide. During the cooler, rainier seasons of the year, the soil becomes saturated, but in late spring and summer, it dries out and develops cracks. In addition, the soil is low in some nutrients, such as phosphorus. The rare plants in the barrens apparently do well under these conditions, perhaps in part because the growth of potential competitors is discouraged.

A woodland of post oak, blackjack oak, and black hickory, with occasional winged elm and water oak, surrounds the barrens. Because of the better soil conditions, the trees may grow 75 feet tall. Common shrubs are wild privet, highbush blueberry, two kinds of hollies, and two kinds of hawthorns. The grasses and wildflowers include narrow-leaved sea oats, woodland buttercup, black snakeroot, small-flowered skullcap, and blue-leaved sage, all species characteristic of shady forest floors. Only a very few of the barrens' prairie species can survive beneath the abundant trees.

Davy Crockett National Forest

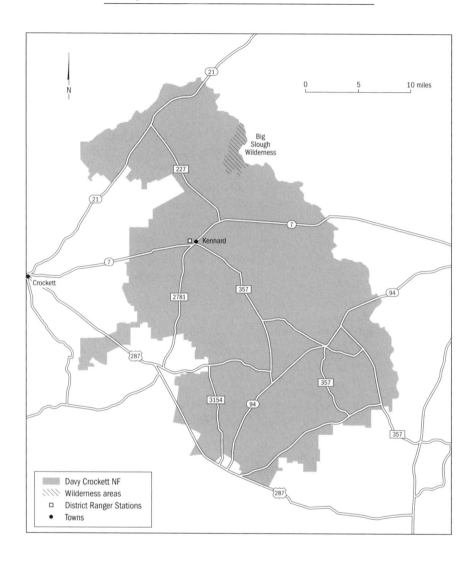

SIZE AND LOCATION: 160,647 acres in eastern Texas, west of Lufkin and east of Crockett. Major access routes are U.S. Highway 283 and State Routes 7, 21, 94, and 103. District Ranger Station: Kennard. Forest Supervisor's Office: 701 N. 1st Street, Lufkin, TX 75901. www.southernregion.fs.fed.us/texas.

SPECIAL FACILITIES: Boat launch areas; swimming areas.

SPECIAL ATTRACTIONS: Ratcliff Lake Recreation Area; Big Slough Canoe Trail.

WILDERNESS AREA: Big Slough (3,455 acres).

Unlike the national forests of Texas that lie to the east, the Davy Crockett National Forest has a preponderance of understory oaks with an overstory of pines. The whole aspect of the national forest is different.

At the center of the national forest is the Ratcliff Lake Recreation Area where you may camp, picnic, fish, swim, and hike. Campsites sit all around the 45-acre lake. The 1.5-mile Tall River Trail is a great one for families as it alternates between forests of pines and hardwood species. During spring, flowering dogwood (fig. 20) is a delight along the trail, and an abundance of spring-flowering wildflowers includes phloxes, Solomon's seal, spring beauty, violets, firecracker honeysuckle (pl. 35), and buttercups on the forest floor. A trail also follows the lake's shoreline.

If you are up to a longer hike, the 20-mile Four C Hiking Trail might be just the one. It begins at the Ratcliff Lake Recreation Area and winds to Neches Bluff at the northern tip of the forest. The trail dips and climbs through wetlands and uplands. The bottomland forests along the way have sweet gum, swamp chestnut oak, willow oak, and red maple as the dominant trees with the gorgeous red buckeye in the shrub layer. Red buckeye flowers in April, often when the plant is barely one foot tall. As the elevation increases slightly, the major trees are loblolly pine, sweet gum, southern red oak, white oak, sugarberry, hop hornbeam, and rusty nannyberry, with smaller trees and shrubs such as Carolina buckthorn, beauty-berry, fringe tree, and yaupon holly common. On the forest floor are three-lobed violet, blue sage (pl. 36), Carolina elephant's-foot, and cut-leaf grape fern. Toward the crest of hills where conditions are drier, red cedar, blackjack oak, and slippery elm occur. Farkleberry is the dominant shrub, and patches of bracken fern occur throughout the dry forests.

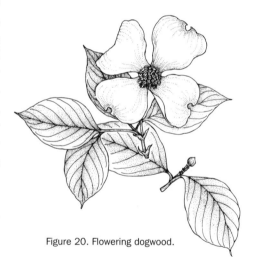

Figure 20. Flowering dogwood.

Neches Bluff overlooks the Neches River and is at the northern terminus of the Four C Hiking Trail. A path around the overlook is through a forest of red oak, blackjack oak, post oak, pignut hickory, sassafras, American holly, and winged elm. A great variety of shrubs and small trees live in this dry community, including devil's-walking-stick, redbud, rough-leaf dogwood, and beauty-berry. Dense entanglements of vines discourage hikers from straying off the path, because the woods teem with poison ivy, several kinds of greenbriers, trumpet vine, crossvine, and muscadine. In addition, the very prickly southern dewberry crawls along the ground. Most of the wildflowers bloom during summer. These include four-leaf bedstraw, Florida wild lettuce, wild bergamot, yellow wood sorrel, and a tiny white-flowered forget-me-not.

The Neches River from the northeastern boundary of the national forest wiggles its way southeastward to just west of Diboll. A few miles south of the Neches Overlook is a large swampy area that is the Big Slough Wilderness. Although this is the smallest wilderness area in Texas, it is the place to study swampy flora and fauna of east Texas. The wettest parts of the wilderness support water hickory, water elm, green ash, swamp red maple, swamp gum, swamp cottonwood, and black willow. On slightly higher ground are short-leaf and loblolly pines. Bird life is incredible, with various species of herons and egrets the showiest. The area is not for the squeamish, however, as it also has cottonmouths, timber rattlesnakes, feral hogs, wood ticks, mosquitoes, chiggers, ground-nesting yellowjackets and, of course, poison ivy. The Big Slough Canoe Trail is perhaps the best way to see the area as it makes a loop using the Neches River and open water in the slough. The trail should be canoed during winter and spring because the water level is usually too low during the hot, dry summer. A primitive campground at Holly Bluff sits above the Neches River a few miles east of Apple Springs in the southeastern corner of the Davy Crockett National Forest.

Sabine National Forest

SIZE AND LOCATION: 160,806 acres in eastern Texas, extending to the Louisiana border. Major access routes are U.S. Highway 96 and State Routes 21, 83, 87, and 147. District Ranger Station: Hemphill. Forest Supervisor's Office: 701 N. 1st Street, Lufkin, TX 75901. www.southernregion.fs.fed.us/texas.

SPECIAL FACILITIES: Boat launch areas; swimming areas.

SPECIAL ATTRACTIONS: Ragtown Recreation Area; Mill Creek Cove.

WILDERNESS AREA: Indian Mounds (12,369 acres).

The Sabine National Forest is in the Pineywoods Province of east Texas, with huge Toledo Bend Reservoir forming the eastern boundary of the national forest. Toledo Bend is the fifth largest artificial reservoir in the country.

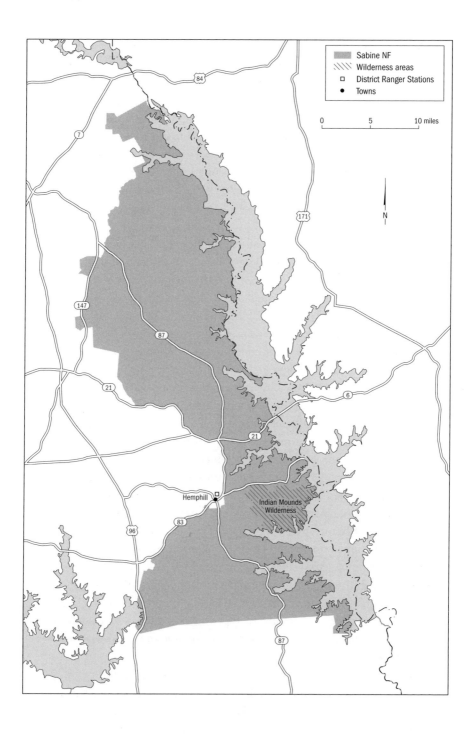

Sabine NF

Wilderness areas

□ District Ranger Stations

● Towns

0   5   10 miles

N

84

7

171

147

87

21

6

21

Hemphill □

Indian Mounds
Wilderness

96

83

87

Although almost all the forest has been logged extensively in the past, there are a few significant and isolated plant communities scattered about. One of the finest is Mill Creek Cove, nestled between two lobes of the Toledo Bend Reservoir shoreline. This old-growth forest has a cathedral-like appearance with the canopy height averaging 120 feet. Beech and southern magnolia dominate the cove, although a good diversity of tree species is present.

The Indian Mounds Wilderness a few miles south of Mill Creek Cove also has some pristine forests, particularly along Indian Creek and Hurricane Bayou. These forests are dominated by beech and white oak, with southern magnolia noticeably absent, apparently because of the low acidic content of the soil. The white-trunked chalk maple is here, as well as the attractive bigleaf snowbell bush. A prominent fall wildflower is blue-stem goldenrod.

State Route 87 from Hemphill goes through a major part of the southern end of the Sabine National Forest. At the settlement of Six Mile, a side road leads to the Lakeview Campground on the shore of the Toledo Bend Reservoir. The campground is the start of the 28-mile Trail Between the Lakes that connects the Toledo Bend Reservoir with the Sam Rayburn Reservoir. The trail passes through various plant communities. A few miles south of Six Mile on State Route 87 is the Willow Oak Campground. Beyond the campground and just before State Route 87 leaves the national forest is Foxhunters Hill, a relatively wild area that possesses some magnificent plant communities. On ridges with deep sandy soil is a community whose major tree species are bluejack oak and shortleaf pine, although other trees such as southern red oak, post oak, blackjack oak, and black hickory are also present. One area with apparently low pH, where longleaf pine is the dominant species, has a well-developed herbaceous layer of little bluestem, pineywoods dropseed, Carolina joint-tail grass, switch grass, and several kinds of sedges. A good shrub layer of beauty-berry, wax myrtle, and farkleberry is also present. Where there are permanently moist soils along creek bottoms, a sweetbay magnolia forest occurs. Gallberry, swamp red bay, titi, evergreen bayberry, red maple, and swamp gum are usually present beneath the trees. Shingle Branch Bog is also in the Foxhunters Hill area. An interesting assemblage of bog species includes cinnamon fern, pipewort, possum haw viburnum, poison sumac, and two rare herbs, the bog coneflower and rough-leaf yellow-eyed grass.

Red Hills Lake is a 19-acre spring-fed lake where fishing and swimming are permitted. A campground is adjacent to the lake, and there is a half-mile trail up Chambers Hill to the site where a lookout tower used to be.

Ragtown Recreation Area is a major facility at Toledo Bend Reservoir. The campground is high on a bluff above the reservoir, where there are spectacular views down into the water. Along the shore there is a boat ramp as well as the mile-long Mother Nature Trail through a beech and white oak woods.

Boat ramps also enter the Toledo Bend Reservoir just north of Ragtown at Haley's Ferry Boat Ramp and just south of Ragtown at East Hamilton Boat Ramp and at Indian Mounds Recreation Area and Willow Oak. Boles Field, about eight miles east of Shelbyville, is the site of a cemetery for fox hounds established around 1900 when fox hunting was popular in the area. Several markers are lined up beneath pine trees. A campground is at the area.

Mill Creek Cove

Although southeastern Texas is commonly referred to as the East Texas Pineywoods, majestic forests composed chiefly of American beech and southern magnolia sometimes prevail in the rolling, southern part of the region. These broad-leaved trees thrive where the soil is moist but well drained, such as near the foot of a hill. During the past century, many such forests, which resemble those of the Appalachian foothills hundreds of miles to the east, have been destroyed or disturbed by human activities, including the creation of huge reservoirs. One 94-acre tract that has been spared is Mill Creek Cove, located in the Sabine National Forest between two arms of the Toledo Bend Reservoir.

Woody plants form three distinct layers, or canopies, in Mill Creek Cove. The upper canopy forms the forest's cathedral-like ceiling and averages about 100 feet high; it is dominated by American beech and southern magnolia, with an occasional tall sweet gum, white oak, water oak, and green ash. Some of the beeches and magnolias are more than 120 feet tall, with diameters of nearly four feet at shoulder height. This upper canopy is completely closed except for an occasional small opening where a large tree has died.

Musclewood, hop hornbeam, black gum, flowering dogwood, and the evergreen American holly grow 25 to 50 feet tall. All these midcanopy species are able to thrive in the heavy shade cast by the taller trees. Beneath the midcanopy is the shrub layer, where shrubs and trees seldom grow more than 15 feet high. The species found there include yaupon holly, beauty-berry, highbush blueberry, styrax, witch hazel, and Carolina buckthorn. These last two are capable of growing twice as tall where light is more abundant.

Because of the deep shade, there are no brushy thickets or entanglements of vines on the forest floor, giving Mill Creek Cove a nearly parklike appearance. What vines exist—such as the southern muscadine grape, yellow jessamine, crossvine, and Virginia creeper—seek sunlight by climbing the nearest tree and do not sprawl across the leaf litter. Except for some scattered plants of jack-in-the-pulpit and hound's-tongue, wildflowers are scarce.

Only partridgeberry, with its deep green, penny-sized leaves, lies in patches on the ground.

The distinct breaks between the different canopies in Mill Creek Cove reflect the dominance of a limited number of species, the result of competition over an extended period of time. This suggests that this is a virgin forest. Botanist E. S. Nixon and his colleagues, who have studied the woody plants in Mill Creek Cove, have also noted an abundance of beech and southern magnolias of every size and age, another clue that this forest has not been cut. Because the growth of the younger trees renew the dominant species, Mill Creek Cove is a climax beech–southern magnolia forest—the only one, according to botanist Steve Orzell, on the West Gulf Coastal Plain, a region that extends around the Gulf of Mexico from Mobile, Alabama, to Galveston, Texas.

Ecologists use the term climax to refer to the final stage in a natural succession of plant communities in a specific environment. In an area such as Mill Creek Cove, for example, the first plants to colonize bare ground are primarily annuals, plants that grow from seeds, reproduce, and then die back each year. As the annuals die, their organic remains contribute to the soil and build up its capacity to retain water. After a year or two, this provides a suitable habitat for perennial herbs to become established. The perennials usually grow taller than the annuals, shading them out.

As the soil becomes richer and more capable of holding water, more robust perennials appear. Shrubs and eventually trees come to dominate. The climax is reached when the forest community consists of plants that continually replace themselves, rather than paving the way for a succeeding stage.

Sam Houston National Forest

SIZE AND LOCATION: 163,037 acres in southeastern Texas south of Huntsville. Major access routes are Interstate 45, U.S. Highways 59, 75, and 190, and State Route 150. District Ranger Station: New Waverly. Forest Supervisor's Office: 701 N. 1st Street, Lufkin, TX 75901. www.southernregion.fs.fed.us/texas.

SPECIAL FACILITIES: Boat ramps; swimming areas; ATV, bicycle, and equestrian trails.

SPECIAL ATTRACTIONS: Big Creek Scenic Area; Big Thicket.

WILDERNESS AREA: Little Lake Creek (3,855 acres).

The Sam Houston National Forest consists of two major tracts, one south-west of Huntsville toward Conroe and one southeast toward Cleveland, and several isolated smaller parcels of land beginning four miles east of Huntsville.

Lake Conroe penetrates the block of land southwest of Huntsville, pro-viding ample opportunity for fishing and boating activities. At the upper reaches of Lake Conroe is an old oxbow now known as Lake Stubblefield. Fishing and canoeing are especially popular on the lake. The campground on the south shore of this lake is nestled beneath southern pines and various kinds of oaks, and flowering dogwood add a touch of color during spring. The Stubblefield Interpretive Trail provides a good cross section of the woods here, although the results of a recent fire that swept through the area may be seen. The long list of trees along the trail includes loblolly pine, slippery elm, American elm, red maple, sweet gum, green ash, hop hornbeam, flowering dogwood, and Carolina buckthorn. Colorful shrubs present are yaupon holly, wax myrtle, arrowwood viburnum, elderberry, beauty-berry, and devil's-walking-stick. Dwarf palmettos are here and there. If you stray off the trail, you encounter hordes of vines—poison ivy, Virginia creeper, pepper-vine, rattan vine, crossvine, moonseed, small yellow passion flower, and three kinds of greenbriers. From the bridge over a muddy creek you may see large sycamores along the banks with deciduous hollies beneath them.

South of Stubblefield with a boat ramp onto Lake Conroe is Cagle Recre-ation Area with new camping facilities. Cagle Recreation Area is one mile south of the road between New Waverly and Richards, and Scott's Ridge boat ramp, also on Lake Conroe, is a few miles north of Montgomery.

Little Lake Creek Wilderness is about eight miles southwest of Stubble-field. Following intensive logging of the area several decades ago, the region has grown back into a nice forest of hardwood species, including cherrybark oak and swamp gum. Little Creek extends through the center of the wilder-ness, with eastern cottonwoods, sycamores, box elders, and swamp privets along its banks. On higher ridges are stands of shortleaf and loblolly pines.

The Lone Star Hiking Trail can be picked up from the wilderness or from several other areas along the 128-mile length of the trail. Five side loops are associated with the main trail. The trail winds through the entire width of the Sam Houston National Forest, at times following roads and sometimes crossing private land. Forest points that it passes are Little Lake Creek Wil-derness, Kelly's Pond Campground, Stubblefield Lake, Double Lake Recre-ation Area, and the Big Creek Scenic Area. Sixty-seven miles of marked trails have been set aside for use by dirt bikes, mountain bikes, equestrians, motor-cycles, and all-terrain vehicles less then 46 inches wide. Kelly's Pond is a good place to join this trail system.

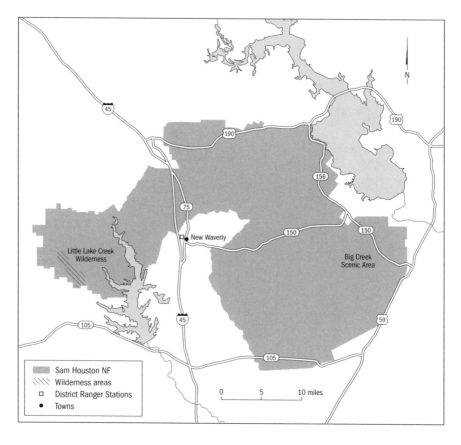

Sam Houston NF
Wilderness areas
District Ranger Stations
Towns

0   5   10 miles

The main features of the Sam Houston National Forest in the unit near Cleveland are the Double Lake Recreation Area and Big Creek Scenic Area. Double Lake is an area ideal for lovers of the outdoors because it offers camping, picnicking, swimming, fishing, hiking, and just plain relaxation. As you wander around the lake, you see laurel cherry, southern red oak, hackberry, post oak, American elm, and loblolly pine. A boat ramp is available and boats also are available for rent. Fishermen enjoy angling for bass, bream, and catfish, and there is a sandy beach adjacent to a swimming area. From the Double Lake Recreation Area you may access the Lone Star Hiking Trail for a five-mile hike to Big Creek Scenic Area. You may also drive to the scenic area.

Big Creek Scenic Area provides a taste of the famous Big Thicket of eastern Texas, because the terrain, flora, and fauna are similar. Double Lake Branch and Henry Lake Branch flow into the scenic area and join to form Big Creek. Hiking trails in the area sometimes parallel the creek, at other times go through wet woods and mesic woods, and sometimes climb onto upland ridges. Along the creeks are sycamores, deciduous hollies, and giant cane, the latter being the only native bamboo in the contiguous United States. Wet woods that border the creeks are home to American elm, sweet gum, green ash, swamp gum, red maple, and the shrubby and very beautiful Virginia sweetspire. In low depressions, often with shallow standing water, are American snowbell bush, cinnamon fern, netted chain

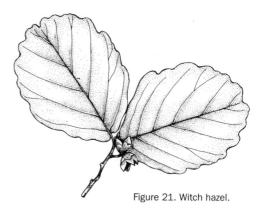

Figure 21. Witch hazel.

fern, sensitive fern, lizard's-tail, broad-leaved arrowhead, swamp dock, crinum lily, smartweeds, and several kinds of sedges. The mesic woods, located between the wet woods near the creek and the ridges, have the most diverse flora. Trees and shrubs in the mesic woods include beech, sugarberry, sweetbay magnolia, hop hornbeam, musclewood tree, silverbell, pawpaw, laurel cherry, wax myrtle, swamp azalea, swamp blueberry, strawberry bush, and beauty-berry. The plants in the upland woods are farther apart, giving this habitat an open appearance. Here are white oak, American holly, black gum, red mulberry, witch hazel (fig. 21), and flowering dogwood.

The greatest population of the endangered red-cockaded woodpecker is in the Sam Houston National Forest. The male of the species is cardinal size with a red patch behind each eye. The national forest has two areas for public viewing of the red-cockaded woodpecker where the birds inhabit a total of 35 cavity trees. One observation site is 5.2 miles west of Interstate 45 on Farm-to-Market (FM) Road 1375 and another is near the Stubblefield Campground. Look for these birds at sunrise or sunset.

Big Thicket

When the Tonkawas, Karankawas, and Caddo Indians roamed freely in what is now eastern Texas, a dense forest stretched for more than 100 miles northeast of the present-day city of Houston. To the Indians it was the Big Woods, but Anglo-Saxon settlers who arrived early in the 1820s called it the Big Thicket because of its nearly impenetrable shrub layer.

Harris B. Parks, who inventoried the plants and animals of the Big Thicket for the Texas Academy of Science during the 1930s, estimated that the forest originally covered 3.5 million acres, a region about the size of the state of Connecticut. After more than a century of intensive lumbering, which began in 1850, and the inroads of oil interests and real-estate developers, the Big

Thicket has been fragmented and reduced to about 300,000 acres. Although exploitation of the forest continues today, the National Park Service has been able to set aside several isolated tracts totaling 84,000 acres as the newly designated Big Thicket National Monument. Another portion falls within Sam Houston National Forest.

The Big Thicket occupies a gigantic basin, which slopes from north to south and consists of low hills, spring-fed streams, and flat, swampy terrain. Biologist Claude McLeod, who studied the Big Thicket for many years, divides the region into the Upper Thicket and Lower Thicket on the basis of topography, soil types, and dominant trees. The terrain of the Upper Thicket, in the northern part of the basin, consists of low hills dissected by clear, sparkling streams. Underlying layers of limestone have made the sandy soil less acidic than that of the Lower Thicket. The latter, which extends nearly to the Gulf Coast, is flat and pocked with swamps and bayous.

The thicket's vegetation is distinctive. Loblolly pine dominates the tree canopy, along with white oak, southern magnolias, and majestic, smooth-barked American beech. In the Lower Thicket the American beech is replaced by swamp chestnut oak, a tree more tolerant of acidic soils. All of these trees grow to immense size because the well-aerated soil holds considerable moisture and abounds with bacteria and fungi that decompose plant and animal residues. Diverse plants such as mesquite and bald cypress, Spanish moss and pitcher plants, as well as animals such as alligators and roadrunners, add to the natural attractions of the region.

The Big Thicket may be sampled within Sam Houston National Forest, where 1,130 acres have been designated the Big Creek Scenic Area, named for a clear stream that empties into the Trinity River. Several trails lead from a parking lot accessible by a forest road off State Route 150. The four characteristic species of the Upper Thicket—loblolly pine, southern magnolia, white oak, and American beech—are common along Big Creek, some attaining giant stature. Tree species beneath the canopy include sugar maple, white ash, and basswood. The dense shrub layer, which gives the Big Thicket its name, is crowded with red bay, sweetleaf, maple-leaved viburnum, strawberry bush, greenbriers, and the magenta-fruited beauty-berry. Wildflowers carpet the forest floor during spring, led by bloodroots, trout lilies, Mayapples, jack-in-the-pulpits, wake-robins, and even several kinds of wild orchids.

Low, sandy hills adjacent to Big Creek provide a drier and more open habitat for post oak, blackjack oak, sassafras, winged sumac, black hickory, and farkleberry. Prickly pear cactus and yucca grow sporadically beneath these scrubby trees.

The plant and animal life of the Big Thicket, a mixture of elements from the northern forests, the southern coastal plain, the northeastern mountains, and the western plains, led one writer to refer to the area as the "biological crossroads of North America." In 1967, a National Park Service study of the Big Thicket identified "forest elements common to the Florida Everglades, the Okefenokee Swamp, the Appalachian region, the Piedmont forests, and the large, open woodlands of the coastal plain. Some large areas resemble tropical jungles in the Mexican states of Tamaulipas and Veracruz."

Biologists have long speculated about the origins of the Big Thicket and the diversity of its flora. Harris B. Parks suggested that the Big Thicket was at one time the bottom of a sea that extended inland from the Gulf of Mexico. As ocean levels repeatedly rose and fell during the last Ice Age, the advancing and retreating water from the gulf deposited sands in the large basin. The low-lying hills we see today are ancient sand dunes, whereas the lower areas toward the Gulf represent prehistoric sand flats. Existing swamps and bogs are relict ponds that were trapped by dunes as the ancient seas finally subsided. Subsequent action by wind and water has molded the thicket into its current configuration.

As the last of the seas retreated, plants sprouted from seeds and other pieces of vegetation that had been deposited by the precursors of the region's present-day rivers. Extensive flooding of the Mississippi River carried in species from farther north, and warm Gulf waters deposited a variety of subtropical plants. Westerly winds sowed seeds from the plains, and other plants colonized the basin in the course of gradually expanding their ranges. Donovan Correll, the late authority on Texas plants, traced the origin of some Big Thicket plants to the Appalachians, noting that at the western edge of their range, these plants tended to differ from their eastern relatives.

Today only a portion of the once vast thicket survives. Despite protection by the National Park Service and the U.S. Forest Service, even this remnant is threatened. The southern pine beetle, which spreads a fungus, is making deep inroads into the native pine populations, destroying pines and forcing foresters to cut down diseased trees in an effort to contain the damage. Ironically, efforts to control fire in southern forests are probably responsible for the pine beetle invasion. This plague is just another in a series of events that has decimated the Big Woods that the Tonkawas, Karankawas, and Caddos knew and appreciated.

NATIONAL FORESTS IN WYOMING

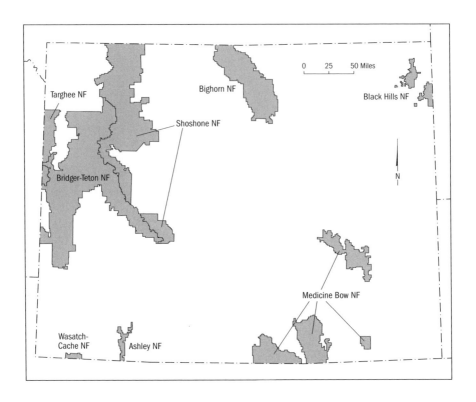

The national forests in Wyoming are in Region 2 of the U.S. Forest Service. The Medicine Bow National Forest also administers Colorado's Routt National Forest. The Bridger-Teton National Forest is treated as a single entry.

Bighorn National Forest

SIZE AND LOCATION: 1,115,073 acres in north-central Wyoming, west of Sheridan and extending northward to Montana. Major access routes are Interstates 25 and 90, U.S. Highways 14 (Bighorn National Scenic Byway), 14A (Medicine Wheel Passage Scenic Byway), and 16 (Cloud Peak Skyway), and State Route 31. District Ranger Stations: Buffalo, Lovell, and Sheridan. Forest Supervisor's Office: 2013 Eastside 2nd Street, Sheridan, WY 82801, www.fs.fed.us/r2/bighorn.

SPECIAL FACILITIES: Winter sports areas.

SPECIAL ATTRACTIONS: Bighorn National Scenic Byway; Medicine Wheel Passage National Scenic Byway; Cloud Peak Skyway; Medicine Wheel.

WILDERNESS AREA: Cloud Peak (189,039 acres).

The Bighorn National Forest is in the heart of the Bighorn Mountains that rise abruptly out of the Great Plains in north-central Wyoming. The Big Horn, Tongue, and Powder Rivers are in the national forest, all emptying into the Yellowstone River. The national forest, mountains, and river are named for bighorn sheep that inhabit the area. The Sioux, Crow, and Cheyenne Indians lived in the Bighorn Mountains prior to the many battles during the 1860s and 1870s when settlers came into the area.

Three scenic byways that cross the Bighorn National Forest and roads off of them allow visitors to reach most parts of the national forest easily. From Ranchester on Interstate 90, U.S. Highway 14 heads westward toward the Bighorn Mountains that you can see rising in the distance. After reaching the eastern border of the Bighorn National Forest, the highway climbs 3,200 feet in just a few miles. U.S. Highway 14 is the Bighorn National Scenic Byway. After you have reached the top of the switchbacks, there is an area south of the highway known as Fallen City. The "city" is a spectacular rock slide in the Madison limestone where chunks of limestone rocks are strewn along the side of a mountain slope. Several scenic pullouts are on the way to the top of the mountain. Forest Highway 16 is a side road south that in four miles comes to a hiking trail that goes to the Black Mountain Lookout. The mountain towers above the surrounding terrain.

After the Bighorn National Scenic Byway passes the side road to the Black Mountain Trail, the highway comes to Sibley Lake where there is an inviting campground. At the Burgess Visitor Center you may get information about the national forest and hike an interpretive nature trail.

At Burgess Junction, U.S. Highway 14, the Bighorn National Scenic Byway, turns abruptly southward, with U.S. Highway 14A branching off to the west. U.S. Highway 14A is the Medicine Wheel Passage National Scenic Byway. By staying on U.S. Highway 14, the route continues to climb to Owen Creek Campground. A side road from the campground goes to the Tie Flume along the South Tongue River and is definitely worth a visit. A campground is near the flume. South of the Tie Flume is Woodrock where there used to be a sawmill, tie flume, and splash dam for logs. The log ties were sent down the flume to the town of Dayton. From the Tie Flume, Forest Highway 26 goes through scenic forested land to Twin Lakes near the edge of the Cloud Peak Wilderness. A pleasant trail to Coney Lake is in the wilderness. Forest Highway 26 continues eastward past Dome Rock to two campgrounds near the Big Goose Forest Service facility.

From the Owen Creek Campground, the Bighorn National Scenic Byway climbs over 8,950-foot Granite Pass to the Antelope Butte Ski Area before descending to spectacular Shell Canyon. The scenic byway is now dwarfed by high bluffs of the canyon. The Shell Canyon Research Natural Area is located on the south side of Shell Creek. The area was established because of its fine community of Rocky Mountain juniper below stands of Douglas fir on the slopes. The area is dominated by steep, north-facing limestone cliffs. Biologists recognize three distinct plant associations in the Research Natural Area. The Douglas fir community is on steep slopes between and just below the limestone cliffs. Plants associated with Douglas fir are mountain ninebark, mountain snowberry, Rocky Mountain juniper, limber pine, white spiraea, creeping Oregon grape, heartleaf arnica, and northern bedstraw. The Rocky Mountain juniper community occurs on moderate to steep lower slopes between the Douglas fir forest above and the sagebrush on the terrace below. Other plants in this community are blue bunch wheat grass, curl-leaf mountain mahogany, Junegrass, Geyer's onion, stemless goldenweed, moss phlox, and a species of fleabane. Between the juniper community and Shell Canyon is the black sagebrush community where stands of sagebrush are joined by blue bunch wheat grass, big sagebrush, winterfat, and Junegrass. Be sure to linger for a moment at picturesque Shell Falls before the scenic byway drops down to Post Creek Campground and then out of the national forest.

By heading west on U.S. Highway 14A, the Medicine Wheel Passage National Scenic Byway, you follow a scenic route along North Tongue River, passing between Little Bald Mountain and Bald Mountain. Near where Gold Creek enters the North Tongue River is the site of Bald Mountain City, an old mining town.

From the Bald Mountain Campground along the scenic byway, there are two options, each of which should be taken. Forest Highway 14, the first road

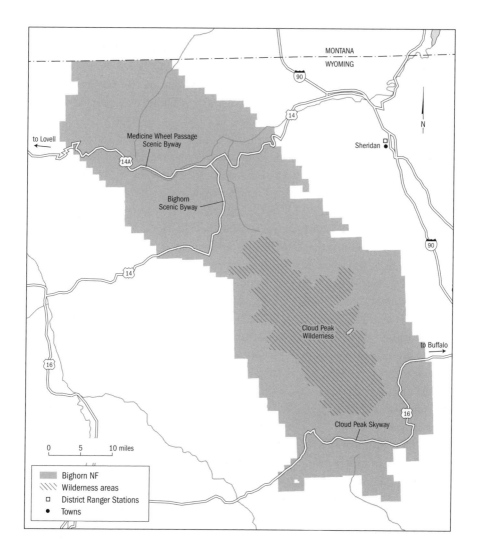

MONTANA
WYOMING

90

14

N

to Lovell

Medicine Wheel Passage
Scenic Byway

Sheridan

14A

Bighorn
Scenic Byway

90

14

Cloud Peak
Wilderness

to Buffalo

16

16

Cloud Peak Skyway

0   5   10 miles

Bighorn NF
Wilderness areas
District Ranger Stations
Towns

to the north past the campground, circles around the Porcupine Guard Station and Campground to Porcupine Falls. You may hike a trail down to a pool at the base of the 200-foot falls. Less than three miles north of Porcupine Falls is the beginning of a three-mile-long trail to Bucking Mule Falls, an even more spectacular waterfall that plummets 500 feet into Porcupine Creek. Porcupine Creek was the site of several gold strikes decades ago. By taking the second road to the north (Forest Highway 12) after leaving the Bald Mountain Campground, you may drive as far as Medicine Mountain. From the road closure you hike a little more than one mile to the mysterious Medicine Wheel, a prehistoric relic on the north side of Medicine Mountain. The wheel is constructed of stones laid side by side, forming an almost perfect circle 80 feet in diameter, and is our country's equivalent to Stonehenge.

The Cloud Peak Wilderness occupies the center of the Bighorn National Forest with the crest of the Bighorn Mountains located down the middle of the wilderness. Cloud Peak (13,175 feet) and one of the Mather Peaks (12,348 feet) are the highest in the wilderness. A hiking trail over Florence Pass goes between these two highest peaks.

U.S. Highway 16 from Buffalo is the Cloud Peak Skyway and it curves around the southern end of the Cloud Peak Wilderness. As soon as the skyway enters the national forest, there is an observation point on Hospital Hill. The highway then passes several campgrounds and picnic areas as it winds its way south to Crazy Woman Campground that is situated along Crazy Woman Creek. Near the campground is the historic Muddy Creek Guard Station, built during the 1930s by the Civilian Conservation Corps. You may rent it from the Forest Service as an overnight cabin. A dirt road from the guard station squeezes through narrow Crazy Woman Canyon.

From the Crazy Woman Campground, the skyway takes a westerly route until it reaches Meadowlark Lake. The route passes over Powder River Pass, the highest elevation on the skyway at 9,677 feet. Around Meadowlark Lake are several campgrounds, picnic areas, and a lodge. Powder River Ski Area is one mile south of the lake. Beyond Meadowlark Lake is Boulder Park Campground, and then the Cloud Peak Skyway drops 2,600 feet in about eight miles where it enters scenic Ten Sleep Canyon, lined by colorful red and white rock walls.

Outside the western boundary of the Cloud Peak Wilderness are the Paint Rock Lakes and the Medicine Lodge Lakes, all picturesque fishing lakes. The region also has several archaeological sites. The lakes may be reached by a series of back roads off of State Route 31.

A paved road from the northern edge of Dayton follows the Tongue River into the Bighorn National Forest. Shortly after entering the forest, the Tongue River Cave is on the south side of the river.

Ten miles northwest of Burgess Junction is the Bull Elk Research Natural Area. The area is not easy to get to because the last six miles to it consist of two miles of a four-wheel-drive road and four miles of hiking. The area is a typical grassland, with occasional limestone outcrops, occupying the top of a high plateau in the Bighorn Mountains. Because of its isolation, the grassy park has never been grazed by domestic livestock but is grazed by elk in spring and summer. The plants are on a grassy bald bisected by a low ridge that extends its entire length. The park is considered to be a disjunct unit of the Palouse Prairie of eastern Washington state. Common grasses that comprise the bald are blue bunch wheat grass, Idaho fescue, spike fescue, smooth brome, mountain brome, Nelson needle grass, timber oat grass, big bluegrass, blue wild rye, slender wheat grass, and bearded wheat grass. Growing among the grasses are false yarrow, lupine, mountain dandelion, yarrow, wild geranium, cinquefoil, northern bedstraw, harebell, asters, fleabanes, snowberry, and sagebrush.

Medicine Mountain

The Bighorn Mountains form a huge arc, extending from southern Montana into central Wyoming. Their slopes are dominated by coniferous species— Douglas and subalpine firs, Engelmann and white spruces, and ponderosa, limber, and lodgepole pines—with quaking aspen and Rocky Mountain maple adding a touch of variation. Below the timberline, roughly 9,700 feet above sea level, daisies, monkey flowers, groundsels, paintbrushes, beard-tongues, and balsamroots provide vivid color, and on the higher peaks grow a number of alpine species. These miniature plants, which bloom in July and August, include a rosy pink shooting-star, Parry's primrose, white-hairy and alpine cinquefoils, tufted saxifrage, paintbrush, gray groundsel, alpine blue-bells, arctic harebell, low blue-eyes, and stemless woollybase.

One of the higher peaks is Medicine Mountain, which rises 9,962 feet in Wyoming's Bighorn National Forest. Winter is so severe there that the mountaintop is often covered with snow as late as mid-June. Despite the inhospitable location, on a shoulder of the mountain, about 340 feet downslope and one mile northwest of the windswept peak, lies a pattern of rocks arranged by unknown native people. Known as the Medicine Wheel, it has been the subject of study and speculation since it was first recorded in 1895. In 1958 archaeologist Don Grey dated it by dendrochronology to A.D. 1760.

The rocks that make up the Medicine Wheel were taken from the surrounding limestone terrain. They form an irregular circle with a diameter of approximately 80 feet. A 2.5-foot opening in the circle faces approximately east, and in the center is a hollow mound of rocks, or cairn, 14 feet in diameter and 2.5 to three feet high. Excavation beneath this cairn has revealed that a hole nearly three feet deep had once been dug there.

Radiating from the central cairn to the outer circle are 28 irregularly spaced rows of stones, resembling the spokes of a wheel. Five other cairns, each an open circle between three and four feet in diameter and standing a few inches tall, are irregularly spaced around the outer edge of the wheel. One is located approximately north, one northeast, one southeast, one nearly south, and one west. Another lies about 12 feet southwest of the wheel, connected by a continuation of one of the 28 spokes.

Since its modern discovery, archaeologists have wondered who built the Medicine Wheel and what it was for. The first published account, in 1895, speculated that the peripheral cairns were occupied by the medicine men of different tribes during religious ceremonies, and that the large central cairn was the abode of the manitou, a supernatural force that gives power to spirits. In 1922, George Grinnell interviewed Plains Indians who still lived in the nearby lowlands. None had ever seen the Medicine Wheel, but many had heard of it.

When Elk River, a revered Cheyenne, was shown a sketch of the Medicine Wheel, he immediately said that it was the plan of an old-time Cheyenne medicine lodge, used for Sun Dance ceremonies. Such a lodge, constructed with a wood framework, consisted of an outside circular wall and a central pole, with 28 rafters radiating to the outer wall. An opening facing east permitted entry. At the western edge (equivalent to the western cairn of the Medicine Wheel) was the sacred altar, "where the thunder came from," according to Elk River.

Grinnell suspected that the Medicine Wheel was visited by a great number of people in prehistoric times, because a well-worn trail runs along the side of the mountain and descends across a narrow saddle before climbing to the shoulder where the Medicine Wheel is built. Grinnell concluded that the Medicine Wheel had been the site of a ceremonial Sun Dance lodge and that stones had been used in its construction because wood was not immediately available above timberline.

In 1953, archaeologist Thomas Kehoe studied a number of similar stone wheels in Alberta, Canada, and adjacent Montana and interviewed Blackfoot Indians about them. Several Indians suggested that the wheels were associated with death or burial. One remarked that "when they buried a real chief, one that the people loved, they would pile rocks around the edge of his lodge and then place rows of rocks out from his burial tipi. The rock lines show that everybody went there to get something to eat." Another noted that "a circle of stones used to mark the place where great chiefs or medicine men died." Others referred to the rock cairns as places that commemorated important events, such as a great battle or a lodge struck by smallpox. Kehoe concluded that certain death rituals of the Blackfoot Indians might have given rise to the strange configurations.

In 1973, astronomer John Eddy studied the Medicine Wheel in the Bighorn Mountains to see if it might instead be a calendric device, much like Stonehenge in England. To visit the site at the summer solstice (around June 22, when the sun is in the sky for the longest time during the year), he had to trudge through the mountain snow. Fortunately, even though the snow was boot-deep on the climb up, the wheel itself had been mostly blown clear.

Before dawn at the summer solstice, Eddy positioned himself at the southwestern cairn, 12 feet outside the wheel, and sighted over the central cairn. When the sun broke the horizon, it was directly in line with the two points! Later, at sunset, he stood at the southeastern cairn and directed his view over the central cairn; again, the sun was nearly on target.

Eddy proceeded to check the alignments of the other cairns to see if they corresponded to other astronomical phenomena. He knew that the rising of bright stars had often been used for calendar references by early peoples, including the Egyptians. At night, Eddy found that if he stood at the western cairn and sighted over the northern cairn, Aldebaran, the second brightest star in the sky during the summer solstice, was in alignment. Aldebaran is the one star, according to Eddy, whose brief appearance near dawn would signal the summer solstice.

Gazing across the northeastern cairn from the western cairn, Eddy found that another bright star, Rigel, was almost in line, and the brightest star, Sirius, was nearly in alignment when sighted over the central cairn from the western cairn. In making further calculations, Eddy noted that Rigel would be in perfect alignment one lunar month after Aldebaran's rise at solstice and Sirius would be in perfect alignment one lunar month after that, a good reminder that winter was just around the corner.

Eddy points out that the comparison of the Medicine Wheel with the floor plan of the Cheyenne medicine lodge is not perfect: there are too many cairns and they are not accurately aligned with the cardinal points. Eddy also believes that because of its inhospitable location, the Medicine Wheel was not used ceremonially by large groups of people. He suggests that it served as a primitive astronomical observatory, visited by only a few.

After Eddy reported his findings, Thomas Kehoe, along with Alice Kehoe, decided to reexamine 11 similar stone patterns in Canada and Montana. At three of these sites, they concurred that the summer solstice and the appearance of Aldebaran and Sirius could be determined from the stones, but the other eight sites did not conform. Carbon-14 dating showed that a medicine wheel on Moose Mountain, in Saskatchewan, which is practically a duplicate of the one on Medicine Mountain, was built in about 750 B.C.

The Kehoes also learned from interviews that the Plains Cree take note of the longest day of the year, regarding it as the beginning of the new year. The Indians once relied on calendar men to tell them when to celebrate. Observation of the solstice sun might have been part of the private knowledge these calendar men held. The Kehoes speculate that when the original function of medicine wheels was no longer remembered by later generations, similar cairns may have been used for the burial of important men, eventually leading to the Blackfoot tradition of erecting memorials to chiefs by placing rock lines that led to the cairns. This would account for some of the wheels not lining up with the sun and the stars.

Similarly, Eddy has suggested that the pattern of the Bighorn Medicine Wheel and the plan of the Cheyenne medicine lodge may be related because the lodge was usually built to celebrate the Sun Dance, traditionally performed near the time of the solstice. Eddy believes that a few observers may have climbed up to the Medicine Wheel in June to ascertain the time to begin the ceremony, which was probably held at a lower site more accessible to the tribe. If the wheel was known to only a select few, this could have added to the mystique of the procedure.

Bridger-Teton National Forest

SIZE AND LOCATION: Nearly 3.5 million acres in western Wyoming, east of Grand Teton National Park and south of Yellowstone National Park. Major access routes are U.S. Highways 26, 87, 89, 189, 191, and 287 and State Routes 22 and 352. District Ranger Stations: Afton, Big Piney, Jackson, Kemmerer, Moran, and Pinedale. Forest Supervisor's Office: 340 N. Cache Street, Jackson, WY 83001, www.fs.fed.us/r2/btnf.

SPECIAL FACILITIES: Winter sports areas; boat ramps; swimming pools.

SPECIAL ATTRACTIONS: Kendall Warm Springs; Gros Ventre Slide Area; Lander Cutoff of Oregon Trail; Periodic Spring.

WILDERNESS AREAS: Gros Ventre (317,874 acres); Teton (585,238 acres); Bridger (427,087 acres).

The Bridger-Teton National Forest is two forests in one. Although the two are contiguous, they were combined into one administrative unit in 1973.

The Teton is the northernmost of the two forests, with its northern border reaching the southern edge of Yellowstone National Park. The Bridger is mostly south of the Teton and is divided into an eastern division and a western division. The Bridger-Teton National Forest is a scenic high mountain area that includes several mountain ranges.

The Teton National Forest occupies a horseshoe-shaped area around Jackson Hole and Grand Teton National Park. The Snake River Range forms the western border of the Teton National Forest, with the Gros Ventre Mountains nearly in the center of the forest. The Snake and Hoback Rivers have caused deep gorges within the national forest.

The northern third of the Teton National Forest and all of it that borders the Yellowstone National Park are in the Teton Wilderness, a vast area bisected by the Continental Divide. The western side of the wilderness has heavily forested ridges, mountain meadows, and grassy slopes with elevations between 7,500 and 9,942 feet. The eastern side, however, has rugged canyons, mountain peaks, high plateaus, and bubbling streams. Elevation in this part of the wilderness goes as high as 12,165 feet on Younts Peak.

Two distinctive plateaus occupy the eastern side of the wilderness. Thorofare Plateau northeast of the Yellowstone River consists of breccia, which is a recemented rock resembling conglomerates, but the rubble of rocks is rough and broken, showing that it has weathered very little. At the southeast corner of the wilderness is Buffalo Plateau, composed of volcanic conglomerates that resemble river-worn rocks and pebbles set in cement. Some of the Jackson Hole elk herd spends summer in the Teton Wilderness. Near the center of the wilderness is the remarkable Two Ocean Creek. When this creek reaches the Continental Divide, it divides into two creeks. Pacific Creek flows west into the Pacific Ocean by way of the Snake and Columbia Rivers. The water of Atlantic Creek eventually reaches the Atlantic Ocean by means of the Yellowstone, Missouri, and Mississippi Rivers. Also in the wilderness are South Fork Falls and North Fork Falls along the South Buffalo Fork and North Buffalo Fork, respectively. The South Fork Falls plunges through a sheer-walled chasm 80 feet deep and six feet wide. North Fork Falls, just one mile south of the Continental Divide, drops several times through a narrow rocky gorge. Grizzly bears live in the wilderness and trumpeter swans may be seen on some of the lakes. The great forest fire in 1988 in Yellowstone National Park burned many acres in the Teton Wilderness, and a tornado in 1987 cut a swath 20 miles long and two miles wide.

From the southern entrance of Yellowstone National Park, U.S. Highway 89/287 follows the western boundary of the Teton Wilderness to Moran Junction. From Moran Junction to the east, U.S. Highway 26/287 stays within two or three miles of the southern border of the Teton Wilderness. Along these highways you can find great things to do, both outside and inside the wilderness.

Only three miles south of Yellowstone National Park is the Sheffield Creek Trailhead. A hike of six miles brings you to the historic Huckleberry Mountain Lookout Tower on the summit of Huckleberry Mountain. If you want to hike beyond the tower, you can easily connect to the Arizona Creek Trail, the Rodent Creek Trail, the Coulter Creek Trail, and several others that go into the heart of the wilderness.

Although U.S. Highway 89/287 to Moran Junction is in Teton National Park, side roads along Pilgrim Creek and Pacific Creek go to trailheads and recreation sites at the edge of the Teton Wilderness. A paved side road north of the main highway ends at the Turpin Meadow Recreation Area where there is a campground and trails into the wilderness. The highway crosses the main Buffalo Fork River and enters the national forest at Hatchet Campground. At several pullouts along the highway you get breathtaking views of the area, including the Togwotee Overlook just past Togwotee Lodge. As the road continues along Blackrock Creek, the impressive Breccia Cliffs appear to the north. The cliffs, at the edge of the Teton Wilderness, are more than 1,000 feet high. The highway climbs over Togwotee Pass and enters the Shoshone National Forest.

Continuing south from Moran Junction takes you on U.S. Highway 26/89/191. Although the highway is in Teton National Park, most of the land

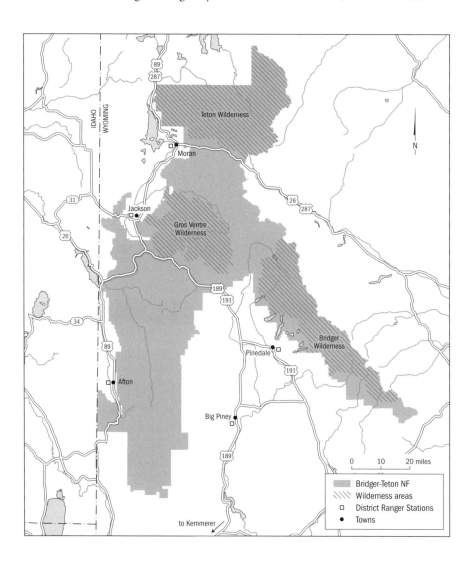

to the east is in the Teton National Forest. Many creeks and small lakes are in this part of the national forest, and a limited number of dirt roads runs through. Mount Leidy, with its conspicuous pyramid-shaped peak at 10,326 feet, may be seen from the highway.

When the highway reaches Gros Ventre Junction, the Gros Ventre side road follows Gros Ventre River to the townsite of Kelly prior to entering the national forest. Just within the boundary of the forest is the Gros Ventre Slide Area. In 1925, the mountain above the Gros Ventre River on the south side consisted of massive sandstone over layers of shale. Heavy rains during spring 1925 saturated and lubricated the shale. On the morning of June 23, 1925, the massive sandstone sheered away from the shale and roared down the mountainside, carrying with it earth, rocks, and trees. Estimates say that the slide consisted of 50 million cubic tons of material as it slid down from an elevation of 9,000 feet, crossing the river, and moving up on the opposite slope for 300 feet before it ground to a halt. The dam it formed across the Gros Ventre River was 225 feet tall and nearly a half-mile wide. The scar on the mountain where the sandstone had been was one mile long, 2,500 feet wide, and several hundred feet deep in places. Water behind the dam began to rise, forming Lower Slide Lake and inundating ranch land and a road up the valley. By July 16, the water was five feet below the top of the dam. Water began seeping through the dam about 30 feet from the top, increasing until the water out of the lake equaled the normal flow of the river. People believed that the dam would be permanent; however, heavy rains during May 1927 caused the dam to be breached in several places and, on May 18, the upper 60 feet of the dam collapsed, sending a wall of mud, rock, and water down the canyon. The rush of material destroyed ranches, drowned hundreds of domestic animals, and wiped out the town of Kelly, 3.5 miles away. Six people were killed during the event that was over the same afternoon. All that remained of Kelly was the post office, housed in a former church. If you look east from Kelly, you should be able to make out Sleeping Indian, a mountain that has the appearance of an Indian in full warbonnet lying on his back. From a parking lot at the Gros Ventre Slide area, hike a four-tenths-mile trail to several vista points. Lower Slide Lake extends today for nearly five miles up the valley. The Gros Ventre Road is paved as far as the Atherton Creek campground along the Lower Slide Lake. Beyond that, the road is dirt and extends for more than 20 miles along the Gros Ventre River. The land south of the Gros Ventre Road is in the wild Gros Ventre Wilderness.

South of Jackson, U.S. Highway 26/89/191 follows the western edge of the Teton National Forest and the Gros Ventre Wilderness. The Gros Ventre Mountains form the backbone of the wilderness, with several mountain peaks above 10,000 feet and a few above 11,000 feet, including the distinctive Doubletop Peak at 11,682 feet. Just outside of Jackson and the Gros Ventre Wilderness is Storm King Mountain Winter Sports Area.

As soon as U.S. Highway 26/89/191 crosses the Snake River, both sides of the highway are in the national forest. The western side includes a part of the Snake River Range with a rather narrow strip of the Teton National Forest on the west side of Jackson Hole.

At Hoback Junction, the highway divides, with U.S. Highway 189/191 following the Hoback River to the east and U.S. Highway 26/89 following the Snake River south and eventually west. You do not want to miss either route. U.S. Highway 189/191 enters scenic Hoback Canyon (pl. 37), at one point passing Stinking Springs, which you can detect before seeing because of the pungent odor of hydrogen sulfide being emitted from the springs. Hoback Campground is squeezed in a very narrow portion of Hoback Canyon. Four miles east of the campground is a side road along Granite Creek to Granite Hot Springs, Granite Falls, and a campground. The therapeutic hot springs has a swimming pool, and a footbridge over Granite Creek permits easy access to Granite Falls. The Gros Ventre Wilderness is immediately north of the hot springs. Also along the southern border of the wilderness are Shoal Falls, West Dell Falls, and Sulphur Springs. U.S. Highway 189/191 continues to follow Hoback River, eventually emerging from the canyon and into Hoback Basin where there is a Forest Service guard station. A hiking trail from the guard station onto Monument Ridge leads to a lookout tower. After the highway passes Clark Butte to the south, the Hobart River abruptly heads south with a dirt road alongside. The main highway takes an easterly course, leaving the national forest in about 10 miles at The Rim.

U.S. Highway 26/89 south of Hoback Junction continues to West and East Elbow campgrounds where it and the Snake River make an abrupt turn to the west. For the next 12 miles, the highway is deep in the Grand Canyon of the Snake River. The land south of the canyon is in the Bridger National Forest. Within the canyon are five campgrounds, and an amphitheater sits at East Table Creek. The Snake River is one of the best whitewater rivers in the world. Less than two miles after leaving the Grand Canyon of the Snake River, the highway comes to Alpine Junction, located between the Targhee National Forest to the north and the Caribou National Forest to the south.

The narrow section of the Teton National Forest north of the Grand Canyon of the Snake River and west of Jackson Hole is crossed by State Route 22. This highway climbs over Teton Pass and into the Targhee National Forest. If you are more adventurous, there is an alternate dirt road up to Teton Pass. Teton Village and the Jackson Hole Ski Area are at the northern end of this area. An aerial tramway runs from the village to Rendezvous Mountain.

The Bridger portion of the Bridger-Teton National Forest is divided into two divisions. The eastern side of the eastern division contains the Wind River Range, the crest of which is on the Continental Divide and in the Bridger Wilderness. The western side of this division includes a small part of the Gros Ventre Wilderness. Between the two wilderness areas is the Green River. This division of the Bridger National Forest may be explored with Pinedale as your base. Just north of Pinedale, extending to the southern boundary of the Bridger Wilderness, are three lakes—Fremont, Willow, and New Fork—all with campgrounds and boating ramps. Skyline Drive is a scenic route that follows the eastern side of Fremont Lake, terminating at the Trail's End Campground.

For a longer scenic route, drive State Route 352 north of Pinedale. After entering the national forest, you arrive at Kendall Warm Springs where the water from the springs rushes over a terrace and delightfully falls into the Green River. Not only is this a highly scenic area, but it is significant in that the warm springs are the only place in the world where the Kendall Warm Springs dace, a federally endangered fish, lives.

As you continue to drive northward along the Green River, you come to the site of Billy Wells Dude Ranch, reputed to be the first dude ranch in the country. The dude ranch was built in 1897 as the Gros Ventre Lodge and was operational until 1906. Dilapidated log buildings and crumbling stone fireplaces are all that remain.

The Green River and the road then make a sharp bend, and the road gradually climbs to the Green River Lakes at an elevation around 8,000 feet. As you stand at the edge of the lakes and look south, you see in the distance the characteristic top of Square Top Mountain. This mountain rises to a height of 11,678 feet. A great hiking trail runs into the Bridger Wilderness from the Green River Lakes to Square Top Mountain, Granite Peak, Beaver Peak, and Three Forks Park before crossing Green River Pass. Eventually this trail passes several lakes and ends at Fremont Lake just north of Pinedale.

The western division of the Bridger National Forest lies south of the Grand Canyon of the Snake River. U.S. Highway 89 through Star Valley is just west of this division. Two parallel mountain ranges are in this division, the Salt River Range and the Wyoming Range. Grey's River flows between these two ranges and Forest Highway 138 parallels the river for several miles. Grey's River is a good river for whitewater rafting. At the north end of the forest highway, near Alpine, is the Alpine elk feedground. A parking lot next to the feedground allows for very close observation of the elk in winter. Just south of the Moose Flat Campground is a trailhead for the Wyoming Range National Recreation Trail. This 75-mile-long trail goes to the heart of the Wyoming Range. The forest highway continues southward, eventually coming to the LaBarge Guard Station. Traces of Conestoga wagon wheels may be seen in the area. These are from the Lander Cutoff of the Oregon Trail, a route used originally by pioneers who were heading for the California gold fields or the rich lands of Oregon.

One mile east of Afton off of U.S. Highway 89 is the Swift Creek Campground situated along one of the most beautiful and turbulent creeks in the national forest. A trail runs from here to Periodic Spring, a cold-water geyser that flows and stops at regular intervals of 18 minutes. East of Periodic Spring is the Swift River Research Natural Area that preserves an alpine scree, coniferous forests of Engelmann spruce, subalpine fir, and Rocky Mountain maple, a riparian community of Engelmann spruce, Booth's willow, and red osier dogwood, and an upland shrub community dominated by sagebrush.

Lake Alice near the southern end of the national forest is 1.5 miles by trail from the Hobble Creek campground. Lake Alice is beautifully situated below heavily wooded slopes. The lake was formed thousands of years ago when a part of Lake Mountain slid into the valley, forming a natural dam for Lake Alice.

Kendall Warm Springs

Although it ends by rushing turbulently into the Colorado River, the Green River rises 730 miles away among the small alpine lakes, springs, glaciers, and streams high in Wyoming's Wind River Range. After a northwestward descent, the river turns abruptly and begins its southward flow through a broad valley. Within a few miles, the clear waters of Kendall Warm Springs feed the river, which cascades over a 10-foot-high ledge of white travertine, a rock made of calcium deposits. Named for an early logger in the region, the springs are part of the Bridger-Teton National Forest.

The springs get both their calcium and their warmth from a low limestone ridge some 1,000 feet east of where the springs enter the river. As surface water in the river circulates downward through subterranean cracks and fissures, it penetrates deeply enough to be warmed by the rock, emerging from the hillside at 85 degrees F. The springs consist of three major and several smaller flows that drop into two main pools at the foot of the ridge. The water then continues in a stream, often fanning out into a series of rills until it pours over the waterfall into the Green River.

A strong odor of sulfur emanates from the springs. Fisheries biologist Niles Allen Binns has analyzed the water and found it also abundant in other dissolved minerals, particularly calcium, magnesium, sodium, and potassium. At the source along the limestone ridge, the warm springwater is very high in carbon dioxide but low in dissolved oxygen. As it percolates downstream, the water gives off some of the carbon dioxide and gains more oxygen from the air.

Although this part of Wyoming has severe winters with recorded temperatures as low as −54 degrees F, the water in the Kendall Warm Springs stream never drops below 78 degrees F. Following a heavy winter storm, mounds of sparkling snow border the green aquatic vegetation in the stream, whereas the water continues to flow. Biologist Galen Boyer has noted that the warm water entering the Green River keeps it free of ice for several miles downriver. Judging from the mink, bobcat, and coyote tracks that are often found on the streamside mud, the springs and the stream appear to be a winter oasis for wildlife.

Kendall Warm Springs and its stream lie in an open landscape. On the limestone ridge, only occasional aspen and rounded, shrubby sagebrush form a canopy above sparse grasses and a few flowering plants. Boggy patches of vegetation and occasional willows grow along the stream, punctuated by yellow monkey flower, silvery cinquefoil, and daisy fleabane. Several plants grow submerged in the water, including large mats of sago pondweed and naiad, which are flowering plants that form small, obscure flowers and minute seeds, whereas stoneworts are algae that reproduce without flowers, sometimes forming egg and sperm cells and sometimes forming asexual spores.

Stoneworts lack the complex cellular structure of leaves, stems, and roots found in flowering plants but may form extensive colonies several feet across. An easy way to distinguish stoneworts from similar-appearing pondweeds and naiads is by their feel: stoneworts are stiff and rough, whereas the others are soft and flaccid. This is because stoneworts (as their name implies) become heavily calcified. At Kendall Warm Springs, calcium from the spring-water precipitates in a crust of tiny calcium carbonate crystals over the body of stoneworts. As the plants die and fall to the bottom of the stream, their limy residue accumulates, eventually forming layers of travertine.

Fisheries biologist Binns also examined the animal life in and near the warm water and recorded an assortment of snails, aquatic beetles, buffalo gnats, dragonflies, damselflies, caddis flies, crane flies, and soldier flies. Although all these organisms are apparently common and widespread, the soldier flies have attracted special interest. Related to, but smaller than, horse-flies, soldier flies are usually brightly colored and shiny and possess short, spinelike projections over parts of their bodies. Fly authority Harold Oldroyd finds these features unusual in an organism that does little but rest, mate, and lay eggs.

The soldier fly's coloration may relate to the way the insects swarm before mating, glistening as they flit through the sunlight. When at rest, the soldier fly crosses its wings in a scissorlike fashion, covering the bright markings so that potential predators are not alerted to its presence. Its spinelike projections may also offer protection. The larvae of the soldier fly lie nestled in decaying, moist vegetation along the stream. Their roughened skin is also due to the deposition of calcium carbonate. Oldroyd postulates that the crust may protect the delicate larvae from desiccation during dry spells.

Only one kind of fish swims in these springs, the Kendall Warm Springs dace, a 1.5-inch-long relative of the minnow. Recorded no other place in the world, it has been considered an endangered species under federal law since 1973. The Kendall Warm Springs dace is related to the speckled dace that lives in the Green River, but the latter species has never been found in the warm springwaters. (Nor do the Kendall Warm Springs dace survive if they enter the Green River, where they encounter not only different water temperature and chemistry but also hungry trout that linger in the vicinity.) Some biologists have questioned the distinctiveness of the dace in the springs, but the special conditions in the springwaters and the fish's probable isolation from the Green River for several thousand years (due to the waterfall at the end of the stream) have given the fish the opportunity to develop its own characteristics.

Discovered by forest ranger Harmon Shannon in 1934, the Kendall Warm Springs dace was described by fisheries biologist Eugene R. Kuhn and his colleague C. L. Hubbs. They called it an odd little fish unlike those seen from any other body of water. The females, which outnumber the males four to one, are a mottled dull green, whereas the males, particularly during the breeding season, are bright purple. The fish live along the entire length of the stream except in the uppermost pools, where the level of carbon dioxide is too high for them to survive. The main channel of the stream contains primarily adult fish, whereas young of all ages remain under the protective cover of aquatic vegetation in small, shallow pools.

Although the number of Kendall Warm Springs dace appears stable today, the survival of the fish was threatened until it received adequate protection under the Endangered Species Act. Grazing livestock used to wander across and into the stream, disturbing the fragile ecosystem, until the U.S. Forest Service fenced off the area. For many years, the warm-water stream was also used for bathing and washing clothes. Because soaps and detergents were potentially harmful to the organisms inhabiting the water, the Forest Service banned all such activities in 1975. And until it was forbidden, fishermen used to take the dace as bait. As a result of these measures, the Kendall Warm Springs dace is now more secure.

Periodic Spring

Near the base of a limestone cliff in Wyoming's Bridger-Teton National Forest, springwater gushes from an opening for several minutes, stops abruptly, then begins a new cycle a short time later. This is Periodic Spring, the intermittent flow of which is a rare geologic phenomenon. The water is cold and clear, an indication that this is not a geyser like Old Faithful; such geysers, of volcanic origin, send forth hot water. Through the years, various observers have timed the flows at anywhere from four to 25 minutes, with similarly varying dry spells. The intermittent flow is especially regular in late summer and fall. During stormy periods or when there is heavy snow melt-off, the flow fluctuates but does not stop entirely.

When geologist William W. Rubey saw the spring in 1931, the water issued from a small cave, forming a stream 9.5 feet wide and 1.25 feet deep. The water cascaded down a talus slope to nearby Swift Creek. By taking measurements of the water velocities, Rubey calculated that the discharge of the spring at full flow was about 285 gallons per second. He also noted that when the flow stopped, the narrow cave, which led downward at an angle of 42 degrees, continued to drain until the water level was 9.5 feet lower than the spillover point. This indicated that water was also leaking through the limestone bedrock into the talus slope.

Ron Shreve, a colleague of Rubey's at the University of California, Los Angeles, constructed a working model in the laboratory that mimicked the stop-and-go flow of the spring. The two concluded that a reservoir existed within the limestone cliff. Drawing water out of the reservoir, they argued, was a curved passage that acted as a siphon. The passage they envisioned began somewhere near the bottom of the reservoir, extended upward several feet, then curved abruptly downward before once again slanting upward to reach the outside. As long as the water level in the reservoir remained lower than the siphon's upper bend, the S-shaped passage contained mostly air. Only the lowest bend was always filled with a little water.

Rubey theorized that the sequence of events leading to the periodic flow goes something like this: First, the reservoir behind the siphon fills with water until the level rises above the upper bend of the siphon. Water then floods the siphon, driving air out of the mouth of the cave, with the water eventually following. The flow ceases when the water level in the reservoir drops below the level of the siphon intake. The next cycle occurs once the reservoir is replenished. Rubey calculated that 100,000 gallons of added water are needed to raise the level of the reservoir high enough for a new cycle to begin.

Recently, geologists Peter Huntoon and James Coogan have attempted to identify the source of the spring. Periodic Spring lies within the Salt River

Range. Huntoon and Coogan note that the reservoir is contained in a type of rock known as Madison limestone. An extensive outcrop of this limestone also lies five miles east of Periodic Spring, at high elevation. Huntoon and Coogan theorize that much of the water that emerges at the spring originates from snow packs on this mountain limestone and flows underground through fissures and caves in the rock layers until it reaches the reservoir. Apparently, during rapid melt-offs or after extreme rainstorms, the supply of water overwhelms the siphon, and the spring pulses rather than turning on and off.

Periodic Spring can be reached by driving east of the town of Afton and following a trail eastward along Swift Creek. The view a visitor to the spring gets today is somewhat different from what Rubey saw in 1931. Fallen rocks now obscure the small cave, and the water issues from beneath a concrete ledge. In 1958, moreover, Periodic Spring became the source of water for Afton, and an intake was constructed at the mouth of the spring to divert some of the water. Nonetheless, enough water still flows from the mouth to cascade down to Swift Creek.

Swift Creek itself arises high in the mountains five miles southeast of Periodic Spring. Its clear but often turbulent waters descend 2,000 feet through rocky terrain and bottom out in a densely shaded gorge some three miles north of their source. After making a sharp turn westward, Swift Creek flows past Periodic Spring, skirts Afton, and eventually empties into the Salt River, just inside the Idaho line.

The trail along Swift Creek east of Periodic Spring follows a willow-lined corridor often narrowed by steep mountain slopes. Intermingled with the willows are diverse wildflowers, including the starry Solomon's-seal, bluebells, sweet-scented bedstraw, and meadow rue. Where low terraces have developed along the stream, the striking red-osier dogwood replaces the willow as the dominant shrub, and a different group of wildflowers, which includes red baneberry and wild geranium, prevails.

After the trail bends southward and begins to climb steeply toward the source of Swift Creek, it enters forests dominated mostly by subalpine fir, with Douglas fir, Engelmann and blue spruces, and lodgepole pine growing in places. Along the way, the forests are frequently punctuated by treeless meadows referred to as forblands (a forb is a nonwoody, broadleaved, flowering plant). Joel Tuhy, of the Utah Natural Heritage Program, has identified a number of different types of these habitats along Swift Creek. One, found more in the bottomlands, is dominated by the mountain bluebell (pl. 38); higher up the prevailing herb is the white-flowered ligusticum, with its carrotlike leaves. Where more rock fragments are exposed on the ground, a herbaceous community of arrowleaf balsamroot, one-headed sunflower, scarlet gilia, and Drummond's aster interrupts the forest.

Medicine Bow National Forest

SIZE AND LOCATION: Approximately 1.1 million acres in south-central and southeastern Wyoming on all sides of Laramie, extending southward to the Colorado state line. Major access routes are Interstates 25 and 80, U.S. Highways 30, and State Routes 10, 11, 15, 16, 61, 62, 70 (Battle Highway), 71, 77, 130 (Snowy Mountain National Scenic Byway), 210, 230, 710, and 721. District Ranger Stations: Douglas, Laramie, and Saratoga. Forest Supervisor's Office: 2468 Jackson Street, Laramie, WY 82070, www.fs.fed.us/r2/mbr.

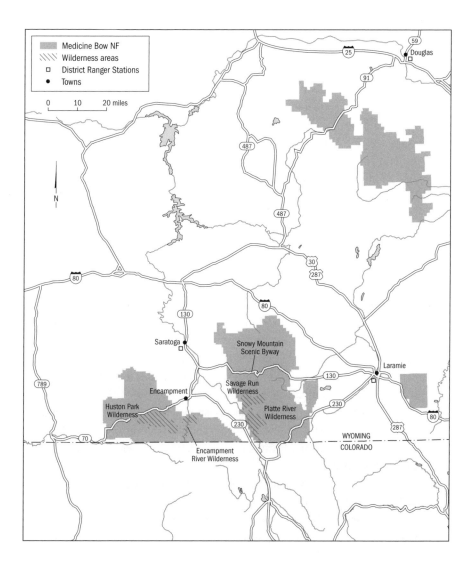

SPECIAL FACILITIES: Winter sports area; boat ramps.

SPECIAL ATTRACTIONS: Snowy Mountain National Scenic Byway; Vedauwoo Rocks; Battle Highway.

WILDERNESS AREAS: Encampment River (10,024 acres); Huston Park (30,588 acres); Platte River (23,492 acres); Savage Run (14,927 acres).

The Medicine Bow National Forest comprises three discreet units along the southern and southeastern border of Wyoming and a fourth unit north of Laramie. Thirty miles west of Laramie are the magnificent Medicine Bow Mountains, popularly known as the Snowy Mountain Range. West of the Medicine Bow Mountains are the Sierra Madres. East of Laramie are the more gentle mountains and rolling hills of the Pole Mountain District. North of Laramie are the Laramie Mountains, part of the Front Range of the Rocky Mountains.

The Snowy Mountains occupy about 40 percent of the Medicine Bow National Forest and extend from the Colorado state line northward until they merge into dense stands of timber, grasslands, and sagebrush at the edge of the Great Plains. Within the Snowy Mountains are dozens of streams and creeks, more than 100 glacial lakes, and steep, rugged slopes densely forested up to the timberline. Several of the peaks are capped by snow, but some of the white peaks are covered not with snow but with deposits of white quartzite. The dominant forest species are Engelmann spruce, ponderosa pine, Douglas fir, alpine fir, and limber pine. Aspens occur throughout the range, and eastern cottonwoods (fig. 22) line many of the streams.

A drive on State Route 130, the Snowy Mountain National Scenic Byway, from Centennial to Saratoga, brings you up close to many features of the Medicine Bow National Forest. Before State Route 130 reaches Centennial, a part of the Medicine Bow National Forest south of the highway encompasses Sheep Mountain. Almost the entire area is in the Sheep Mountain Game Refuge. State Route 11 parallels the western side of this region.

The small but interesting community of Centennial is the eastern terminus of the Snowy Mountain National Scenic Byway. The Wyoming Department of Transportation tries to clear the snow from the highway around Memorial Day. Drive this highway shortly after its cleaning and you feel like you are on a bobsled course because the snowbanks on either side of the road are several feet deep.

As soon as the Snowy Mountain National Scenic Byway enters the national forest, it starts an immediate ascent. In less than one mile, Forest Highway 351 branches off of the scenic byway, rejoining the scenic byway a few miles later. Along the side road is the Libby Creek Recreation Area. In the vicinity is the Snowy Mountain Range Lodge, a large log structure on the National Register of Historic Places.

Back on the scenic byway, the highway comes to the Snowy Range Ski Area where all the winter sports amenities are available. Forest Highway 101 to the north is a lengthy route through beautiful scenery past several small lakes. The North Fork Campground, only 1.5 miles north of the scenic byway, is nestled along the creek, with the Deep Creek campground nearly 20 miles farther north. The Deep Creek Hiking Trail may be hiked from the campground.

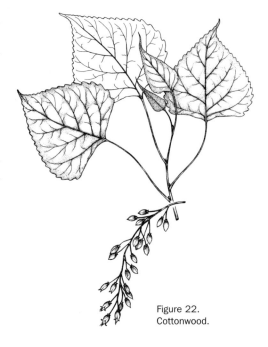

Figure 22.
Cottonwood.

Just beyond the Green Rock Picnic Area, the scenic byway comes to the most beautiful part of the Snowy Mountains. Dozens of sky blue glacial lakes are north of the highway for the next eight miles, with much of the area above 10,000 feet. A pretty side road connects North Fork Campground with the Brooklyn Lake Campground. The North Fork and Glacier Lakes Trail may be hiked in this area. South of the scenic byway, from the North Fork Campground, is the Snowy Range Research Natural Area. This area is one of the few timbered areas in the Medicine Bow Mountain Range that has not been logged since the arrival of white settlers. Two tributaries to the Little Laramie River flow through the natural area. Elevations within the natural area range between 9,500 and 10,500 feet. The forest contains a virgin stand of Engelmann spruce, alpine fir, and lodgepole pine.

After passing a side road to Sugarloaf Campground and Lewis Lake, the scenic byway crawls over the 10,847-foot Snowy Range Pass and then to Libby Flats where there is an interpretive nature trail. At a stone observation you may survey the entire area. The vegetation here is rocky tundra with carpets of dwarf alpine wildflowers such as moss campion and alpine cinquefoil. Any trees at this elevation are dwarfed and gnarly. Shortly, one of the most awe-inspiring views looms before you with sparkling Lake Marie nestled below scenic Medicine Bow Peak. Trails lead to several lakes and to Medicine Bow Peak.

The scenic byway stays above 10,000 feet as far as the Silver Lake Campground before it begins its descent to Ryan Park Campground. Between these two campgrounds is an interpretive trail along North French Creek. The Brush Creek Visitor Center is just before the scenic byway leaves the national forest.

A large area of the Medicine Bow District of the national forest lies south of the Snowy Mountain National Scenic Byway and extends to the Colorado state line. Within this district are the Savage Run Wilderness and the Platte River Wilderness. These two wilderness areas are separated by less than one mile, and in this narrow corridor is Forest Highway 511. The forest highway branches off of State Route 230, the highway that connects Laramie with Encampment. Where Forest Highway 511 enters the national forest are two campgrounds nestled in a loop of the North Platte River. After passing between the two wildernesses, the forest highway crosses the Medicine Bow National Forest south of Rob Roy Reservoir and north of Lake Owen, both of which have Forest Service campgrounds.

Savage Run Wilderness has steep-sided, forested slopes at the western edge of the Medicine Bow Range. Savage Run Creek crosses the wilderness, with a hiking trail along nine miles of it. This is an area of high plateaus with elevations between 8,000 and 10,000 feet.

The North Platte River twists along the western edge of the Platte River Wilderness, going through spectacular Laramie Canyon. Six-mile Gap Campground is just outside the western edge of the wilderness's boundary, and Pelton Creek Campground is along the eastern edge. Fishing for trout is popular in the wilderness. State Route 230 cuts across a corner of the Medicine Bow District, providing access to Miller Lake and Campground, several hiking trails, and a few four-wheel-drive roads.

The community of Encampment is near the northeastern edge of the Hayden District of the Medicine Bow National Forest that is centered around the Sierra Madres. This district extends from the Colorado state line to the Great Plains, some 25 miles to the north. The Continental Divide (pl. 39) winds through the district. The Encampment River Wilderness and the Huston Park Wilderness are in the southern part of the district.

Much of the Huston Park Wilderness is near or above 10,000 feet, with alpine vegetation and bogs at the upper elevations and dense forests of lodgepole pine, Engelmann spruce, and subalpine fir on mid and lower slopes. A few abandoned mines are on the western side of the wilderness. The Continental Divide National Recreation Trail is the best in the wilderness, but it is marked only by rock cairns and blazed trees in places.

The Encampment River Wilderness follows the Encampment River for the entire eight-mile length of the wilderness. A hiking trail follows the river, beginning from the Lakeview Campground. Forest Highway 550 is located between the two wildernesses and provides access to Lakeview Campground. The forest highway then turns abruptly westward along the north shore of Hog Park Reservoir where there is a picnic area, campground, and a boat ramp. You may access the Continental Divide National Recreation Trail two miles west of Hog Park Reservoir.

The Battle Highway, State Route 70, from Encampment to Boggs crosses the Sierra Madres and is the best way to get an overview of this district. Shortly after the highway enters the national forest is Battle Creek Campground, where Forest Highway 550 originates. The highway crosses the Continental Divide and follows the northern edge of Huston Park Wilderness, passing the sites of several towns—Battle (on the Continental Divide), Rambler, and Copperton. Along the route is the Edison Memorial where Thomas Edison reputedly got the idea for electricity when he saw the frayed ends of his bamboo fishing pole. The national forest part of the Sierra Madres lies north of the Battle Highway. The only good road to this section of the Medicine Bow National Forest is Forest Highway 801, which branches off the Battle Highway about six miles west of the Lost Creek Campground. Forest Highway 801 crosses several creeks and the Continental Divide 1.5 miles south of the Jack Creek Campground.

Laramie Peak District is located in the Laramie Mountains and comprises 177,000 acres of national forest land. These mountains rise above the Great Plains and are characterized by mountains parks, deep valleys, and high, rocky peaks. Major trees in this district are ponderosa pine, limber pine, lodgepole pine, and Douglas fir. Because the Laramie Peak District is a checkerboard of public and private land, visitors should make sure that they do not trespass on the private parcels of land. State Routes 15, 16, 61, and 62 are the major access routes into this part of the Medicine Bow National Forest. La Bonte Creek has carved scenic La Bonte Canyon near the Curtis Gulch Campground. Laramie Peak, at 10,272 feet, is the highest in the district and is located near the southern part of the district. A lookout tower on Black Mountain provides fine views of the area.

The Pole Mountain District southeast of Laramie shows a completely different aspect of the Medicine Bow National Forest. It consists of rolling terrain and low rugged mountains. Much of the timber in the area has been harvested repeatedly. Interstate 80 cuts across the western edge of this district and provides great access to a group of fantastic rock formations known as the Vedauwoo Rocks (pl. 40). These rocks are a favorite of photographers and rock climbers. State Route 210 crosses the Pole Mountain District. From the rest area on Interstate 80 is a hiking trail into the Sherman Mountains.

Shoshone National Forest

LOCATION AND SIZE: Nearly 2.5 million acres in northwestern Wyoming. Major access routes are U.S. Highways 14, 16, 20, 26, 212 (Beartooth Highway), and 287, and State Routes 28, 131, 290, and 296 (Chief Joseph Scenic Highway). District Ranger Stations: Cody, Dubois, and Lander. Forest Supervisor's Office: 225 W. Yellowstone Avenue, Cody, WY 82414, www.fs.fed.us/r2/shoshone.

SPECIAL FACILITIES: Winter sports areas; boat ramps; wilderness horsepacking.

SPECIAL ATTRACTIONS: Beartooth Highway; Chief Joseph Scenic Highway; North Fork Highway; Wyoming Centennial National Scenic Byway.

WILDERNESS AREAS: Absaroka-Beartooth (943,626 acres, partly in the Custer and Gallatin National Forests); Fitzpatrick (198,525 acres); North Absaroka (350,488 acres); Popo Agie (101,870 acres); Washakie (704,274 acres).

The Yellowstone Park Timberland Reserve was set aside by proclamation by President Benjamin Harrison in 1891. Later, the major part of that reserve became the Shoshone National Forest. Thus the Shoshone National Forest and the Teton National Forest, which was set aside the same day, were the country's first national forests. The Shoshone is sometimes referred to as the horse forest because so much of it can be reached only by horseback because of the lack of roads in the very rugged terrain. Fifty-eight percent of the national forest is in designated wilderness, and some of the land that is not in wilderness is still fairly inaccessible. The highways that do cross the national forest are spectacular and do give the visitor a good cross section of the area.

The Beartooth, Wind River, and Absaroka mountain ranges form the backbone of the national forest. Several national forests are contiguous with the Shoshone National Forest, and Yellowstone National Park is on the western side.

The northern part of the national forest is crossed by the Beartooth Highway, one of the most scenic routes in the country as it passes between the southern end of the Absaroka-Beartooth Wilderness and the northern end of the North Absaroka Wilderness. After leaving Red Lodge, Montana, the Beartooth Highway (U.S. Highway 212) enters the Shoshone National Forest when it crosses into Wyoming. Ahead of you on the east side of Beartooth Pass is the Twin Lakes Headwall, a snowfield where the Red Lodge International Ski Race Camp is located. Twin Lakes west of U.S. Highway 212 are

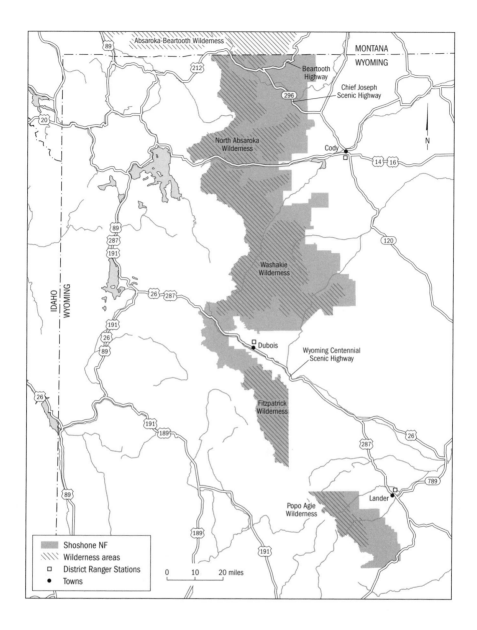

just two of numerous alpine lakes in the Beartooth Mountains. As the highway approaches Beartooth Pass, you may want to stretch your legs by hiking a part of the Beartooth Loop National Recreation Trail that begins at Gardner Lake. If you hike the entire 10 miles of the loop, you circle Tibbs Butte and pass Losekamp and Stockade Lakes. Take time at Beartooth Pass to walk to the overlook and admire the breathtaking surroundings. The highway descends from the pass by a series of sharp switchbacks, passing Long Lake and

Little Bear Lake to the north. A half-mile side road brings you to Island Lake and Campground.

A half-mile west of Long Lake is Morrison Trail, a four-wheel-drive road that provides access to Rainbow Lake, Chain Lakes, Dollar Lake, and Top Lake, passes Sawtooth Meadows at the foot of Sawtooth Mountain, and comes to the Sawtooth Lake Peatlands. The unusual feature of the peatland is that it has the only known palsa in the lower 48 states. A palsa is a peat-covered mound over a core of permafrost. Botanists from the Rocky Mountain Research Station note that the overlying peat acts as an insulating blanket over the frozen core. The permafrost core exists up to 18 inches below the surface. The palsa is about five acres in size and is raised three to six feet above the surrounding fen that is dominated by sedges. The surface of the palsa has little vegetation except for clumps of hairgrass and occasional plants of mountain sorrel.

The Top of the World Store provides respite from the tenseness of driving the highway, and you can stock up on snacks and drinks here. In two more miles is Beartooth Lake, nestled at the foot of 10,514-foot Beartooth Butte with its red summit. A campground is on the eastern side of the lake. Across the highway from Beartooth Lake is the 100-foot drop of Beartooth Falls, reached by a scenic half-mile trail. Take time to drive the three-mile side road to the Clay Butte Fire Tower where there is a visitor information center during summer. From the tower you can see Montana's highest peak, Granite Peak to the north, in the heart of the Absaroka-Beartooth Wilderness, the Bighorn Mountains to the east, Clarks Fork Valley to the south, and Yellowstone National Park to the west.

The area around Beartooth Pass is alpine tundra that is above the growth limit of trees. The rocky ground is carpeted with miniature wildflowers during the six- to 12-week growing season. In moist areas are Indian paintbrushes, monkey-flowers, and buttercups, whereas somewhat drier areas support lupines, beardtongues, forget-me-nots, and balsamroot. Where the highway descends below timberline, lodgepole pine, limber pine, whitebark pine, alpine fir, Douglas fir, Engelmann spruce, and patches of quaking aspen occur. The pinkish tint that may appear on some of the snowfields comes from microscopic algae growing on the surface of the snow. During winter, some of the snowbanks may be as much as 30 feet deep.

West of Clay Butte, the Beartooth Highway follows the southern end of the Absaroka-Beartooth Wilderness. Although a few hiking trails lead into the wilderness from the highway, only a small part of the wilderness is in the Shoshone National Forest. A couple of fine overlooks pop up as the highway descends to the Clarks Fork of the Yellowstone River. As you drive the highway along the river as it heads northwest, the two prominent peaks that loom in front of you in the distance are Pilot Peak and Index Peak, both located at the eastern edge of the North Absaroka Wilderness.

Before the Beartooth Highway reaches the Clarks Fork, there is a road junction with State Route 296, the Chief Joseph Scenic Highway. This is one of the most scenic routes that goes into the heart of Clarks Fork Valley. Campgrounds and picnic areas are available at various places along the highway as are two Shoshone National Forest ranger stations. After about three miles on this road, the Chief Joseph Scenic Highway comes alongside the Clarks Fork, which it follows for several miles into Clarks Fork Canyon. The highway is always within a few miles of the eastern side of the North Absaroka Wilderness, and there are occasional trails you may hike into the wilderness, some of them crossing the wilderness all the way to Yellowstone National Park. Several mountain peaks in the wilderness jut above 10,000 feet, with Dead Indian Peak the highest at 12,216 feet. Because of the region's volcanic origin, it is extremely rugged and often inaccessible.

Back on the Chief Joseph Scenic Highway, you come to the Crandall Ranger Station a few miles south of Hunter Peak Campground. East of the ranger station, at the base of the spectacular Cathedral Cliffs, is Swamp Lake, which is unlike other lakes in the area because of the presence of three types of peatlands. In a calcareous marl fen insoluble deposits of calcium carbonate have influenced the kinds of wetland plants that can live there. Common dominants of the marl fen are three dwarf sedges. Another area of peat has gnarly specimens of blue spruce, with twinflower, horsetail, bladder sedge, and blue joint grass found here and there. A third type of fen is dominated by sedges and shrubs, the most common being bog birch, shrubby cinquefoil, arrow-grass, and several sedges. Botanists have found other plants that generally do not occur in this region here, among them alpine manzanita, round-leaved orchis, bird's-eye primrose, blueberry willow, and several species of sedges and mosses.

At the eastern end of Sunlight Basin, the Chief Joseph Scenic Highway crosses the Sunlight Bridge where there is a splendid overlook before reaching Dead Indian Campground. From the campground it is three miles by trail to the edge of the North Absaroka Wilderness and the tranquil Dead Indian Meadows. Beyond the campground, the scenic highway climbs a series of extremely sharp switchbacks to Dead Indian Pass before eventually leaving the Shoshone National Forest about 25 miles west of Cody.

When you reach the Sunlight Bridge, instead of continuing on the Chief Joseph Scenic Highway, you may choose to take the side road to the west that goes through Sunlight Basin for 20 miles to the edge of the North Absaroka Wilderness. This road passes several ranches, a waterfall, and the Sunlight Ranger Station. After the first 17 miles, the road becomes a jeep road, which you may take to the abandoned town site of Lee City and several abandoned mines.

Another drivable access through a part of the Shoshone National Forest is on U.S. Highway 14/16/20, the North Fork Highway, from Cody to the east entrance to Yellowstone National Park. The highway follows a narrow corridor between the North Absaroka Wilderness to the north and the Washakie Wilderness to the south. Before the highway gets to the national forest, it comes to Buffalo Bill Reservoir five miles west of Cody. About halfway along the northern shore of the reservoir, there is a side road northward along Rattlesnake Creek that enters Rattlesnake Canyon where the road reaches the national forest boundary. The sheer cliffs that form the eastern side of the canyon are The Palisades. The side trip is well worth your time. U.S. Highway 14/16/20 enters the national forest 11 miles west of the Buffalo Bill Reservoir.

This scenic highway between the two wildernesses follows the North Fork of the Shoshone River. Several fanciful rock formations are adjacent to the highway—Flag Peak, Camel Rock, Goose Rock, Anvil Rock, Slipper Rock, Holy City, Elephant Head Rock, Chimney Rock, and Henry Ford Rock, which to some resembles a person behind a steering wheel. Nine campgrounds are available before reaching Yellowstone National Park. The Wapiti Ranger Station that you pass dates back to 1903 and is now on the National Register of Historic Places. The Clearwater Campground is on the site of a Civilian Conservation Corps camp, which was in use until 1941. A trail from the campground along Clearwater Creek reaches the North Absaroka Wilderness in about one mile. Near the Rex Hale Campground is Mummy Cave, which hunters used between 7200 B.C. and A.D. 1580. A mummified human body was found inside the cave dating back to A.D. 734. Just beyond the cave, a firefighter's monument commemorates 27 firefighters killed in 1927 during the Blackwater Fire. From this monument is a trail to the south that splits in about two miles. The left fork is the Fire Memorial National Recreation Trail and leads to another firefighters memorial in two miles and to Clayton Mountain in another mile. The right fork is the Natural Bridge Trail that heads toward the spectacular Blackwater Natural Bridge just below Coxcomb Mountain. The natural bridge is on the northern border of the Washakie Wilderness.

After the North Fork Highway passes the Newton Creek Campground, the spectacular Palisades line the north side of the river. Just before the scenic highway reaches Yellowstone National Park is the Sleeping Giant Winter Sports Area. Nearby is a trail along Grinnell Creek into the North Absaroka Wilderness. Pahaska Tepee just north of the highway was Buffalo Bill Cody's hunting lodge at one time.

The Washakie Wilderness is a huge roadless area occupying much of the southwestern side of the Shoshone National Forest. The wilderness includes the South Absaroka Mountains, which consist mostly of flat-topped mountains and plateaus separated by deep, narrow valleys. Major ridges are the Wapiti and Boulder Ridges. Numerous waterfalls are dispersed throughout the wilderness, and there are many long, difficult trails to a few of them.

Some nonwilderness area exists between the southern end of Washakie Wilderness and the north end of the Popo Agie Wilderness. U.S. Highway 26/287 passes through this area and is part of the Wyoming Centennial National Scenic Byway. Coming from the west and the Teton National Forest, the highway enters the Shoshone National Forest at Togwotee Pass, at 9,544 feet, on the Continental Divide. The curvy highway follows Wind River, coming to Falls Campground. From the campground is a very short trail to lovely Brooks Lake Creek Falls. From near the campground is a road north to Brooks Lake (pl. 41) where there is a lodge and two more campgrounds. On the east side of Brooks Lake are The Pinnacles, a unique rock formation. Within the area are petrified forests and a great abundance of fossils. You may hike the Dunoir Trail along Bonneville Creek, climbing over Bonneville Pass. The trail continues along West Dunoir Creek.

The highway follows Wind River to Dubois, with the Shoshone National Forest south of the highway. Three miles south of Dubois is the northern edge of the Fitzpatrick Wilderness in the Wind River Range. The Continental Divide forms the western edge of the wilderness where there are 13 peaks with elevations greater than 12,000 feet. Gannett Peak, the highest peak in Wyoming at 13,804 feet, is here as are 44 active glaciers, numerous lakes, alpine meadows, rocky plateaus, and several waterfalls. The only road access to the wilderness is a forest road off of U.S. Highway 26/287 five miles from Dubois.

The southernmost district of the Shoshone National Forest lies west of Lander. The western half of this district is in the Popo Agie Wilderness. The Continental Divide on the crest of the Wind River Range forms the western edge of this wilderness. This is a high-elevation wilderness, ranging from 8,400 feet along the Middle Fork of the Popo Agie to 13,255-foot Wind River Peak. More than 300 alpine lakes are in the wilderness, with numerous streams and waterfalls. The mountain peaks are high and jagged, and the narrow canyons have sheer granite walls. The eastern half of the district, which is not designated wilderness, is traversed by Forest Highway 300, the Louis Lake Road. State Route 131 leaves Lander and becomes Forest Highway 300 as the highway enters the Shoshone National Forest. At the entrance to the forest is Sinks Canyon where the Middle Fork of the Popo Agie disappears beneath the ground, reappearing in a trout-filled pool later on. Once the forest highway passes the University of Missouri Geology Camp, there is a trail along the creek to Popo Agie Falls. The road then climbs by a series of switchbacks to Frye Lake. Just past the lake is a side road westward to Worthen Meadows Reservoir in the middle of Worthen Meadows. A trail from the western end of the reservoir enters the Popo Agie Wilderness at Roaring Fork Lake.

By staying on Forest Highway 300, you travel through the southern end of the Shoshone National Forest, past Fiddlers Lake and Louis Lake, each with a pleasant campground. The highway then crosses Grannier Meadow and leaves the national forest when it comes to State Route 28.

ART CREDITS

We wish to express our gratitude to the Forest Service staff who put us in touch with photographers, helped to locate good images, or donated their images. Peter Keller, the wilderness program manager of the Pacific Southwest Region, directed us to Roxane Scales, public affairs officer, and Anne Bradley, regional botanist. They in turn spread the word about this project among their counterparts in other regional offices. Elton (Sonny) Cudabac of the Southern Region helped us to locate images in their Digital Image Library.

Plates

Jennifer Anderson, USDA-NRCS PLANTS Database, 33

Gerald & Buff Corsi, California Academy of Sciences, 26

William Follette, 23, 30

John Game, 13

William Grenfell, 29

Tanya Harvey, 5

Kirk Keogh, www.first2lastlight.com, 24

Peter Knapp, 8, 14

Robert Mohlenbrock, 11, 20, 25, 34, 35

Robert Mohlenbrock, USDA-NRCS PLANTS Database / USDA NRCS. 1995. Northeast wetland flora: Field office guide to plant species. Northeast National Technical Center, Chester, PA, 18

Clarence A. Rechenthin, USDA-NRCS PLANTS Database, 36

Lynn & Donna Rogers, bear.org, bearstudy.org, 16, 17

Scott T. Smith, ScottSmithPhoto.com, 6, 9, 10, 21, 32, 38, 39, 40, 41

John Mark Stewart, 7

Bill Summers, USDA-NRCS PLANTS Database / USDA SCS. 1989. Midwest wetland flora: Field office illustrated guide to plant species. Midwest National Technical Center, Lincoln, NE, 19

T. Taylor, 22

USDA Forest Service, Intermountain Region, Susan Marsh, 37

USDA Forest Service, Intermountain Region, Teresa Prendusi, 15

USDA Forest Service, Pacific Southwest Region, 27, 28

USDA Forest Service, Pacific Southwest Region, Ken DeCamp, 12, 31

USDA Forest Service, Southern Region, Bill Lea, 2, 3

USDA Forest Service, Southern Region, Barry Nehr, 1, 4

Figures

Tanya Harvey, 5, 7

Illinois Department of Natural Resources, Division of Forestry, MW Meyer & FJ Burton, 1, 3, 11, 12, 15, 20, 21, 22

Robert Mohlenbrock, 4, 10

Hans Peeters, CA Mammals, 16

Andrea J. Pickart, 6, 8, 9, 13, 17, 18, 19

Lynn & Donna Rogers, 14

USDA Forest Service, Southern Region, Bill Lea, 2

INDEX OF PLANT NAMES

coreopsis, palmate-leaved *(Coreopsis palmata)*, 164
cotton grass
    Chamisso's *(Eriophorum chamissoni)*, 238
    green-keel *(Eriophorum viridicarinatum)*, 207
    narrow-leaved *(Eriophorum angustifolium)*, 91
cottonwood *(Populus deltoidea)*, 78, 120, 273, 294
    black *(Populus trichocarpa)*, 196
    eastern *(Populus deltoides)*, 203, 300, 317, 342
    Fremont's *(Populus fremontii)*, 269, 285
    narrowleaf *(Populus angustifolia)*, 51, 134, 268, 269
    swamp *(Populus heterophylla)*, 34, 312
creeper, Virginia *(Parthenocissus quinquefolia)*, 285, 315
cress
    golden glade *(Lesquerella aurea)*, 13
    heart-leaved bitter *(Cardamine cordifolia)*, 49
    white bitter *(Cardamine cordifolia)*, 259, 289
    white bulbous *(Cardamine bulbosa)*, 178
crossvine *(Bignonia capreolata)*, 312, 315, 317
crown
    King's *(Rhodiola integrifolia)*, 49, 52
    rose *(Sedum rhodanthum)*, 49, 52, 54
crownbeard, yellow *(Verbesina alternifolia)*, 14
Culver's root *(Veronicastrum virginicum)*, 173
Currant
    black *(Ribes americanum)*, 300
    Colorado *(Ribes coloradense)*, 133, 134
    prickly *(Ribes missouriense)*, 134
    wax *(Ribes cereum)*, 134, 284
    wild *(Ribes coloradense)*, 61, 226
cyrilla, swamp *(Cyrilla racemiflora)*, 304
cypress, bald *(Taxodium distichum)*, 13, 34, 136, 173, 305, 320

daisy
    Easter *(Townsendia hookeri)*, 108
    five-flowered rock *(Perityle quinqueflora)*, 275
    Hershey's cliff *(Chaetopappa hersheyi)*, 275
    least *(Chaetopappa asteroids)*, 309
    one-flowered *(Erigeron simplex)*, 40
    ox-eye *(Leucanthemum vulgare)*, 119, 133
dandelion
    mountain *(Agoseris glauca)*, 327
    western *(Agoseris glauca)*, 309
danthonia, Parry *(Danthonia parryi)*, 286

deerberry *(Vaccinium stamineum)*, 306
deer's ears *(Swertia perennis)*, 49
devil's walking-stick *(Aralia spinosa)*, 311, 317
dewberry, southern *(Rubus trivialis)*, 312
dock
    prairie *(Silphium terebinathaceum)*, 165, 173
    swamp *(Rumex verticillatus)*, 319
dogwood
    flowering *(Cornus florida)*, 10, 18, 140, 162, 178, 181, 311
    rough-leaved *(Cornus drummondii)*, 14, 311
    swamp *(Cornus foemina)*, 167, 172
draba, rockcress *(Draba grayana)*, 126
dragonhead, false *(Physostegia virginiana)*, 15, 172
dropseed
    pineywoods *(Sporobolus junceus)*, 306, 308, 314
    prairie *(Sporobolus heterolepis)*, 176

elderberry *(Sambucus canadensis)*, 148, 317
    blue *(Sambucus cerulea)*, 238
    red *(Sambucus racemosa)*, 61, 134
elephant's foot, Carolina *(Elephantopus carolinianus)*, 311
elephant's head *(Pedicularis groenlandica)*, 54, 89, 107
elm
    American *(Ulmus americana)*, 21, 148, 169, 300, 317
    cedar *(Ulmus serotina)*, 14
    slippery *(Ulmus rubra)*, 148, 162, 173, 178, 311, 317
    water *(Planera aquatica)*, 14, 34, 305, 312
    winged *(Ulmus alata)*, 11, 22, 166, 171, 309, 311
everlasting, pearly *(Anaphalis margaritacea)*, 108

farkleberry *(Vaccinium arboreum)*, 22, 162, 177, 311, 314, 320
Fendler bush *(Fendlera rupicola)*, 270, 284
fern
    bracken *(Pteridium aquilinum)*, 149, 300, 306, 311
    broad beech *(Thelypteris hexagonoptera)*, 11
    bulblet bladder *(Cystopteris bulbifera)*, 182
    Christmas *(Polystichum acrostichoides)*, 11
    cinnamon *(Osmunda cinnamomea)*, 307, 314, 319
    crested *(Dryopteris cristata)*, 196
    cutleaf grape *(Botrychium dissectum)*, 311

Engelmann's milk *(Astragalus engelmannii)*, 11
Nuttall's milk *(Astragalus nuttallianus)*, 309
viburnum
  arrow-wood *(Viburnum dentatum)*, 140, 317
  maple-leaved *(Viburnum acerifolium)*, 320
  possumhaw *(Viburnum nudum)*, 142, 314
violet
  bog *(Viola nephrophylla)*, 307
  kidney-leaved *(Viola reniformis)*, 225
  marsh blue *(Viola cucullata)*, 167
  round-leaved *(Viola orbiculata)*, 226
  three-lobed *(Viola triloba)*, 311
  yellow *(Viola nuttallii)*, 53
virgin's-bower, western *(Clematis 259*

wait-a-minute bush *(Mimosa binucifera)*, 273
wake-robin *(Trillium recurvatum)*, 320
wallflower, western *(Erysimum capitatum)*, 286
walnut
  Arizona *(Juglans major)*, 268, 269
  black *(Juglans nigra)*, 5, 6, 21, 29, 165, 171, 182, 294
watercress *(Nasturtium officinale)*, 163, 168, 179, 264
waterleaf *(Hydrophyllum fendleri)*, 53
  blue *(Hydrophyllum appendiculatum)*, 15
whitlow grass *(Draba albertina)*, 52, 266
willow
  alpine fen *(Salix planifolia)*, 89
  barren ground *(Salix brachycarpa)*, 47
  black *(Salix nigra)*, 182, 312
  blueberry *(Salix myrtillifolia)*, 349
  blue-leaf *(Salix myricoides)*, 148
  bluestem *(Salix irrorata)*, 259

Booth's *(Salix boothii)*, 336
Carolina *(Salix caroliniana)*, 179
flat-leaved *(Salix planifolia)*, 48, 91
Geyer's *(Salix geyeriana)*, 134, 221
heart-leaved *(Salix rigida)*, 167
hoary *(Salix candida 206, 225*
mountain *(Salix monticola)*, 89
purple-twig *(Salix wolfii)*, 89
seep *(Baccharis salicifolia)*, 273
short-fruited *(Salix brachycarpa)*, 48
silky *(Salix sericea)*, 172
snowy *(Salix nivalis)*, 189
Ward's *(Salix caroliniana)*, 178
water *(Justicia americana)*, 163
Wolf's *(Salix wolfii)*, 134, 189
willow-weed *(Epilobium ciliatum)*, 285
winterfat *(Eurotia lanata)*, 324
wintergreen *(Gaultheria procumbens)*, 151, 191, 226, 293
  one-sided *(Orthilia secunda)*, 238
  smooth *(Gaultheria humifusa)*, 238
witch hazel *(Hamamelis virginiana)*, 11, 179, 315, 319
  Ozark *(Hamamelis vernalis)*, 178
wolfberry *(Lycium berlandieri)*, 273
woodsia, western *(Woodsia scopulina)*, 238
woollybase, slender *(Hymenoxys acaulis)*, 327

yarrow *(Achillea millefolium)*, 39, 119, 133, 327
  false *(Chaenactis douglasii)*, 327
yellowbells *(Disporum trachycarpum)*, 197
yellow puff *(Neptunia lutea)*, 140
yellowwood *(Cladrastis kentuckea)*, 29, 170, 171
yew, western *(Taxus brevifolia)*, 196, 215
yucca, banana *(Yucca baccata)*, 265

zornia *(Zornia bracteata)*, 140

GENERAL INDEX

Mosquito View, 98
Mountain
  Absaroka, 213, 346
  Aeneas, 207
  Aetna, 102
  Anaconda, 190, 191
  Antero, 101
  Arrowhead, 76
  Audubon, 86, 91
  Bald, 54, 102, 324
  Baldy, 61, 247
  Bear, 257, 258
  Bear Lodge, 293, 295, 298
  Beartooth, 346, 347
  Beaverhead, 186
  Bee, 7
  Bierstadt, 40
  Big Agnes, 93
  Big Belt, 218, 219
  Big Burro, 272
  Big Creek Baldy, 225
  Big Horn, 293
  Big Snowy, 231
  Bilk, 220
  Bitterroot, 184, 192, 199,
    233
  Black, 96, 271, 345
  Black Fork, 6
  Blowout, 7
  Blue, 72, 165, 233
  Blue Bouncer, 5
  Boston, 16, 19, 20, 22, 23,
    24
  Boulder, 101, 220
  Boy Scout, 280
  Brazos, 250
  Bridger, 209
  Bristol Head, 79
  Brokeoff, 276
  Bross, 73
  Brush Heap, 8
  Bullet Nose, 206
  Bullion, 111
  Bull of the Woods, 248
  Burro, 131
  Cabinet, 223, 226, 233
  Caddo, 7, 8
  Canjilon, 250
  Capitan, 274, 278, 279,
    280
  Carrizo, 279
  Castle, 230
  Cave, 189, 229
  Chickalah, 22
  Chittenden, 85
  Cimarron, 248
  Cinnamon, 211
  Clayton, 350

Coeur d'Alene, 233, 234
Cook, 202
Copper, 45
Cossatot, 7, 8
Cowan, 213
Cow Creek, 5
Coxcomb, 350
Crane, 165
Crazy, 209, 231
Crystal, 7, 9
Datil, 258
Dross, 74
Eagle, 152, 153
Eddy, 234
Elbert, 98, 100
Elk, 225, 272, 298
Elkhead, 93, 95
Elkhorn, 190, 220
Emmons, 62
Eolus, 110
Ethel, 94
Evans, 38, 40, 76
Fairview, 44
Fall, 88
Flint Creek, 191
Fourche, 7
Gallinas, 255
Gallo, 272
Garfield, 111
Garnet, 211
Goliath, 38, 39
Gothic, 61
Grassy, 165
Graybark, 80
Gray Wolf, 38
Green, 223
Greenhorn, 104, 105
Greyrock, 87
Grizzly, 101
Gros Ventre, 330, 333
Guadalupe, 274, 275, 276
Guyot, 74
Harvard, 100
Hells Half Acre, 194
Henderson, 17
Herman, 70
Highland, 190, 229
Hollister, 190
Horse, 229
Huckleberry, 22, 331
Jackfork, 5
Jemez, 283, 284, 285, 286,
    288
Jicarilla, 274, 278, 279
John Long, 191, 236
Jupiter, 111
Keller, 165
Kelly, 266

Kevan, 227
La Mosca, 256
La Plata, 113
Laramie, 89, 342, 345
Leidy, 333
Leveaux, 153
Lewis and Clark, 221
Little Bald, 324
Little Belt, 229
Little Snowy, 231
London, 73
Lone, 204
Lookout, 156
Lynn, 5
Madre, 258
Magazine, 15, 16, 22, 29,
    30, 31
Magdalena, 257
Mamma, 101
Mangas, 266, 272
Manzanita, 255
Manzano, 255, 260
Marble, 103
Marcelline, 62
Massive, 99
Maverick, 188
McConnel, 87
Medicine, 325, 327, 329
Medicine Bow, 90, 342,
    343, 344
Middle, 116
Mill Creek, 7
Mimbres, 270
Missouri, 7, 8, 30
Mogollon, 266
Monarch, 229
Monument, 212
Moon, 218
Moose, 153, 329
Morrell, 236
Mount of the Holy Cross,
    128
Needle, 110
Negrito, 272
North Eolus, 110
Notch, 128
Oak, 9
O'Brien, 224
Oliphant, 96
Ouachita, 3, 7, 10
Ouray, 103
Oxford, 100
Ozark, 3, 29, 33, 158, 159,
    164, 168, 170, 173, 177,
    179, 181
Patos, 279
Patterson, 165
Pine, 21

ABOUT THE AUTHOR

Robert H. Mohlenbrock is Distinguished Professor Emeritus at Southern Illinois University, Carbondale, where he taught botany for 34 years. He is also Senior Scientist for Biotic Consultants, Inc. He is the author of 50 books, most of them field guides, and is a contributing editor to *Natural History* magazine, which has just published the 201st article in his *This Land* series. He served for 16 years as Chairman of the North American Plant Specialists Group of the Species Survival Commission of the International Union for the Conservation of Nature. Since his retirement, he has taught over 180 week-long wetland plant identification classes in 29 states.

Series Design:	Barbara Jellow
Design Development:	Jane Tenenbaum
Cartographer:	Bill Nelson
Compositor:	Jane Tenenbaum
Text:	10/13.5 Minion
Display:	Franklin Gothic Book and Demi
Printer:	Friesens
Binder:	Friesens